D1708837

SCIENCEWISE 12

Succeeding in Today's World

ScienceWise 12

Succeeding in Today's World

Mirella Agusta-Palmisano
Sarah Barrett
Alan Davis
Deborah Fairchild
Kandis Thompson

Toronto, Ontario

ScienceWise 12:
Succeeding in Today's World

Authors
Mirella Agusta-Palmisano
Sarah Barrett
Allan Davis
Deborah Fairchild
Kandis Thompson

Director of Publishing
David Steele

Publisher
Kevin Martindale

Project Developer
Doug Panasis

Senior Developmental Editors
Jackie Dulson
Lee Geller

Editorial Assistant
Matthew Roberts

Senior Production Editor
Linh Vu

Production Editors
Patricia Ciardullo
Julia Hubble

Copy Editors
Allyson Latta
Denyse O'Leary
Judith O'Leary

Production Coordinator
Sharon Latta Paterson

ArtPlus Ltd. Production Coordinator
Dana Lloyd

Composition
Alicia Countryman/ArtPlus Ltd.

ArtPlus Ltd. Illustrators
Nancy Cato
Joelle Cottle
Donna Guilfoyle
Sandy Sled

Interior Design
Dave Murphy/ArtPlus Ltd.

Cover Design
Peter Papayanakis

Photo Research and Permissions
Nancy Belle Cook
Lisa Brant

Photo Shoot Coordinators
Trent Photographics
Jackie Dulson
Joe Nizio

Printer
Transcontinental Printing Inc.

Printed and bound in Canada
1 2 3 4 05 04 03 02

For more information contact Nelson, 1120 Birchmount Road, Toronto, Ontario, M1K 5G4.
Or you can visit our Internet site at http://www.nelson.com

National Library of Canada Cataloguing in Publication

ScienceWise 12 : succeeding in today's world / Mirella Agusta-Palmisano ... [et al.].

For use in grade 12 in Ontario.

Includes index.
ISBN 0-7725-2930-2

1. Science—Textbooks.
I. Agusta-Palmisano, Mirella
II. Title: Science wise 12.
III. Title: ScienceWise twelve.

Q161.2.S393 2002 500
C2002-905491-5

Acknowledgements

The authors and publisher would like to thank the following reviewers for their insights and suggestions:

Dave Arthur, former Teacher,
Waterloo Region District School Board

Kyn Barker, Consultant,
York Region District School Board

John Bottos, Teacher,
Toronto Catholic District School Board

Robert Brown, Teacher,
Toronto Catholic District School Board

Robert Callott, former Head of Science,
York Region District School Board

Ann Harrison, Teacher,
Niagara Catholic District School Board

Elizabeth Jarman, Teacher,
Simcoe County District School Board

Igor Nowikow, Teacher,
York Region District School Board

The authors and publisher would like to thank the staff and students at Pine Ridge Secondary School and Loyola Catholic Secondary School for use of their facilities and resources in the production of this book.

Table of Contents

UNIT 2 *Communications— Sounds and Pictures* 67

UNIT 3 *Medical Technology* 141

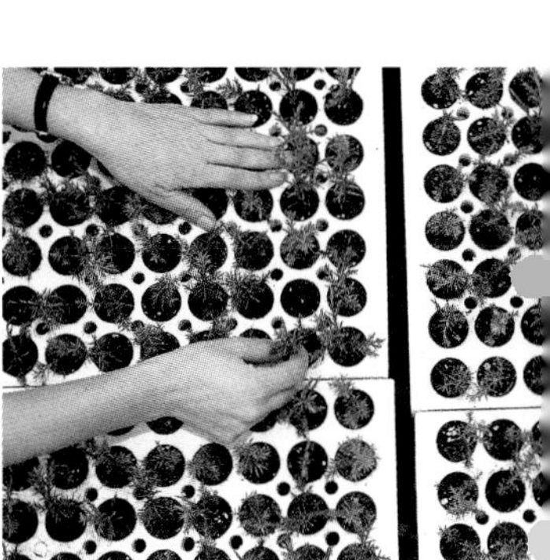

UNIT 4 *Gardening, Horticulture, Landscaping, and Forestry* 199

UNIT 5 *Alternative Environments* 263

Chemistry at Work and Home

UNIT 1

CHAPTER 1

Chemistry Around Us

Chemistry is everywhere. Chemicals make up all the clothes we wear, the food we eat, and the objects we use. Look at Figure 1.1. List as many chemicals being used as you can. What does a pencil have in common with a hair roller? In this chapter, you will learn about the structures, properties, and reactions of common organic substances encountered in the home and workplace. You will also discover health, safety, economic, and environmental issues related to using organic substances.

FIGURE 1.1

What You Will Learn

After completing this chapter, you will be able to:

- Draw and explain the formation of covalent and ionic bonds (1.1, 1.2)
- Identify and collect information on science-and-technology careers (Job Link)
- Describe the process of polymerization in terms of one or two simple molecules (1.3)
- Describe the importance of common organic substances used in the home and workplace (1.3)
- Use library and Internet resources to locate, select, and analyze information (1.3)

What You Will Do

- Determine the miscibility of a variety of organic liquids (Lab 1A)
- Predict the solubility of common organic substances in aqueous and non-aqueous solvents (Lab 1A)
- Prepare and present a report on the social, environmental, and economic consequences of using and discarding organic products (1.3)
- Safely prepare organic products by the process of polymerization (Activity 1B)
- Compare the properties of naturally occurring polymers with synthetic polymers (Lab 1C)
- Carry out experiments to identify physical and chemical properties of recycled plastics (Lab 1D)
- Demonstrate an understanding of WHMIS safety practices (Labs 1C, 1D)
- Select and use appropriate vocabulary and terms (1.1, 1.2, 1.3, Labs 1A, 1C, 1D, Activity 1B)
- Compile, organize, and interpret data (Labs 1A, 1C, 1D)

Words to Know

atom
chemical bond
compound
covalent bond
electronegativity
immiscible
ionic bond
insoluble
miscible
molecule
monomer
natural
non-polar molecule
organic compound
petrochemicals
polar covalent bond
polar molecule
polymerization
polymer
soluble
solute
solvent
synthetic

A puzzle piece indicates knowledge or a skill that you will need for your project, Making a Consumer Product, at the end of Unit 1.

1.1 Bonding

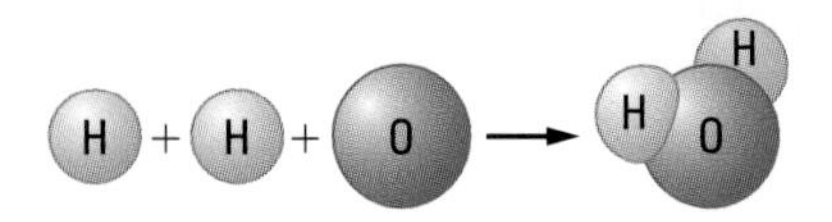

FIGURE 1.2 When atoms bond, what do they form?

Atoms, molecules, and compounds—how do they all relate? What does it mean when atoms are bonded together? How is a compound different from an atom?

In this section, we will answer these questions and more. **Atoms** are the smallest units of matter found on Earth. They have many properties. Atoms have the ability to unite with one another, forming compounds. **Compounds** are substances made by bonding atoms together in a particular way. Most compounds are made up of **molecules**, a group of atoms held together that could be considered one particle. For example, the water that we drink—H_2O—is a combination of hydrogen (H) and oxygen (O) atoms. See Figure 1.2. These atoms bond together to form a molecule of water. A glass of water contains billions of molecules, and these molecules together form the compound water (Figure 1.3).

Types of Bonds

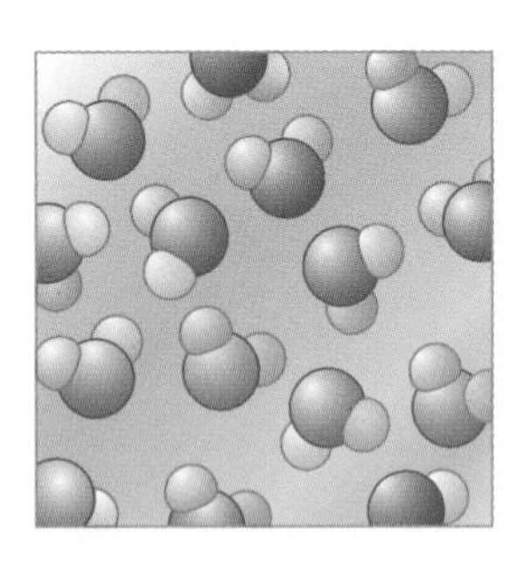

FIGURE 1.3 The compound water

The attraction that holds atoms together to make a molecule is called a **chemical bond**. When atoms come together to form bonds, the outer electrons of each atom are involved in the chemical bond. Not all atoms attract in the same way. Some atoms have a very strong ability to pull the electrons from another atom toward themselves. This is an **ionic bond**, which involves the complete transfer of electrons. In other bonds, the atoms share electrons. The electrons can be shared to form single, double, or triple bonds. See Figure 1.4. If two atoms have an equal sharing of electrons, they form a **covalent bond**. If the two atoms have an unequal sharing of electrons, they form a **polar covalent bond**.

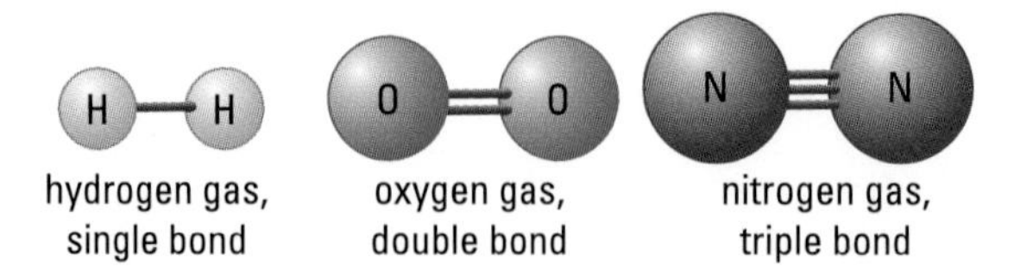

hydrogen gas, single bond
oxygen gas, double bond
nitrogen gas, triple bond

FIGURE 1.4

H 2.1																
Li 1.0	**Be** 1.5											**B** 2.0	**C** 2.5	**N** 3.0	**O** 3.5	**F** 4.0
Na 0.9	**Mg** 1.2											**Al** 1.5	**Si** 1.8	**P** 2.1	**S** 2.5	**Cl** 3.0
K 0.8	**Ca** 1.0	**Sc** 1.3	**Ti** 1.5	**V** 1.6	**Cr** 1.6	**Mn** 1.5	**Fe** 1.8	**Co** 1.9	**Ni** 1.9	**Cu** 1.9	**Zn** 1.6	**Ga** 1.6	**Ge** 1.8	**As** 2.0	**Se** 2.4	**Br** 2.8

FIGURE 1.5 Electronegativity values for some elements of the periodic table

Determining Bond Type: Electronegativity

Figure 1.5 shows the electron-attracting ability, or **electronegativity**, of some elements. The higher the value, the greater the electron-attracting ability. These electronegativities can be used to determine the type of bond formed in a compound, and in some cases, to determine the overall polarity of a molecule. To do this, calculate the difference in electronegativity between elements in the compound. A difference of between 0 and 0.3 represents a covalent bond. If the difference is between 0.3 and 1.7, this represents a polar covalent bond. If the difference is greater than 1.7, an ionic bond is formed.

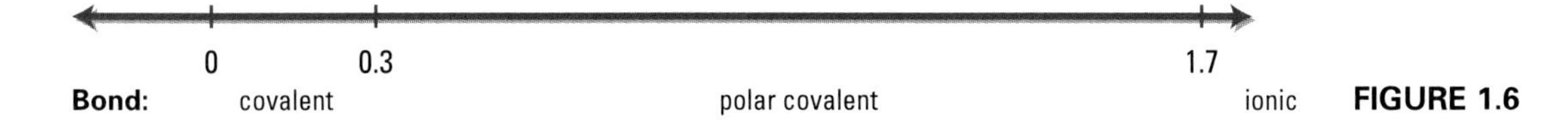

FIGURE 1.6

Look at Figure 1.7 for some examples of the types of bonds formed between elements.

Electronegativity, Bonds, and Polarity

Examples	Difference in electronegativity	Bond type
Chlorine Cl_2	Cl = 3.0 Cl = −3.0 0.0	Covalent bond
Ammonia NH_3	N = 3.0 H = −2.1 0.9	Polar covalent bond
Table salt NaCl	Cl = 3.0 Na = −0.9 2.1	Ionic bond

FIGURE 1.7 Determine the difference in electronegativity and bond type for H–H, H–F, and Mg–O. To calculate the difference in electronegativity, the smaller number should always be subtracted from the larger number.

Polar and Non-Polar Molecules

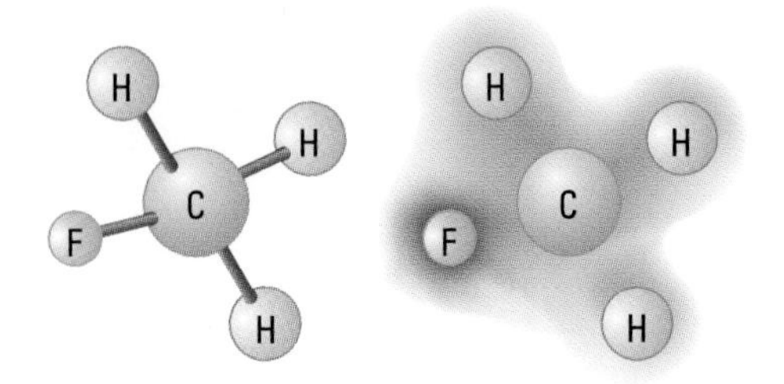

example: CH_3F
bond: polar covalent
molecule: polar

FIGURE 1.8 The electronegativity between the C–H bonds is 0.4 making it slightly polar, but the electronegativity between the C–F bonds is 1.5, making the bonds very polar and allowing the electrons to be more concentrated around fluorine, making the overall shape of the molecule asymmetrical. This is a polar molecule.

There are two main factors that determine if a molecule is polar or non-polar: 1) the degree of polarity in the bonds and 2) the shape or structure of the molecule.

In some molecules the electrons tend to concentrate on one side more than the other. This causes one side of the molecule to be more negative than the other. This is called a **polar molecule**. There is a negative end (the end where the electrons are more concentrated) and a positive end (the end where the electrons are less concentrated). A molecule is polar when the bonds are polar and it has an asymmetrical shape. See Figure 1.8.

In some molecules, the bonding electrons are evenly distributed so that both ends of the molecule have the same amount of negative or positive charge. There is no difference between the two ends—neither one is more negative or positive than the other. This is called a **non-polar molecule**. A molecule is non-polar when the bonds are not very polar and/or it has a symmetrical shape, see Figure 1.9A–B.

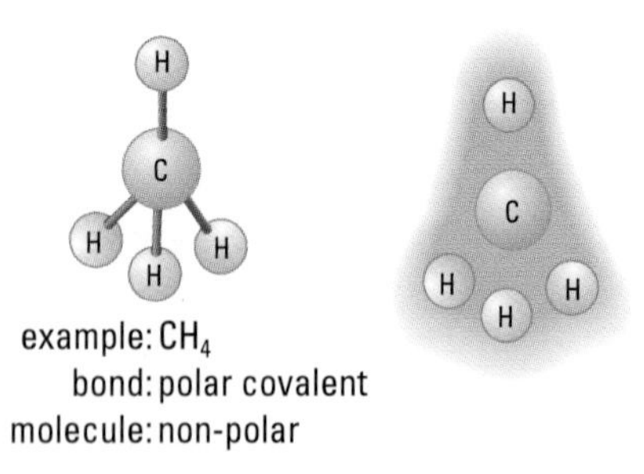

example: CH_4
bond: polar covalent
molecule: non-polar

FIGURE 1.9A The difference in electronegativity in the C–H bonds is 0.4, and the bonds are polar. In this case, the electrons are distributed evenly within the molecule, making the overall shape of the molecule symmetric and non-polar.

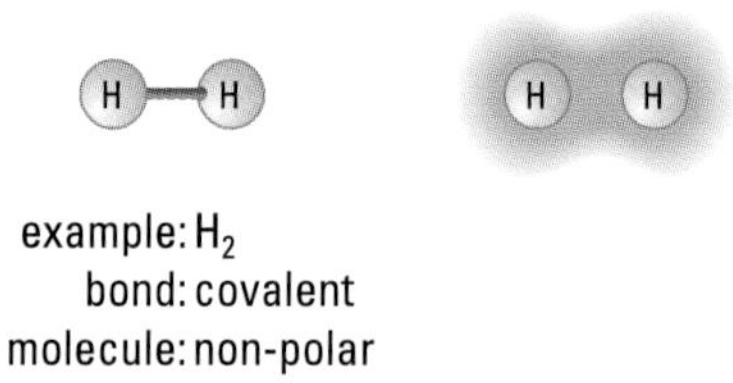

example: H_2
bond: covalent
molecule: non-polar

FIGURE 1.9B The difference in electronegativity in the H–H bond is 0, and the bond is covalent. In this case the electrons are evenly distributed within the molecule making the overall shape of the molecule symmetric and non-polar.

Solubility

What is a solution? A solution is a uniform mixture of two or more substances. It is usually composed of a **solvent**, the substance that does the dissolving, and a **solute**, the substance being dissolved. For example, with drink crystals and water, the water is the solvent and the drink crystals being dissolved are the solute.

Why do you think oil does not dissolve in water, but drink crystals do? When a solute and solvent do not mix, the solute is said to be **insoluble**. It is called **immiscible**. When a solute and solvent mix, the solute is said to be **soluble**. It is called **miscible**. See Figure 1.10.

FIGURE 1.10 Oil and water are immiscible.

FIGURE 1.11 Drink crystals and water are miscible.

ScienceWise Fact

Many of the liquids we drink and the fluids in our body are solutions. We are also surrounded by solutions such as the air (consisting of mostly nitrogen and oxygen) and water (in rivers, lakes, and oceans).

To describe the physical property of a substance as soluble or insoluble is to determine its solubility. The solubility of a substance depends on the answers to two questions:

1. Is the solute that is being dissolved polar or non-polar?
2. Is the solvent polar or non-polar?

Chemists use a general rule—"Like dissolves like"—meaning a polar molecule will be soluble (or miscible) in a polar solvent, and a non-polar molecule will be soluble (or miscible) in a non-polar solvent. A polar molecule will *not* be soluble or miscible in a non-polar solvent.

Now you can answer the question why oil and water do not mix. Water is a polar molecule and oil is a non-polar molecule; they are not like molecules. Drink crystals and water will mix because drink crystals are a polar molecule and water is a polar molecule; they are like molecules.

Does your consumer product dissolve in water? Is this a pro or con for the environment? Find out for your project, Making a Consumer Product, at the end of Unit 1.

Surf the Web

In the early days of the dry-cleaning industry, a variety of solvents were used to clean clothes and fabrics, including gasoline and kerosene. Now one of the solvents used is perchloroethylene (perc). Visit www.science.nelson.com and follow the links for ScienceWise Grade 12, Chapter 1, Section 1.1 to find other solvents used to remove stains in dry cleaning. Classify them as polar or non-polar. Should you dry clean clothing made of microfibre materials? Why or why not?

Review and Apply

1. Make a cartoon strip to show the difference between ionic, covalent, and polar covalent bonds.

2. Why does a painter
   a) need to ask if the walls in a room are painted with latex (water-based) or acrylic (oil-based) paint?
   b) clean paintbrushes with either water or paint thinner (acetone)?

3. Determine if each of the following types of bonds are covalent, polar covalent, or ionic: HCl, LiF, H_2O, NO, K_2O, KCl, H_2, O_2, CO, N_2, CH_4, CCl_4, NH_3, PH_3. Give reasons for your answers.

4. Food colouring is often added to foods such as candies, ice cream, and icing. Are food colouring dyes more likely to be polar or non-polar molecules? Explain your answer.

5. In a graphic organizer, organize the concepts you have learned about in this section.

Molecules and Dyes

What if the world had no colour? There would be no sunrises, autumn leaves, or blue skies. Every day we are surrounded by colour in nature and in objects we use. In manufactured products, colours are added as dyes. There are hundreds of different types of dyes; however they all share the same basic molecular structure. The dye compound consists of molecules that contain C, H, O, and N.

FIGURE 1.12 Dyes add colour to manufactured products.

Clothing can be dyed using artificial or natural processes. For example, red cabbage is considered a naturally occurring dye because its colour can be extracted and then used to dye other products. List as many naturally occurring dyes as you can.

1.2 Organic Compounds

In Section 1.1, you learned about the types of bonds that can form based on their electronegativities. In this section, you will learn about the types of bonds that carbon can form and how these bonds are involved in creating products used at home, at school, and in the workplace.

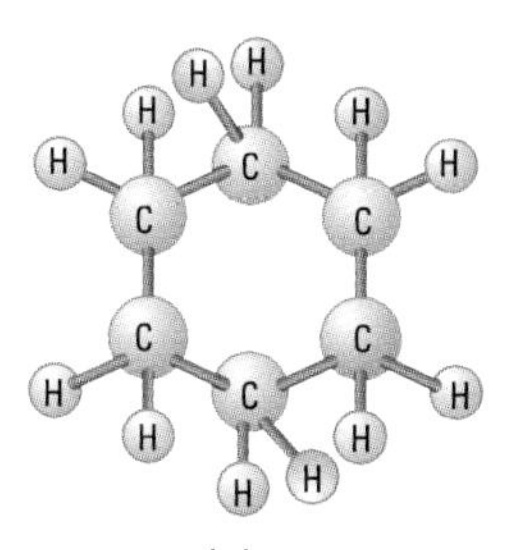

FIGURE 1.13A Carbon atoms can form rings.

Carbon

Carbon makes up less than one percent of Earth's crust, oceans, and atmosphere, but about 20 percent of the human body. **Organic compounds** are all carbon-containing compounds except carbon monoxide, carbon dioxide, and ionic carbon compounds. Carbon atoms have unique bonding properties, such as producing long chains or rings of carbon atoms. See Figure 1.13A–B. This is because the carbon atom forms the most covalent bonds of any main group element of the periodic table. Carbon has many uses, some of which are identified in Figure 1.14A.

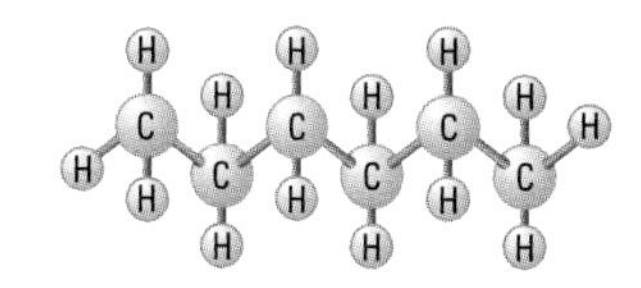

FIGURE 1.13B Carbon atoms can form long chains.

Carbon at Home and at Work

Carbon compound	Use
Carbohydrates	Wood, cotton, paper, food
Proteins	Wool, silk, food, hair
Hydrocarbons	Rubber, gasoline, natural gas
Vinyls	Synthetic fabrics
Polyamides	Nylon

FIGURE 1.14A Types and uses of some carbon compounds

FIGURE 1.14B Identify four types of carbon compounds in this picture.

Find out the structure of your consumer product for your project at the end of Unit 1.

Carbon also forms four chemical bonds. It can form a covalent bond with other carbon atoms, and may also form polar covalent bonds with hydrogen, nitrogen, or oxygen. Carbon's ability to form a variety of bonds allows many different carbon-based compounds to be made for society's use. Refer to Figure 1.15.

Examples of Organic Compounds

Carbon compound containing H, N, or O	Examples	Structure	Workplace example
Hydrocarbons—contain only carbon and hydrogen. They can contain single, double, or triple carbon–carbon bonds or the carbon atoms can be joined in a ring.	acetylene C_2H_2 methane CH_4 butane C_4H_{10} 1-butene C_4H_8 1-butyne C_4H_6 cyclobutane C_4H_8	H–C≡C–H acetylene	Using an acetylene (C_2H_2) torch to cut through sheet steel.
Amides—contain atoms of carbon, nitrogen, and hydrogen.	1,6-diaminohexane $N_2H_4C_6H_{12}$ (used to make nylon)	1,6-diaminohexane	This tennis racket is made with nylon strings.
Alcohol, acids, ketones, aldehydes—contain atoms of carbon, hydrogen, and oxygen.	acetic acid CH_3COOH (vinegar) propanone C_3H_6O (acetone) methanal CH_2O (formaldehyde) 2-propanol C_3H_8O (rubbing alcohol)	acetone	Nail polish remover contains acetone.

FIGURE 1.15

Review and Apply

1. Use Figure 1.16 to do the following:
   a) Identify five organic compounds.
   b) Classify each compound as a carbohydrate, protein, hydrocarbon, vinyl, or polyamide.
   c) List two jobs in which organic compounds are used.

FIGURE 1.16

2. Lily has a part-time job as a mechanic. She is performing an oil change on a car. What organic compounds is she using? Identify all safety precautions and Workplace Hazardous Materials Information System (WHMIS) symbols involved in the oil change.

3. You need to clean a stain of grease (a mixture of hydrocarbons). Would you choose a non-polar solvent or water to remove the stain? Explain your answer.

4. Add the new concepts from this section to the graphic organizer you started in Section 1.1.

5. **Creating Models**

Chemists use molecular models to help them visualize the shapes and bonding angles of molecules that are too small to see, just as a model car is a visual representation of an actual car. In this activity, you will use a molecular model kit or modelling software to better understand the structure of different organic molecules. Using your modelling kit, create molecular structures for each of the following molecules:

C_5H_{12}, C_5H_{10}, C_5H_8, CH_3COOH, $N_2H_4C_6H_{12}$

a) In your notebook, draw a sketch of each of the structures.

b) Compare your structures with those of a classmate and describe the similarities and differences among the structures.

LAB
1A

Miscibility of Organic Liquids

Finding out about the solubility of a common organic compound can help you determine its best uses. Is water the best choice to remove suntan oil? Could you use rubbing alcohol? Why or why not?

Purpose

To determine the solubility of different solutes with water, mineral oil, rubbing alcohol, olive oil, sesame oil, vinegar, and air.

Materials

- micro-tray
- pipette dropper
- scoopula
- electronic balance or scale
- solutes: petroleum jelly, baby oil, suntan oil, butter, sugar, acetylsalicylic acid, solid air freshener
- solvents: water, mineral oil, rubbing alcohol, olive oil, sesame oil, vinegar, air
- dropper bottles, one for each solvent
- dropper bottles, one for each solute: baby oil, suntan oil
- petri dishes, one for each remaining solute: petroleum jelly, butter, sugar, acetylsalicylic acid, solid air freshener

Procedure

1. Create a data table like Figure 1.17 to record your observations.
2. Add five drops of water to seven wells of the micro-tray. Label the wells 1, 2, 3, 4, 5, 6, and 7.
3. Add 1 g of petroleum jelly into well 1, five drops of baby oil into well 2, five drops of suntan oil into well 3, 1 g of butter into well 4, 1 g of sugar into well 5, 1 g of acetylsalicylic acid into well 6, and 1 g of solid air freshener into well 7.

Solvents	Solutes						
	Petroleum jelly	Baby oil	Suntan oil	Butter	Sugar	Acetylsalicylic acid	Solid air freshener
Water							
Mineral oil							
Rubbing alcohol							
Olive oil							
Sesame oil							
Vinegar							
Air							

FIGURE 1.17

4. Stir each well with the pipette dropper and record your observations in your data table. In each box write if the solution is miscible or not miscible.
5. Clean out the used wells of your micro-tray and dispose of the solutions as directed by your teacher
6. Repeat steps 2–5 with each of the remaining solvents.

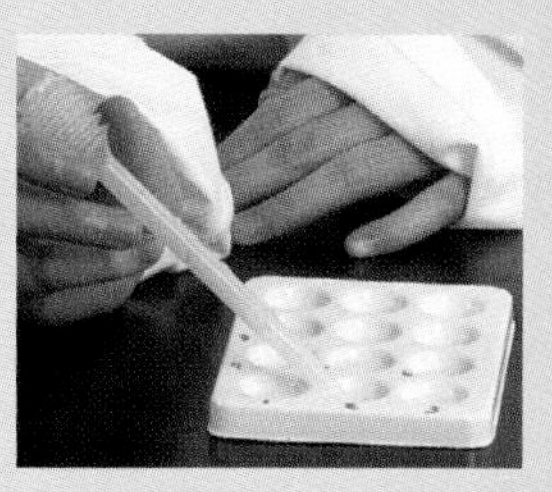

FIGURE 1.18

Analysis and Conclusion

1. Summarize the general properties of organic compounds you observed in this lab.
2. Is the rubbing alcohol a polar or non-polar molecule? Explain.
3. Which solutes are polar molecules? Which solutes are non-polar molecules? Justify your answers.

Extension and Connection

4. You are a waiter in a diner. Before the breakfast rush, you have to clean and refill all the sugar containers. One of the lids is stuck.
   a) Which solvent from this lab would best be used to help you?
   b) Why would a solvent from this lab that may work, not be the best choice?

1.3 Polymerization

Natural products are those that occur in nature. Some examples of natural products are leather, cotton, rubber, hydrocarbon fuels, and metals. **Synthetic**, or artificially made, products do not exist in nature. They are manufactured by the chemical industry. Synthetic substances are made mostly from petroleum (crude oil), and are referred to as **petrochemicals**. In this section, you will look at how some synthetic materials are created, used, and recycled, and their impact on society and the environment.

FIGURE 1.19 List each product as either natural or synthetic.

Polymers

Polymers are extremely big molecules made by combining hundreds or thousands of identical simple molecules. These simple molecules are called **monomers**. **Polymerization** is the process that forms polymers from monomers. Some polymers, such as starch and cellulose, are naturally produced in living organisms (plants). Some are produced by the chemical industry from petrochemicals. They include plastics such as polyethylene (thermoplastics—plastics that harden when cooled but soften when heated), polypropylene, and synthetic fibres. You would better recognize these polymers as plastic chairs, tables, grocery bags, pens, pipes, gas tanks, fibres in carpets, ropes, soft drink bottles, and so on. Refer to Figure 1.20.

Process of Polymerization from Simple Molecules

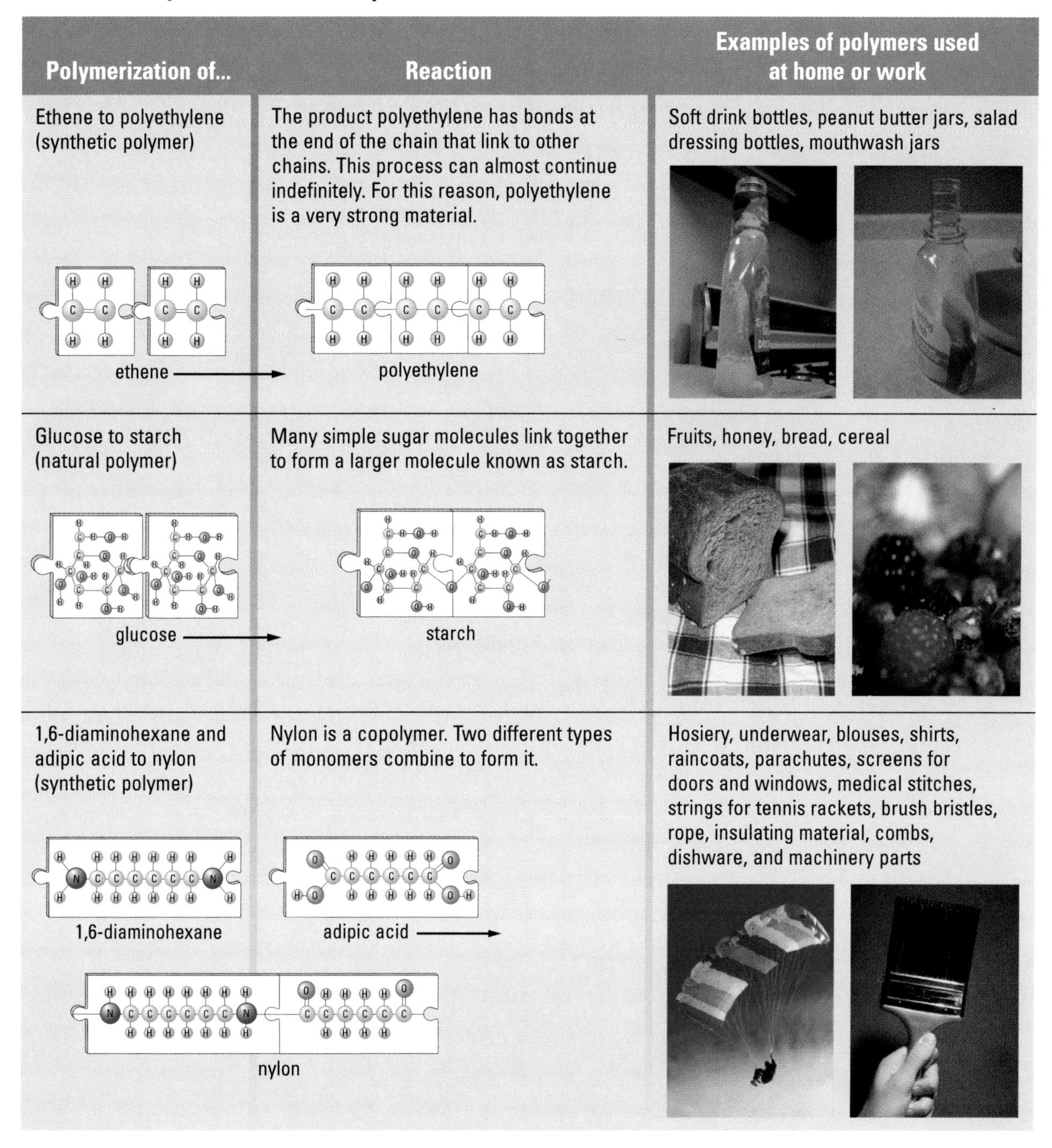

Polymerization of...	Reaction	Examples of polymers used at home or work
Ethene to polyethylene (synthetic polymer)	The product polyethylene has bonds at the end of the chain that link to other chains. This process can almost continue indefinitely. For this reason, polyethylene is a very strong material.	Soft drink bottles, peanut butter jars, salad dressing bottles, mouthwash jars
ethene →	polyethylene	
Glucose to starch (natural polymer)	Many simple sugar molecules link together to form a larger molecule known as starch.	Fruits, honey, bread, cereal
glucose →	starch	
1,6-diaminohexane and adipic acid to nylon (synthetic polymer)	Nylon is a copolymer. Two different types of monomers combine to form it.	Hosiery, underwear, blouses, shirts, raincoats, parachutes, screens for doors and windows, medical stitches, strings for tennis rackets, brush bristles, rope, insulating material, combs, dishware, and machinery parts
1,6-diaminohexane	adipic acid →	
nylon		

FIGURE 1.20

ScienceWise Fact

Synthetic polymers play an important role in medicine. Look at Figure 1.21A–B. Natural polymers, like DNA and protein, are the major molecules of your body.

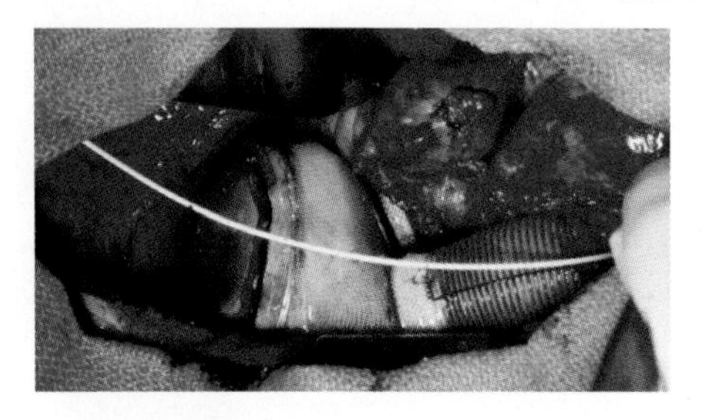

FIGURE 1.21A This was the first artificial heart to be implanted into a human and it is made of polymers: rubber and polyurethane.

FIGURE 1.21B This is artificial skin made from the polymer silicone.

Recycling Synthetic Polymers

Plastics are polymers that are produced in one of two ways. They are shaped either by melting the material, pouring it into a mould, and letting it cool in the desired shape, or by placing the material in a mould and baking it. Recycling plastics is very important because they originate from crude oil, a non-renewable resource. Natural polymers break down easily, but synthetic polymers (plastics) are designed to be very strong and durable. This means that they take a long time to decompose in a landfill site (Figure 1.22).

FIGURE 1.22 Disposing of polymers and other wastes in a landfill site

Recycling plastics is a complex process. For one thing, not all types of plastics are recyclable, and those that can be recycled cannot all be processed in the same way (Figure 1.23). When you place all your different plastics in the recycling bin for pick-up or take them to the recycling drop-off, they will not all be recycled in the same manner. For this reason, in 1988 the Society of the Plastics Industry, Inc. (SPI) introduced its resin identification coding system, which has been adopted by the Environment and Plastics Industry Council Canada. It is a consistent, uniform system that identifies the resin (the type of polymer) used in bottles and containers. When plastic material enters the recycling plant, everything is first separated according to its resin polymer symbol, and then each type is recycled accordingly (Figure 1.24).

FIGURE 1.23 Each of these symbols on the plastic containers indicate the type of polymer they contain.

Generally in Canada, all plastics are identified with a code number inside arrows. Code 1 products are recyclable: the process is widely available and economically practical. Code 2 products have limited collection in only a few communities. Codes 3 to 7 are currently non-recyclable in most areas of Canada. Recycling can become a very expensive process. What are some alternative ways of making use of your plastics? List two ideas.

ScienceWise Fact

Currently, more than 3.8 million households across Ontario have access to recycling. This represents 99 percent of the Ontario population in more than 500 municipalities!

Major Plastic Resins, Resin Symbols, and Examples of Recycled Products

Resin name / Resin symbol	Common uses	Examples of products made from recycled resins
Polyethylene terephthalate (PET or PETE) 1	Soft drink bottles, peanut butter jars, salad dressing bottles, mouthwash containers	Liquid soap bottles, fibrefill for winter coats, surfboards, paint brushes, fuzz on tennis balls, soft-drink bottles, film
High-density polyethylene (HDPE) 2	Milk, water, and juice containers, grocery bags, toys, liquid detergent bottles	Soft-drink cups, flower pots, drain pipes, signs, stadium seats, trash cans, recycling bins, traffic barrier cones, golf bag liners, toys
Polyvinyl chloride or vinyl (PVC-V) 3	Clear food packaging, shampoo bottles	Floor mats, pipes, hoses, mud flaps
Low-density polyethylene (LDPE) 4	Bread bags, frozen food bags, grocery bags	Garbage can liners, grocery bags, multipurpose bags
Polypropylene (PP) 5	Ketchup bottles, yogurt containers, margarine tubs, medicine bottles	Paint buckets, video cassette cases, ice scrapers, fast-food trays, lawn mower wheels, automobile battery parts
Polystyrene (PS) 6	Video cassette cases, compact disk jackets, coffee cups, cutlery, cafeteria trays, grocery store meat trays, fast-food sandwich containers	License plate holders, golf course and septic tank drainage systems, desktop accessories, hanging files, flower pots, trash cans
Other—products made of a resin not listed above, or a combination of resins 7	Certain ketchup bottles	Plastic lumber

FIGURE 1.24

Is the packaging of your consumer product recyclable or is it made from recycled material? You'll need to find this out for your project, Making a Consumer Product, at the end of Unit 1.

Review and Apply

❶ List five natural materials found in the home or workplace.
List two products that contain each material you listed.

❷ You have a summer job working at the local recycling plant. You received the products shown in Figure 1.25. Identify each resin name and give an example of a recycled product that can be produced from each resin.

❸ Add the new concepts from this section to the graphic organizer you started in Section 1.1.

FIGURE 1.25

❹ **Reporting on Plastic**

With a partner, choose one of the following topics:

- common addition plastics
- thermosetting plastics
- copolymer products
- vulcanized products
- natural or synthetic fibres

a) Research your chosen topic using the library and/or Internet resources, and prepare a report to do the following:

   i. Explain about the product using diagrams and simple word equations.

   ii. Identify the pros and cons to society, the environment, and the economy of using this type of organic product.

   iii. Describe the social, environmental, and economical consequences of discarding this type of organic product.

   iv. Write a concluding paragraph explaining why you would or would not use your chosen type of product. Make sure to support your viewpoint with your researched information.

b) Present your report to the class. Make notes during each presentation. Be prepared to ask questions.

Job Link

Plastics Processing Machine Operator

Plastics processing machine operators set up and operate machines that mix, press, and mould plastics into products.

Responsibilities of a Plastics Processing Machine Operator

- keeping the machines clean and in good repair
- lifting and moving products with or without the use of lifting aids
- checking for defects through visual and weight checks
- monitoring that the temperature and pressure of the equipment is correct

Where Do They Work?

- chemical plastic and rubber manufacturing companies
- recycling companies
- factories that have machines

Skills for the Job

- knowledge of machine operations and procedures
- ability to identify, troubleshoot, and correct equipment problems
- proper handling, storage, and packaging of plastic materials
- knowledge of safety regulations such as plant safety procedures, accident reporting, and WHMIS

Education

- completion of secondary school may be required
- an interest in mechanics is a definite asset
- a knowledge of computers is also an asset

FIGURE 1.26 Proper safety protection is important when drilling plastic.

Surf the Web

To find four related careers, visit **www.science.nelson.com** and follow the links for ScienceWise Grade 12, Chapter 1, Section 1.3.

ACTIVITY 1B

Making Polymer Putty

Putty is actually a polymer and it has many uses. Different types of putty are used in the workplace. Plumbers use putty to make a watertight seal around toilet bases and sinks. Tool and die makers also use putty in some machines. Moulders and sculptors work with different types of putty to create art.

Be Safe!

Avoid inhaling the powdered borax. It may cause an allergic reaction. Wash your hands after working with the putty.

What You Will Need

- safety goggles
- lab coat
- two 250-mL beakers
- 45 mL white glue
- 105 mL of water
- 15 mL solid borax
- stirring stick
- graduated cylinder

What You Will Do

1. Put on your safety goggles and lab coat.
2. Pour all the glue into a beaker and add 45 mL of warm water.
3. In the second beaker, add the borax and 60 mL of warm water. Stir for about 2–3 minutes. **Note:** Do not worry if you find that not all the borax solution dissolves.
4. Using a graduated cylinder, measure 30 mL of the borax solution and add it to the glue mixture. Quickly stir the mixture.
5. Remove the material from the beaker.

What Did You Find Out?

1. How does the putty feel?
2. What happens when you pull the putty apart slowly or quickly?
3. Can you bounce the putty?

Making Connections

4. Identify at least one workplace application not listed here for this product.
5. Predict what would happen if you changed the amount of borax used. How do you think this would change the use of the putty?

FIGURE 1.27 Creating polymer putty

LAB 1C

Physical and Chemical Properties of Natural and Synthetic Fibres: Mystery Swatches

Have you ever wondered why clothes are rough or soft, easy to clean or easily stained, easily wrinkled or able to stay neatly pressed? The fibres of your clothing have a chemical structure that gives them these physical properties. Since ancient times, people of all civilizations have made fabric from plant fibres (like flax and cotton), from animal hair (like sheep's wool), and from insects (like silk from silkworm cocoons). In recent times, natural fibres have been replaced with synthetic fibres, which are made from coal, tar, and water using chemical processes. The fibres available today are either natural or synthetic. These fibres, whether natural or synthetic, are polymers, or long chains of matter.

Purpose

To determine the properties of natural and synthetic fibres by performing various tests on fabric samples.

Be Safe!

The following chemicals may cause burning and irritation: hydrochloric acid, sodium hydroxide, chlorine bleach, stain remover solvent. Be careful when pouring them. Bleach can cause damage to clothes.

Prediction

Predict how you think swatch 1 and swatch 2 will react to the solutions listed in the Materials list below.

Materials

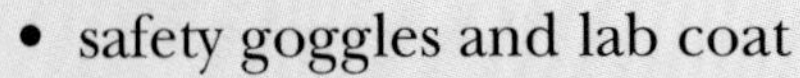

- safety goggles and lab coat
- 10 small squares of swatch 1
- 10 small squares of swatch 2
- red litmus paper
- 100 mL distilled water
- 10 mL hydrochloric acid
- 10 mL sodium hydroxide
- 10 mL chlorine bleach
- 10 mL stain remover solvent
- 10 mL liquid detergent solution
- 10 mL liquid soap solution
- 16 test tubes
- two 1-hole stoppers
- 14 no-hole stoppers
- 100-mL beaker
- Bunsen burner
- test tube tongs
- stirring rod
- 250-mL beaker, labelled "Waste"

Procedure

1. Create a data table like Figure 1.28 to record your observations for this experiment.

Physical and Chemical Properties of Fibres

	Swatch 1	Swatch 2
Does the fabric produce a flame? **FLAME**		
Is there an odour when you burn the fabric? **ODOUR**		
What does the residue look like after the fabric is burned? **RESIDUE**		
Does the fabric contain protein? Does it turn red litmus paper blue? **PROTEIN**		
How does the fabric react to: **ACID** (hydrochloric acid)		
How does the fabric react to: **BASE** (sodium hydroxide) (bleach)		
How does the fabric react to: **SOLVENT (POLAR or NON-POLAR)**		
How does the fabric react to: **DETERGENT**		
How does the fabric react to: **SOAP**		
How does the fabric react to: **WATER**		

FIGURE 1.28

Part One: The Flame Test

1. Put on your safety goggles and lab coat.
2. Set aside one piece of swatch 1 and one piece of swatch 2 as control samples.

CONTINUED

Be Safe!

If the swatch catches fire, put it in the beaker of water.

3. Fill a 100-mL beaker with 80 mL of water.
4. Light the Bunsen burner and using test tube tongs, hold a piece of swatch 1 over the flame for 30 seconds. Record your observations.
5. Repeat step 4 with swatch 2.

Part Two: The Protein Test

Note: Animal fibres containing protein will give off nitrogen when heated. The nitrogen will react with the water, producing ammonium hydroxide. The ammonium hydroxide will turn red litmus paper blue.

1. Place a piece of each swatch in separate test tubes.
2. Moisten two pieces of red litmus paper with distilled water.
3. Using test tube tongs, place one piece of the litmus paper in the opening of each test tube. See Figure 1.29.
4. Cover each test tube with a 1-hole stopper. Carefully heat one test tube, holding it over the flame with the test tube tongs. Hold the test tube over the flame for 30 seconds.
5. Record your observations.
6. Repeat steps 4–5 with the second test tube.

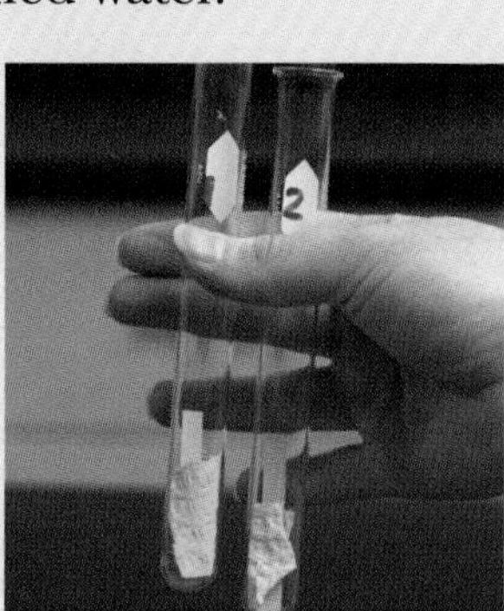

FIGURE 1.29

Be Safe!

Be sure to use a 1-hole stopper to avoid any build-up of pressure that could blow out the stopper.

Part Three: Other Chemical Tests

1. Place seven pieces of swatch 1 in seven test tubes, and seven pieces of swatch 2 in seven different test tubes.
2. Label each group of seven tubes as follows:

#1 – water	#2 – water
#1 – hydrochloric acid	#2 – hydrochloric acid
#1 – sodium hydroxide	#2 – sodium hydroxide
#1 – bleach	#2 – bleach
#1 – solvent	#2 – solvent
#1 – detergent	#2 – detergent
#1 – soap	#2 – soap

3. Place 10 mL of the appropriate solution into each tube.

Be Safe!

Be careful when pouring these solutions because they may cause burns or skin irritation.

4. Place a no-hole stopper on each tube and shake constantly for 1 minute. Allow the fabrics and solutions to stand for 20 minutes. Shake once more after 10 minutes.

5. Carefully pour off the solution first into the waste beaker. Then use a stirring rod to remove each piece of the fabric, one at a time. See Figure 1.30. Rinse each test tube well. Dispose of the solutions in the waste beaker as directed by your teacher.

6. Blot each piece of fabric using the stirring rod and allow to dry.

7. Check the appearance of the swatches for changes in strength and elasticity by comparing each experimentally treated swatch with your observations of the untreated (control) swatch and with the swatch treated with water only. Record your results in your data table.

FIGURE 1.30

Analysis and Conclusion

1. Based on your experiments, how can you distinguish between swatch 1 and swatch 2? Summarize the physical and chemical properties of each fibre. Identify swatch 1 and swatch 2 as either silk or nylon.

2. Why was it important to keep a piece of each swatch aside as the control pieces?

3. Which fibre would make a better shirt? Better stockings? Which tests gave you information to answer this question? What else would you need to know?

4. What problems do you see with fibres that melt at a high temperature?

Extension and Connection

5. Some clothes contain tags that describe their make-up as 35 percent silk and 65 percent nylon. Suggest reasons why this blend of fibres may be used.

6. Research and report on synthetic fibres used in athletic clothing. Investigate the advantages and disadvantages of these fibres. Consider how athletic clothing may be different from everyday clothing.

LAB 1D

Physical and Chemical Properties of Recycled Plastics

The plastics industry uses resin pellets to make recycled plastic containers. In this experiment, each of the resins is colour coded or cut into a specific shape to help you keep them separate, but in real life each of the resins may be made in any colour or shape. The resin samples used in this experiment have come from recycled material. Refer to Figure 1.24 for examples of recycled material and the six main types of resins.

Be Safe!

Acetone is very flammable and should not be inhaled. Work in a fume hood or well-ventilated area.

Purpose

To identify the six main kinds of recycled plastic resins by identifying their physical and chemical properties.
The six resins to identify are as follows:

1. PET—polyethylene terephthalate
2. HDPE—high-density polyethylene
3. PVC—polyvinyl chloride
4. LDPE—low-density polyethylene
5. PP—polypropylene
6. PS—polystyrene

Materials

- masks, safety goggles, lab coat
- 18 samples of resins, three of each colour or shape
- three 250-mL beakers
- four 100-mL beakers
- one 50-mL beaker
- Bunsen burner
- test tube tongs
- hot plate
- glass stirring rod
- 5-cm piece of copper wire
- glass petri dish to use as a waste container
- 100 mL corn oil
- 65 mL isopropyl alcohol
- 10 mL acetone

Procedure

1. Each test is to be done at the station indicated by your teacher. Before starting any experiment, put on your safety goggles and lab coat.

2. Create a data table like Figure 1.31 to record your observations for each station.

	Type of resin					
Test	**1**	**2**	**3**	**4**	**5**	**6**
Water test						
Metal heat test						
Acetone test						
Heat test						
Isopropyl alcohol test						
Oil test						

FIGURE 1.31

Station 1: Water Test

1. Fill a 100-mL beaker with tap water.

2. Place one resin sample into the beaker and stir with the glass stirring rod. Record in your data table whether it sinks or floats. What does each result tell you about each type of plastic?

3. Remove the resin and set it aside for later use.

4. Repeat steps 2–3 with the remaining resins.

Note: Stations 2–4 are to be done **ONLY** for the resins that sank. Resins that sank are more dense than water.

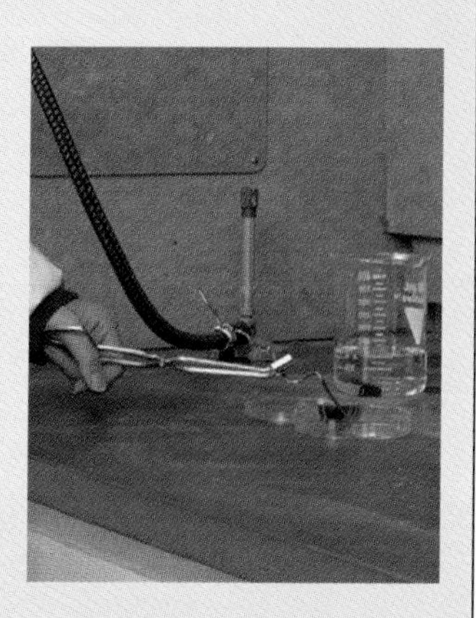

FIGURE 1.32

Station 2: Metal Heat Test (To be done in a fume hood)

1. Fill a 100-mL beaker with water.

2. Using test tube tongs, hold the 5-cm piece of copper wire over the hottest part of the Bunsen burner flame until it is red hot.

3. Remove the wire from the flame and carefully touch it with one of the resins that sank in the water test. (If the resin sticks to the wire, use test tube tongs to pull it apart.) See Figure 1.32.

Be Safe!

This test must be done in a fume hood.

CONTINUED

4. If the wire has some plastic stuck to it, place it back into the flame.
5. Observe the colour of the flame. Record your observations in your data table.
6. Place the wire with the remaining plastic stuck to it into the beaker of water to stop the burning and cool the wire.
7. Repeat steps 2–6 for the other resins that sank in the water test.

Be Safe!

This test must be done in the fume hood.

Station 3: Acetone Test (To be done in a fume hood)

1. Put on your mask.
2. Place 10 mL of acetone into a 50-mL beaker.
3. Place a resin sample into the acetone for 20 seconds.
4. Remove the resin using test tube tongs. Try to scrape off some plastic with the test tube tongs. See Figure 1.33. Record your observations in your data table.

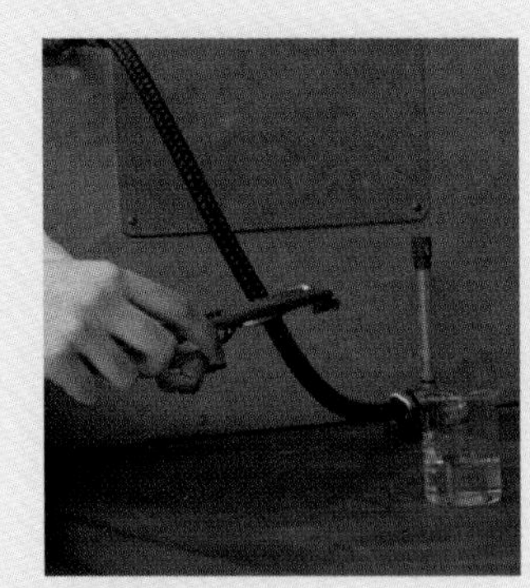

FIGURE 1.33

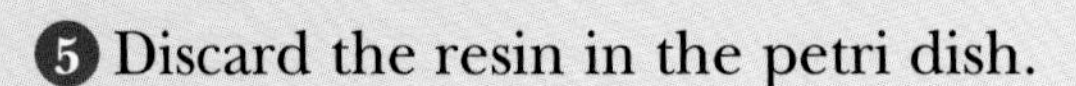

5. Discard the resin in the petri dish.
6. Repeat steps 3–5 with the other resins that sank in the water test.

Be Safe!

Use care around hot water.

Station 4: Heat Test

1. Boil 125 mL of water in a 250-mL beaker on a hot plate.
2. Take a resin that sank in the water test. Using tongs, hold the resin in the boiling water for 30 seconds.
3. Remove the resin from the water, allow it to cool a little, and press it with your fingers. Record your observations.
4. Repeat steps 2–3 for the other resins that sank in the water test.

Note: Stations 5 and 6 are to be done **ONLY** for the resins that floated in the water test. Resins that floated are less dense than water.

Station 5: Isopropyl Alcohol Test

1. Place 65 mL of isopropyl alcohol into a 250-mL beaker and add 35 mL of water to make 100 mL of solution.
2. Place all the resin samples that floated during the water test into the beaker with the isopropyl alcohol. Use the glass rod to stir the solution. Record which resin samples float and which sink in your data table.
3. Remove the resins using the test tube tongs.

Station 6: Oil Test

1. Place 100 mL of corn oil in a 250-mL beaker.
2. Place all the resin samples that floated during the water test into the beaker with the corn oil. Stir and record which ones float and which sink.
3. Remove the resins using the test tube tongs.

At the end of the lab, recycle the resins in the appropriate containers.

Analysis and Conclusion

1. Identify the resins by colour or shape.
2. Copy the flow chart in Figure 1.34, on page 30, into your notebook. By number, identify the resin samples observed in each case. If there were no resin sample results, write N/A.
3. From what you have observed, and using the density table in Figure 1.35 on page 30, what is the approximate density of the isopropyl alcohol and water solution? Explain your answer.
4. Based on your observations, identify the physical and chemical properties of each resin sample.

CONTINUED

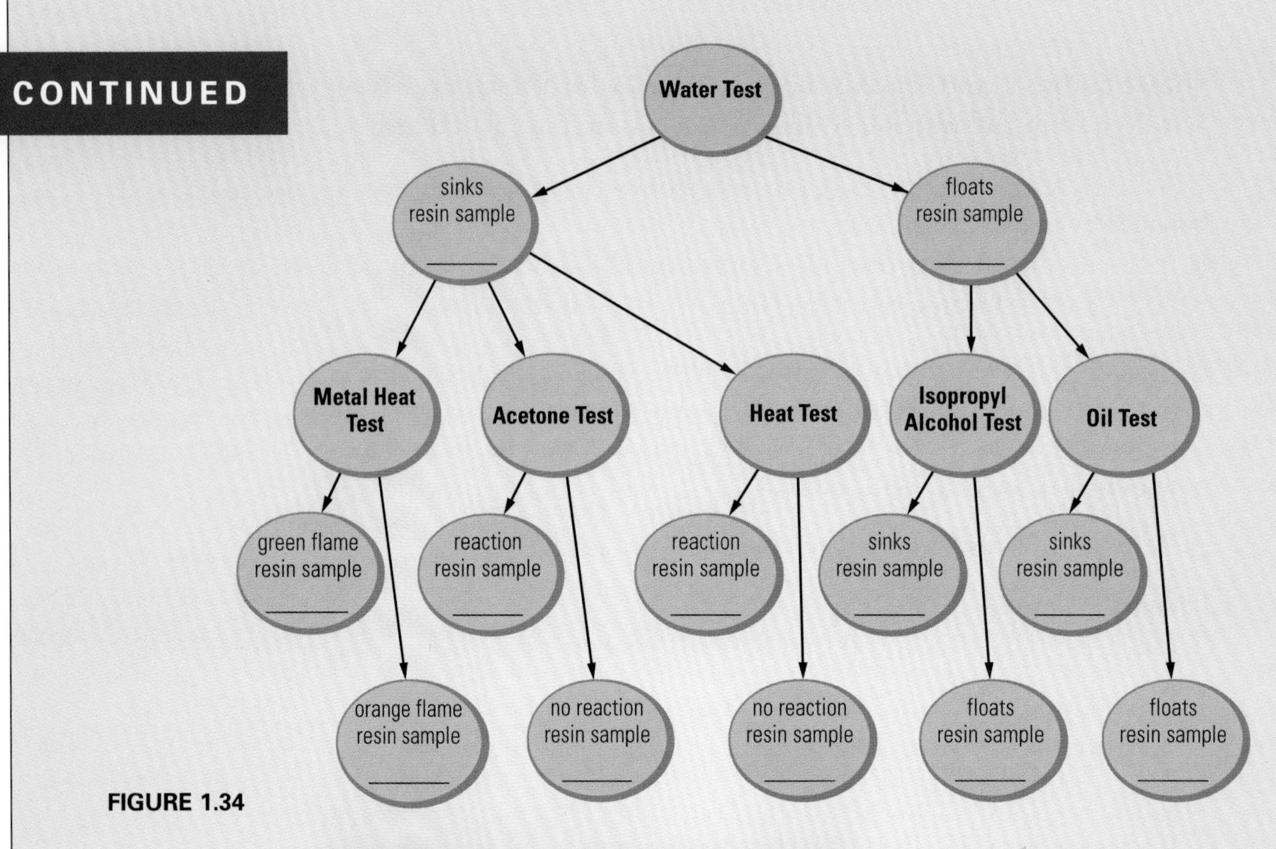

FIGURE 1.34

Density Table	
Substance	**Density**
Water	1.00
1—PET	1.38–1.39
2—HDPE	0.95–0.97
3—PVC	1.16–1.35
4—LDPE	0.92–0.94
5—PP	0.90–0.91
6—PS	1.05–1.07

FIGURE 1.35

ScienceWise Fact

Density is a physical property. It is a measure of how much mass is contained in a unit of volume.

$$\text{Density} = \frac{\text{Mass}}{\text{Volume}}$$

Extension and Connection

5. You want to make a plastic handle for a cooking pan out of recycled plastic. Which plastic should you avoid? Why?

6. Choose two resin samples. What type(s) of products is each one suited for? Explain your answers.

1.4 Chapter Summary

Now You Can...

- Illustrate and explain the formation of covalent bonds (1.1, 1.2)
- Identify and collect information on science-and-technology careers (Job Link)
- Describe the process of polymerization in terms of one or two simple molecules (1.3)
- Describe the importance of common organic substances used in the home and workplace (1.3)
- Report on the social, environmental, and economic consequences of using and discarding organic products (1.3)
- Use library and Internet resources to locate, select, and analyze information (1.3)
- Determine the miscibility of a variety of organic liquids (Lab 1A)
- Predict the solubility of common organic substances in aqueous and non-aqueous solvents (Lab 1A)
- Safely prepare organic products by the process of polymerization (Activity 1B)
- Compare the properties of naturally occurring polymers with synthetic polymers (Lab 1C)
- Carry out experiments to identify some physical and chemical properties of recycled plastics (Lab 1D)
- Demonstrate an understanding of WHMIS safety practices (Activity 1B, Labs 1C, 1D)
- Select and use appropriate vocabulary and terms (1.1, 1.2, 1.3, Labs 1A, 1C, 1D, Activity 1B)
- Compile, organize, and interpret data (Labs 1A, 1C, 1D, Activity 1B)

Concept Connections

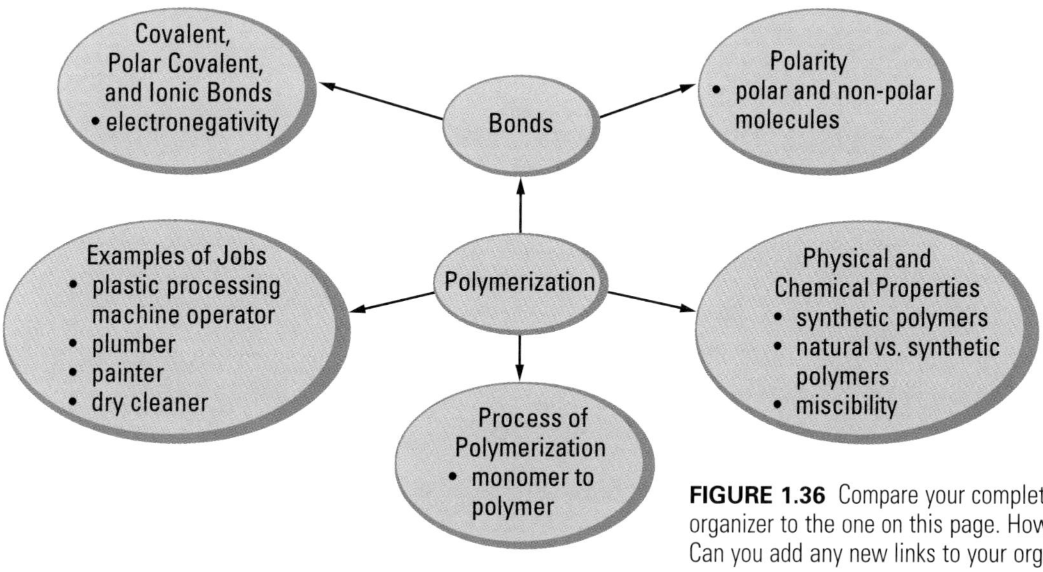

FIGURE 1.36 Compare your completed graphic organizer to the one on this page. How did you do? Can you add any new links to your organizer?

CHAPTER 1 *review*

Knowledge and Understanding

1. For each of the following statements indicate True or False. If False, rewrite the statement to make it true.

a) Compounds are made up of molecules.
b) A covalent bond is the equal sharing of electrons between two different atoms.
c) Organic compounds contain all the elements of the periodic table.
d) Hydrocarbons contain only hydrogen and carbon elements.
e) Polar solutes will dissolve in non-polar solvents.
f) Polymers are extremely small molecules made up of only one element.
g) Synthetic compounds appear in nature.

2. One of your chores is to do the laundry. As you begin to separate the clothes you notice that there is a grease stain on a pair of jeans. What type of substance is grease—polar or non-polar? What kind of product would you need to use to eliminate the stain? Explain your answer.

3. Copy Figure 1.37 into your notebook and complete it for all six resins.

Resin polymer name	Two common uses	Two examples of recycled material made from the resin

FIGURE 1.37

Inquiry

4. Write an investigation to determine if each of the following solutes will dissolve in the solvents. Justify your answers.
a) salt in vegetable oil
b) baby oil in vegetable oil
c) petroleum jelly in water
d) motor oil in rubbing alcohol

5. Margaret is a plastics technician. She is working with two resins, PET (polyethylene terephthalate) and PP (polypropylene), however she is not sure which is which. Help Margaret identify the plastic resins by giving her step-by-step instructions and explanations.

6. Jack works for a textile company. He was asked to perform tests to determine if a material was a synthetic or organic fabric. Describe the tests Jack would need to perform and the results he would obtain to distinguish between the synthetic and organic fabric.

Making Connections

7. Paula works in a paint store. A customer wants to know what type of paint he should use to paint his kitchen. Use your new knowledge and vocabulary to make a list of the questions Paula should ask to determine the type of paint to sell to the customer.

8. Kevlar™ is used to make bullet-proof vests and sails of modern racing sailboats. What chemical and physical properties must Kevlar™ have to be used for these products?

9. Synthetic material such as fibreglass is replacing natural material such as wood in some sports equipment. Discuss whether you believe the new material gives a competitor an unfair advantage. Justify your position.

Communication

10. As a plumber you are asked to use PVC piping instead of copper piping. Identify the advantages and disadvantages, both environmentally and economically, of using PVC piping instead of traditional copper piping.

11. Write a letter to the editor of a newspaper, make a web page, or create a poster to explain the consequences to the environment and society of using and discarding thermosetting plastics.

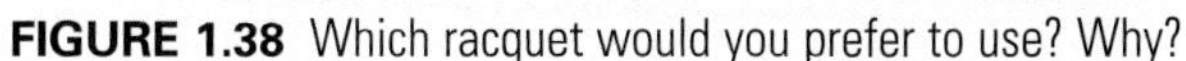

FIGURE 1.38 Which racquet would you prefer to use? Why?

CHAPTER

2

Applying Chemistry

Consumers need to be well informed in order to use products properly and safely. Common organic substances such as soap and detergent each perform similar functions, but the chemical make-up of each product is different. In this chapter, you will learn about the structures, properties, and reactions of common organic substances. Look at Figure 2.1. Why do you wash your hands with soap? What does detergent do? Why doesn't mayonnaise separate the way oil and vinegar does?

FIGURE 2.1

What You Will Learn

After completing this chapter, you will be able to:

- Explain the behaviour of emulsifying agents (2.1, Lab 2A, 2.3)
- Identify and collect information on science and technology related careers (2.1)
- Write word equations for condensation and hydrolysis reactions (2.2)
- Explain how the hydrophobic, hydrophilic, or amphiphilic character of organic molecules is related to the presence of O, N, or ions in the molecule (2.3)

What You Will Do

- Use lab equipment safely and accurately to investigate emulsions (Lab 2A)
- Research and report on an application of condensation, hydrolysis, or emulsification processes (2.2)
- Safely prepare a common organic product (Lab 2B)
- Investigate and compare the relative quantities of soap and detergent required to form emulsions in hard and soft water (Lab 2C)
- Communicate ideas and results of laboratory experiments (Labs 2A, 2B, 2C)
- Demonstrate an understanding of safe laboratory practices (Labs 2A, 2B, 2C)
- Use data tables to present information (Labs 2A, 2B, 2C)
- Select and use appropriate vocabulary (2.1, 2.2, 2.3, Labs 2A, 2B, 2C)

Words to Know

addition reaction
amphiphilic
colloid
condensation reaction
emulsifying agent
emulsion
hard water
heterogeneous
homogeneous
hydrolysis reaction
hydrophilic
hydrophobic
mixture
saponification
scale
soft water
suspension

A puzzle piece indicates knowledge or a skill that you will need for your project, Making a Consumer Product, at the end of Unit 1.

2.1 What Is Inside a Mixture?

What do orange juice, milk, cement, and paint have in common? They are all mixtures. Mixtures are combinations of substances. Have you ever noticed that in some mixtures, you can see the different layers, but in other mixtures you cannot tell that they are made of more than one substance? In this section, you will find out what the different types of mixtures are and how these mixtures are used.

Types of Mixtures

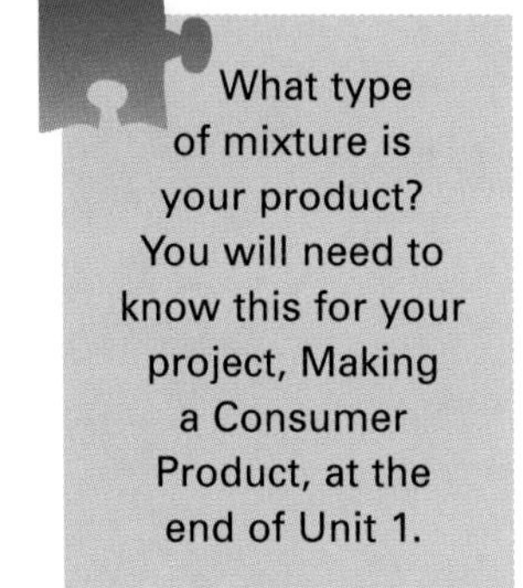

What type of mixture is your product? You will need to know this for your project, Making a Consumer Product, at the end of Unit 1.

When two or more different types of particles are combined, this is a **mixture**. A mixture can be seen as homogeneous or heterogeneous. **Homogeneous** mixtures look as if they are just one substance. For example, look at a small amount of sugar dissolved in water. In a **heterogeneous** mixture, you can see the different substances that make up the mixture. For example, a bag of mixed nuts is a heterogeneous mixture.

What about mixtures that have a bit of both? If you look at 2-percent milk through a microscope you will see that it has two parts—a liquid part and tiny particles of cream. The particles are not mixed completely enough to be a homogeneous mixture, but they are mixed better than a heterogeneous mixture. These "in-between" mixtures are called suspensions, colloids, or emulsions. See Figure 2.2.

A)

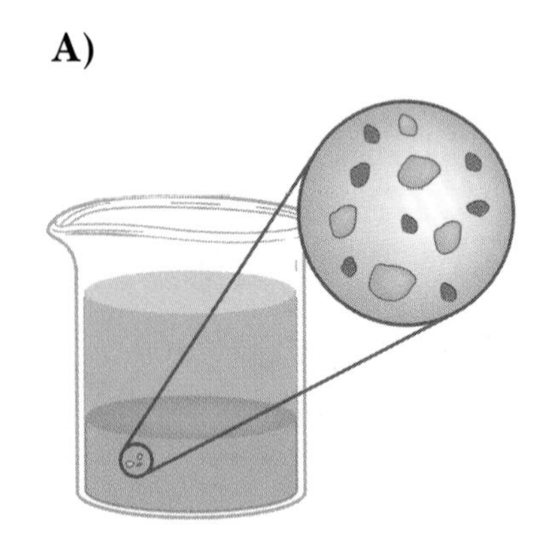

B)

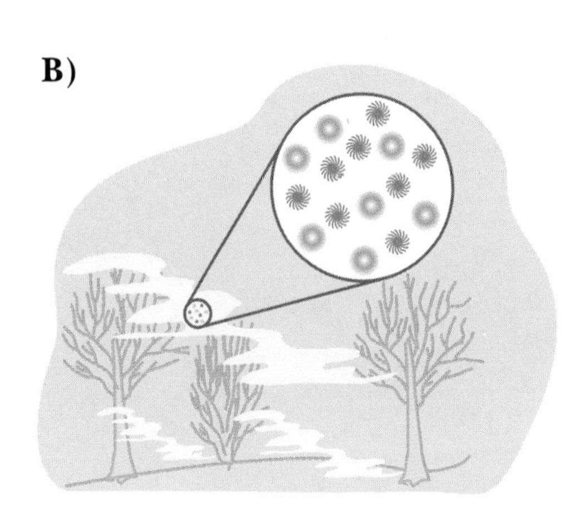

C)

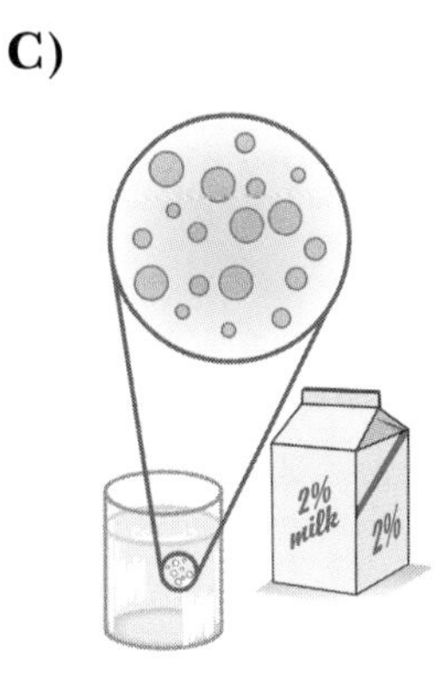

FIGURE 2.2 Three types of heterogeneous mixtures

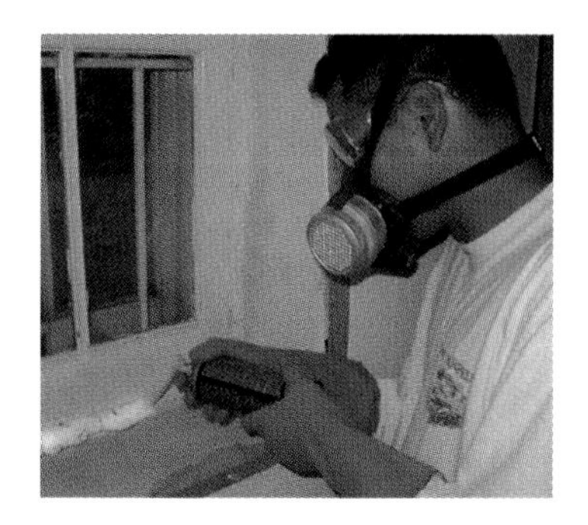

FIGURE 2.3 This insulation foam is a colloid. It is used to seal cracks around windows.

A **suspension** (Figure 2.2A) is a mixture containing suspended solids that cause a cloudy appearance. The pieces of solid are large enough that they will eventually settle out. Sand in water is an example of a suspension. Some of the particles of sand settle quickly, while others are so small that they make the water appear cloudy. List two suspensions used in the home or workplace. Why do you think some medicine labels read "Shake well before using"?

A **colloid** (Figure 2.2B) is a mixture in which pieces of solid or drops of liquid are suspended. The pieces or drops are so small that they are not visible and they will not settle out. An example of a colloid is fog; tiny drops of water are mixed with air. The drops are too small to fall as rain, but large enough to feel as water on your skin.

An **emulsion** (Figure 2.2C) is a type of colloid in which tiny liquid droplets are mixed in another liquid. It is the mixture of two immiscible liquids. Recall from Chapter 1 that immiscible liquids are liquids that do not mix. Homogenized milk is an example of an emulsion in which drops of cream are mixed with water. An **emulsifying agent** is any substance that helps keep other substances from separating out of the mixture. An example of an emulsifying agent is an egg yolk. In mayonnaise, the egg yolk prevents the separation of oil and water because molecules of the yolk form a "skin" around each droplet of the oil, which keeps the oil droplets from coming together and forming a separate layer. See Figure 2.4.

FIGURE. 2.4 Mayonnaise is an emulsion of oil droplets in water. The egg yolk (emulsifying agent) has coated each micro-droplet of oil. What would happen to mayonnaise if egg yolks were not added?

Try This at Home

Identifying Suspensions, Colloids, and Emulsions

Look around your home or the grocery store for the following products: gel toothpaste, shaving cream, mayonnaise, pudding, gelatin, whipping cream, homogenized milk, Italian salad dressing, jam, and fresh-squeezed orange juice.

Create a data table identifying each as a suspension, colloid, or emulsion. Explain your choices.

LAB 2A

Emulsions

Mayonnaise is classified as an oil-in-water emulsion whereas margarine is a water-in-oil emulsion. What is the difference? In an oil-in-water emulsion oil droplets are surrounded by water, but in a water-in-oil emulsion water molecules are surrounded by oil. What makes this happen? In this activity, you will determine the formation and stability of emulsions.

Purpose

To investigate and determine which substances are emulsion-forming and emulsion-breaking agents.

Materials

- 2 scoopulas
- 10 test tubes
- 10 rubber stoppers
- one test tube rack
- three 50-mL graduated cylinders
- 140 mL vegetable oil
- 50 mL vinegar
- 5 mL egg white
- 5 mL egg yolk
- 10 mL mustard
- 1 mL salt
- 1 mL pepper
- 1 mL sugar
- 1 mL paprika
- 5 mL dishwashing detergent

Procedure

1. Create a data table with the following headings to record your observations: Test tube; Immediately after shaking; After 10 min.; Temporary/Permanent Emulsion.
2. Label each test tube from 1 to 10 and place in the test tube rack.
3. Use a clean graduated cylinder to measure vinegar. To each of the labelled test tubes, add 5 mL vinegar. See Figure 2.5.
4. Use a clean graduated cylinder to measure oil. Add 5 mL of oil to test tube 1. To each of the remaining test tubes, add 15 mL of oil.

5. Measure and add 5 mL of mustard to test tube 3.
6. Measure and add 5 mL of mustard and 1 mL of salt to test tube 4.
7. Measure and add 1 mL of pepper to test tube 5.
8. Measure and add 1 mL of sugar to test tube 6.
9. Measure and add 5 mL of egg yolk to test tube 7.
10. Measure and add 5 mL of egg white to test tube 8.
11. Measure and add 1 mL of paprika to test tube 9.
12. Measure and add 5 mL of dishwashing detergent to test tube 10.
13. Place rubber stoppers on each test tube and shake each tube 100 times.
14. Record in your data table your observations of each test tube immediately after shaking 100 times.
15. Record your observations again after the test tubes have been left standing in the rack for 10 minutes.

FIGURE 2.5

Analysis and Conclusion

1. Which substances had a stabilizing (emulsion forming) influence on the emulsion?
2. Which substances had a de-stabilizing (emulsion breaking) influence on the emulsion?
3. Why do certain substances help to form a stable emulsion?
4. Why are stabilizers incorporated into emulsions and how do they act?

Extension and Connection

5. Why does mayonnaise or sour cream last longer than fresh cream?
6. Why is it important for a cook to know which foods are good emulsifying agents?

Review and Apply

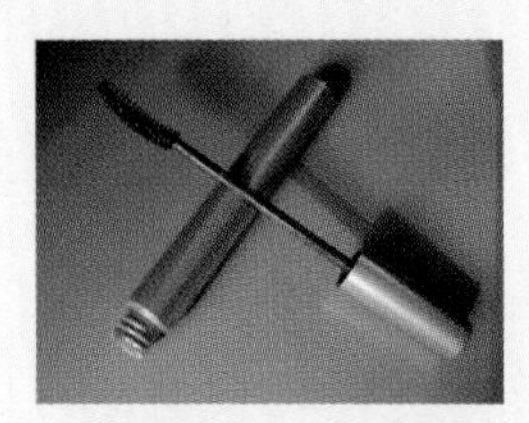

FIGURE 2.6 Cosmetics

1. As a cosmetic quality control technician your job is to ensure that each product meets the standards set by the industry and the Canadian government. What are three factors that would influence the quality of a cosmetic that is applied to the face? Explain your answer.

2. Your cousin is a chef and you are helping with the menu for an important event. With the information learned in this section, give suggestions on making a creamy Caesar-salad dressing that will not separate over time.

3. Draw sketches of consumer products to represent the different types of mixtures: suspensions, colloids, and emulsions.

4. In a graphic organizer, organize the concepts you have learned about in this section.

5. **Classifying Consumer Products**

Cosmetics are used to change or enhance the appearance of the person wearing them. An actor may wear cosmetics to change appearance for a new role.

Classification of Mixtures

Powder: a mixture of minute solid particles with a talcum–powder base
Wax/oil: a mixture of dyes and oils with a wax base
Gel: a colloidal suspension that appears to be solid

Cosmetics can be thought of as emulsions, suspensions, powder, wax/oil, or gel.

a) Classify each of the cosmetic products in Figure 2.6. Give reasons for your choice.
b) Why should you read the label on a product before using it?

Job Link

Chef

Chefs plan and direct food preparation and cooking activities, and prepare and cook meals and other speciality foods.

Responsibilities of a Chef

There are several different levels of responsibility within the food preparation industry:

- *Executive chefs:* Direct food preparation in one or more restaurants or other establishments. They plan menus, hire and supervise chefs and cooks, and may prepare meals.
- *Sous-chefs:* Supervise chefs, cooks, and other kitchen workers. They may plan menus and prepare meals.
- *Working chefs:* Prepare meals or specialize in pastry, sauces, salads, meats, and other foods. They may also supervise kitchen staff and plan menus.
- *Cooks:* Prepare a variety of foods and supervise kitchen workers. They may plan menus and specialize in particular dishes.

Where Do They Work?

- Chefs and cooks work in restaurants, hotels, hospitals, educational institutions, central food commissaries, ships, construction and logging camps, and other organizations.

Skills for the Job

- attention to detail and ability to meet high standards of cleanliness
- ability to problem-solve—for example, if a shipment of food does not arrive
- good organizational skills, to ensure that scheduled tasks are completed on time
- ability to work as part of a team

Education

- a secondary school diploma is usually required
- cooks complete a three-year apprenticeship, attend a college cooking program or other cooking course, or gain experience through several years of commercial cooking
- sous-chefs and working chefs require a cooking apprenticeship or formal training abroad or equivalent training and experience
- executive chefs usually require several years' experience including two years' supervisory experience in commercial food preparation and as a chef

FIGURE 2.7 Chefs must regulate the temperature for ovens, broilers, and grills so that food cooks properly.

2.2 Organic Reactions

Organic compounds are a part of your daily life. Much of your body is composed of organic compounds. The food you eat, the clothes you wear, fuels that heat your home or power cars—these all are made from organic compounds.

Look at Figure 2.8. What do you think these substances have in common? They are both examples of products of organic reactions.

FIGURE 2.8 Examples of organic substances formed by addition reactions

ScienceWise Fact

Perfumes and fragrances that smell like bananas, pineapples, apricots, oranges, apples, or grapes smell this way because they contain esters.

Types of Organic Reactions

Addition reactions occur when compounds react when they are combined. These reactions often occur when a double or triple bond breaks so that new single bonds are formed.

alkene + H_2 → alkane

alkene + Br_2 → bromoalkane

alkene + water → alcohol

alkene + haloacid → alkylhalide

alkyne + H_2 → alkene

In **condensation reactions**, two molecules combine to form a single organic molecule. A small molecule, usually water, is produced as a second product during the reaction. For example, when a carboxylic acid reacts with an alcohol it can condense to form an ester (which has a very pleasant odour) and water.

general word equation:	carboxylic acid	+	alcohol	→	ester	+	water
example:	butanoic acid	+	ethanol	→	ethyl butanoate (fragrance of pineapple)	+	water

Hydrolysis reactions use the addition of a molecule of water to break apart large molecules. Hydrolysis means water-splitting. This reaction is the reverse of a condensation reaction. Heat and sulfuric acid are used in the process.

general word equation: ester + water $\xrightarrow[\text{sulfuric acid}]{\text{heat}}$ carboxylic acid + alcohol

example: ethyl butanoate + water $\xrightarrow[\text{sulfuric acid}]{\text{heat}}$ butanoic acid + ethanol

Making Polymers

Polymers are large organic molecules made of many smaller molecules. Polymerization can occur through either addition or condensation reactions.

1. Addition Method

Addition reactions: Examples					Uses for the polymer
1. ethylene	+	ethylene	→	polyethylene	Packaging, bottles
2. vinyl chloride	+	vinyl chloride	→	polyvinyl chloride	Insulation, plastic pipes
3. propylene	+	propylene	→	polypropylene	Bottles, ropes, pails, medical tubing

FIGURE 2.9

2. Condensation Method

Condensation reactions: Examples							Uses for the polymer
amino acid	+	carboxylic acid	→	nylon	+	water	Clothing, mountain climbing gear, ropes, carpet
terephthalic acid	+	ethylene glycol	→	polyester	+	water	This polyester is known by the trade name Dacron, used to make clothing fibre, and under the trade name Mylar, used to make plastic film and recording tape.
glycylalanine	+	phenylalanine	→	polypeptide	+	water	Aspartame used as an artificial sweetener

FIGURE 2.10

Review and Apply

1. Brainstorm with a partner. Find a simple word equation that describes each of the following organic reactions. Give an example of each.
   a) addition
   b) condensation
   c) hydrolysis

2. Look at Figures 2.11 and 2.12. Which polymer was made by an addition reaction or condensation reaction? How are plastics that are formed by addition reactions different from plastics formed by condensation reactions?

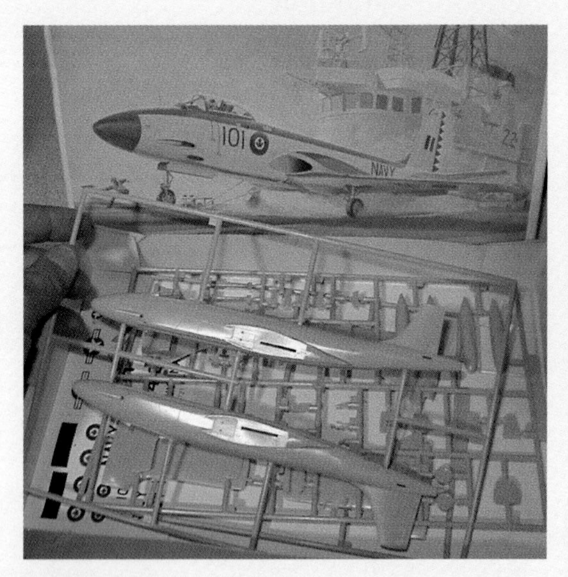

FIGURE 2.11

FIGURE 2.12

3. "A condensation reaction is the reverse of a hydrolysis reaction." Janet is having a difficult time seeing how this is possible. Use word equations to help Janet understand this idea.

4. As you and a friend walk by a perfume counter, you notice an intense smell of bananas. Your friend wonders how they can make perfumes with such realistic smells. What would you tell your friend?

5. Add the new concepts you learned in this section to the graphic organizer you started in Section 2.1.

❻ The Chemistry of Diapers

One special group of polymers is called super absorbent polymers. These polymers absorb many hundreds times their own mass. One common type of this polymer is sodium polyacrylate. It is formed by a condensation reaction. It is specially formulated to swell very rapidly in water but does not dissolve. Such polymers can be found in disposable diapers (largest use), nursery potting soil (provides a water source for plants during shipment), fuel filtration material (removes small amounts of moisture from auto, diesel, and jet fuels), and the toy creatures that grow when placed in water. In this activity, you will determine what is in a disposal diaper that keeps a baby dry.

Materials:

- safety goggles
- lab coat
- thin disposable diaper
- scissors
- plastic grocery bag
- two paper cups or beakers
- graduated cylinder
- water
- sodium chloride (table salt)
- sucrose (table sugar)

Be Safe!

Do not ingest the crystals or touch your face or eyes after coming in contact with them; the crystals may cause irritation and dehydration.

Put on your safety goggles and lab coat. Carefully dissect a diaper. Holding the diaper inside a plastic grocery bag, peel back its outer lining. Pull the filling out of the diaper into the bag. Discard the outer lining. Shred the filling into small bits. Hold the bag tightly shut and shake vigorously. Repeat three times, then discard the filling. Shake the granules into a corner of the bag. Cut off the corner and empty the granules into a paper cup or beaker.

Add 100 mL of water to the granules and stir.

a) What happens? Would you consider the granules in the diaper a polymer? Explain.

After a few minutes, divide the material formed into two equal portions in two cups. To one portion add about 2 mL of table salt and to the second, add 2 mL of table sugar.

b) What happens to the granules?

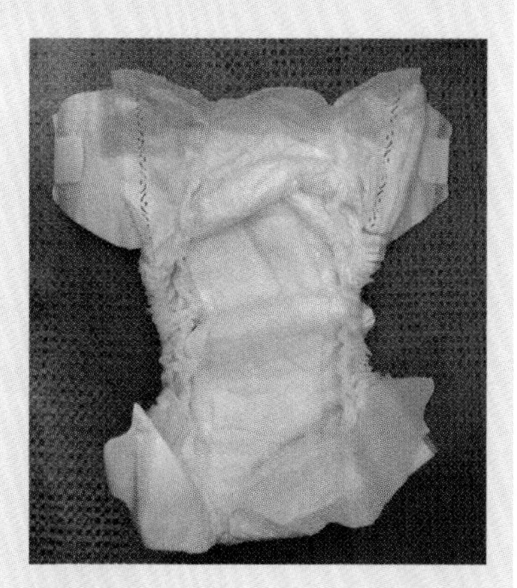

FIGURE 2.13

What organic compounds is your consumer product made of? Write a word equation for it. You will need to know how to write word equations for your project, Making a Consumer Product, at the end of Unit 1.

CASE STUDY

Organic Processes

The food and cosmetic industries contribute millions of dollars each year to the Canadian economy. Many people are employed in both these industries as chemical technicians, cosmetologists, food analysts, dietitians, salespeople, tool and die makers, and so on. Both of these industries deal with chemistry, especially organic chemistry. Many applications of condensation, hydrolysis, or emulsification processes contribute to these industries.

Application of Organic Processes

FIGURE 2.14

Cosmetics Industry

Cold cream has been around a lot longer than you may think. A Greek physician named Galen invented it in the second century. His preparation consisted of one part melted white wax and three parts olive oil. To add a scent to this cream, he soaked rosebuds in the oil. He then blended in as much water as he could. This produced the first cold cream.

a) Which of the organic processes is involved in the production of cold cream? How do you know?

FIGURE 2.15

Food Industry

Margarine was developed in 1869 by a French chemist named Mége-Mouriés in response to a request by Napoleon III for a substitute for butter. There were not enough animal fats (butter and lard) to feed the ever-increasing population of Europe during the Industrial Revolution. The first margarine was made of a mixture of animal fat churned with milk and salt. It was originally called oleomargarine from the Latin word *oleum,* which means "oil," and the Greek word *margaron,* which means "pearl" (because it had a pearl-like shine). After the Second World War, it began to be called margarine. Now, most margarines are made using vegetable oils such as coconut oil, olive oil, and cottonseed oil.

b) Why is the making of margarine considered an organic process?

Food and Your Body

Carbohydrates are organic compounds that contain the elements carbon, hydrogen, and oxygen. We find carbohydrates in starchy foods such as bread, potatoes, pasta, and fruit. Our bodies, through the process of digestion, break down carbohydrates into simple sugar molecules called glucose. The breakdown of glucose through the process of cellular respiration provides the body almost all the energy for life's activities. The word equation for cellular respiration is:

FIGURE 2.16 Pasta is a form of carbohydrates.

glucose + oxygen → carbon dioxide + water + energy

c) What type of reaction is cellular respiration? How do you know?

Some foods contain more complex sugar molecules such as sucrose, found in sugar cane and sugar beets. The sucrose is produced by a condensation reaction between glucose and fructose.

d) What is the common name for sucrose?

e) Are carbohydrates considered polymers? Explain your answer.

Analysis and Communication

1. Research further one application of condensation, hydrolysis, or emulsification for an everyday product such as lip gloss, cold cream, mayonnaise, salad dressing, mustard, or lipids (fats).

a) For the product chosen include the following steps: type of process, word equation, and the steps used in the manufacturing process (an industrial method, home preparation, or in your body). Present your information as a brochure, pamphlet, or some form of electronic media.

Making Connections

2. What health or safety issues are related to the product you chose? How could you eliminate or reduce them?

2.3 Getting Rid of Dirt

ScienceWise Fact

If the base in a saponification reaction is sodium hydroxide, a solid soap that can be moulded into the desired shape is produced. If the base is potassium hydroxide, a softer, liquid soap is produced. The type of fat used also makes a difference. Solid fats such as lard or tallow tend to produce solid soaps, while oils tend to produce liquid soaps.

Using water is traditionally thought of as the best way to get things clean. Water dissolves many liquids and solids. But what happens if you wash your greasy hands with only water? In this section, you will find out how soap is made, how detergent can remove dirt, and how the type of water used can make a difference. In Section 2.2 you learned about hydrolysis reactions. The hydrolysis of esters is also called a **saponification** reaction. It gets this name because the hydrolysis of a special type of ester (fats and oil) produces soap. Soap is actually a special type of salt that can remove grease and dirt. Salts are formed when acids react with bases. When a carboxylic acid called a fatty acid is reacted with a strong base such as sodium hydroxide or potassium hydroxide, you get a salt called soap.

fat or oil + strong base → salts of fatty acid (soap) + glycerol

In Lab 2B, you will make soap. The process you will use is similar to the kettle process. The kettle process was the original way that soap was made, in large kettles. Some independent soap makers still use this process, however the continuous process is now preferred by most large companies as it ensures a more consistent batch of soap and takes less time to complete. The kettle process involves adding salt to separate the soap and the glycerol. Adding a salt (sodium chloride) solution causes the solid soap to separate out as a crust on the mixture. At the end of the process, the soap is washed with plenty of water to eliminate any excess base.

How Soaps and Detergents Work

In Chapter 1, you learned that like dissolves like. If water is a polar molecule and it cannot dissolve dirt that has a thin layer of oil around it, then the oil must be a non-polar molecule. We need to find some way for the oil and water to mix.

Soaps and synthetic detergents are **amphiphilic** molecules. They consist of two distinct parts—a head and a long tail section. Look at Figures 2.17 and 2.18. The head is water-soluble, or **hydrophilic** (the "water loving" polar end). The head is made of a carboxyl group (COO^-) in soaps and sulfate group (SO_3^-) in synthetic detergents. The tail is oil-soluble, or **hydrophobic** (the "water hating" non-polar end). The tail is made of a long non-polar hydrocarbon chain in both soaps and synthetic detergents.

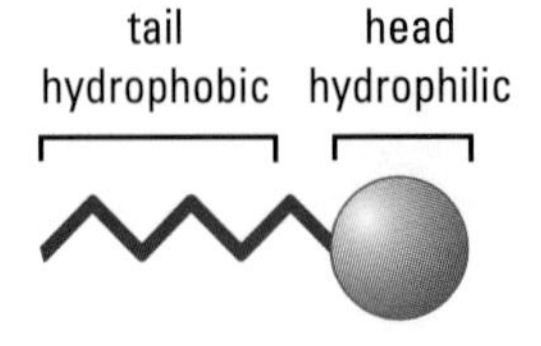

FIGURE 2.17 A simple diagram of a soap molecule. Which end would attach itself to the greasy dirt?

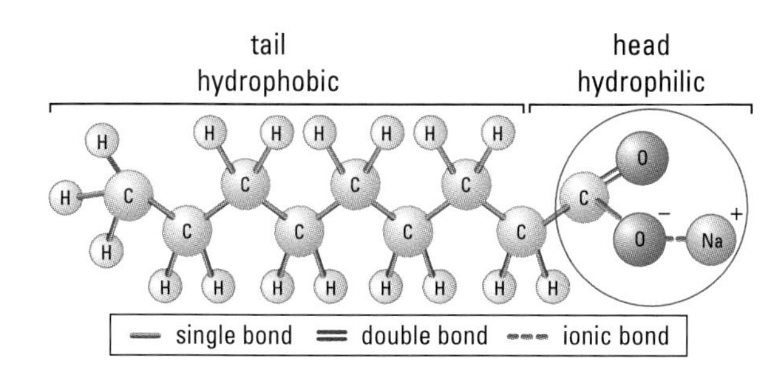

FIGURE 2.18 This is a soap molecule. In soap, there is an ionic bond between the sodium ion and the carboxyl group in the hydrophilic head. Which part is different in a detergent molecule?

When the amphiphilic molecules of soap or detergent form a droplet in water, all their water-loving heads point toward the water and their water-hating tails point away from the water—where they may surround a glob of greasy dirt. See Figure 2.19.

Removing Dirt from Fabric in a Washing Machine

Step	Process	Observation
1	The oil-covered dirt does not mix with the water.	dirt particle; water; fabric
2	The added detergent molecules surround the dirt particles. The hydrophobic tail (non-polar end) attaches to the film of oil and the hydrophilic head (polar end) attaches to the water.	detergent molecules
3	The detergent molecules attach to the dirt particle on the surface of the spot to break the dirt into smaller particles. Agitation (shaking) causes the dirt to loosen from the fabric because the polar ends of the soap and water molecules attract.	dirt surrounded by detergent molecules
4	The water solution contains the oily dirt and detergent, until it is removed in the washing cycle.	detergent molecules; dirt particles

FIGURE 2.19 Based on this model, explain how soap removes grease from your hands.

Surf the Web

Besides the saponification of fats and oils to make soap, there is another soap making process. It involves the neutralization of fatty acids with a base. Find out how this process works. Start your search by visiting www.science.nelson.com and follow the links for ScienceWise Grade 12, Chapter 2, Section 2.3. Create a table comparing and contrasting the two methods of soap making.

Water temperature is a factor when washing clothes. It is important to learn how to read fabric care labels. Your special-occasion sweater cannot be treated in the same way as T-shirts or socks. Hot water is recommended for whites and heavily soiled clothes because dirt comes out better with hot water. Cold water is recommended for dark colours and dyes that run.

Soap Lather and Water Hardness

Lather is the suds made from soap and water. It is a result of air and water mixing with the soap. But not all water has the same result. Do you get the same amount of soap lather if you take a bath in the Muskoka area or in the Niagara region? It all depends on the type of rock found in the area and the length of time that the water is in contact with the rocks.

The Rocky Mountains, the Canadian Shield, and most of the Maritimes have very insoluble bedrock. This means the water in these regions contains few or no dissolved minerals/ions. This type of water is called **soft water**. The rocks of the Niagara Peninsula and the Prairies are more soluble. They are usually limestone (calcium carbonate) and the result is water containing dissolved minerals/ions such as calcium, magnesium, and iron. This type of water is called **hard water**. The amount of soap lather depends on the ions found in the water. The harder the water, the less the soap will lather.

A)

B)

FIGURE 2.20 A) Scale deposit in a humidifier. B) Soap film or scum.

There are two types of water hardness. Hardness caused by hydrogen carbonates is called temporary hardness and can be removed by boiling the water. Kettles that have been used to boil hard water produce a deposit of calcium carbonate known as **scale**. Scale buildup is also found inside pipes, hot water heaters, humidifiers (Figure 2.20A), faucets, and shower heads. To remove scale, you can use a solution of vinegar and water. Hardness caused by sulfate and chloride salts is called permanent hardness because it cannot be removed by boiling the water. To remove permanent hardness, you need to use a water softener such as washing soda (sodium carbonate) or an ion exchange column (Figure 2.21A–B). An ion exchange column is a glass tube filled with a usually solid material called a resin. Calcium (or magnesium) ions are exchanged for sodium ions as water flows through the column.

Which Is Better in Hard Water: Soap or Detergent?

When soap is used in hard water, insoluble salts of calcium and magnesium form. They appear as a soap film or sticky scum (Figure 2.20B). This scum leaves deposits on clothes, skin, and hair. The scum will not form in detergents, because the calcium and magnesium salts of the detergent are soluble in hard water. Even though both detergents and soaps have similar cleansing action, detergents are made using alcohols instead of fat molecules. Detergents, in products such as shampoos and toothpastes, have replaced most soap products. Which is the better emulsifying agent, soap or detergent? Why?

A)

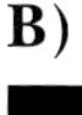

B)

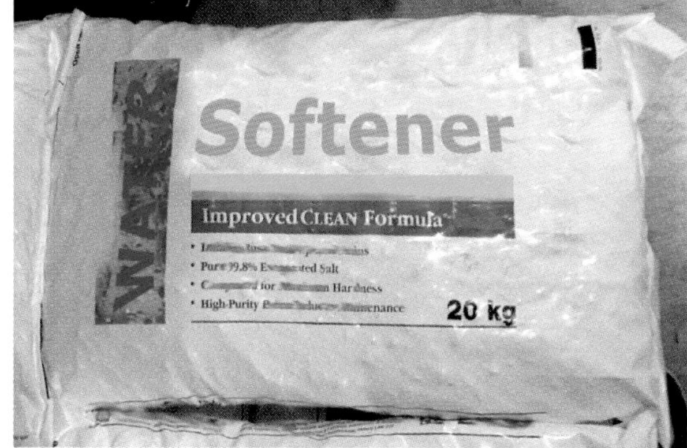

FIGURE 2.21 A) An ion exchange tank. B) Water softener.

Review and Apply

1. Make a chart comparing the properties of soaps and detergents, including reactants used and a description of the reaction with hard water.

2. In what way does soap behave as an emulsion?

3. Early settlers in Canada often made their own soap using animal fats and lye. What type of chemical would you expect lye to be? Explain your reasoning.

4. Doreen is changing the oil in the lawn mower and her hands get very dirty with grease. She rinses them off with water. Will the grease be removed with only water? If not, what should she use? Draw a cartoon strip showing how to remove the grease from her hands.

5. What is scale build-up? What can be done to remove it?

6. You want to start your own detergent factory. Which areas of your province would benefit most from using a water softener? Explain why.

7. Add the new concepts you learned in this section to the graphic organizer you started in Section 2.1.

8. **Look After Our Waters**

   Some cleaning products are toxic, are non-biodegradable (not broken down by bacteria or other micro-organisms), or contain harmful chemicals. Other cleaners are phosphate-based and cause problems for marine organisms. They may contribute to unwanted algae growth and decreased dissolved oxygen levels in water, and even kill fish.

   People have a responsibility to help keep the water clean and protect marine organisms. They should clean with processes and products that have little or no environmental impact on water quality.

Before cleaning, consider the following:

- Use all cleaners sparingly—minimize the use of soaps and detergents by washing more frequently with plain water.
- Do not use cleaners that contain ammonia, sodium, chlorinated solvents, petroleum distillates, or lye.
- Substitute natural cleansers such as vinegar, citric juices, borax, and baking soda for chemical ones.
- Look for the "Ecologo" symbol (Figure 2.22). Use only non-toxic, phosphate-free, biodegradable (able to be broken down by bacteria or other micro-organisms) cleaners.
- Do not use high-pressure washers if paint particles can be washed into bodies of water.

a) Do you know of someone who is not following the rules? What would you tell them?

b) What can you do to prevent pollution in the waters at home or in the workplace?

c) Create a poster, web page, or video skit explaining the importance to the environment of using biodegradable products.

FIGURE 2.22

Are there any health or safety concerns with your product? You will need to find out this information for your project, Making a Consumer Product, at the end of Unit 1.

FIGURE 2.23

LAB 2B

Making Soap

Now that you understand how soap works, it is time to make some! The soap that you make in the activity will be used in Lab 2C.

Purpose

To make a common product, soap.

Be Safe!

Sodium hydroxide is corrosive. Handle with care. If you spill any on yourself, rinse it off immediately with large amounts of cold water. Inform your teacher of any spills.

Materials

- 5 g lard
- 5 g sodium chloride
- two 250-mL beakers
- one 500-mL beaker
- three ring clamps
- Bunsen burner
- stirring rod
- tongs
- lab coat
- paper towels
- 20 mL sodium hydroxide
- graduated cylinder
- support stand
- wire gauze
- boiling chips
- 15 mL cool tap water
- safety goggles
- funnel
- filter paper

Procedure

1. Put on your safety goggles and lab coat.
2. Add 5 g of lard and 20 mL of sodium hydroxide solution to a 250-mL beaker. Stir well and let stand.
3. Set up the apparatus as shown in Figure 2.24.
4. Add 100 mL of water and 4–5 boiling chips to a 500-mL beaker and bring it to a boil over a Bunsen burner.

Be Safe!

Use care around boiling water.

FIGURE 2.24

5. Place the 250-mL beaker from step 2 into the boiling water bath and heat for 10 minutes. Stir the mixture during heating.

6. Add 5 g of sodium chloride and stir well. Continue heating and stirring for another 3 minutes.

7. Turn off the Bunsen burner. Use tongs to carefully remove the beaker from the steam bath. Allow to cool for a few minutes on the wire gauze.

8. Use beaker tongs to hold the beaker. Filter the contents of the beaker using a funnel and filter paper. Wash the soap by pouring 15 mL of cool tap water over the soap and funnel. See Figure 2.25.

9. Dry the soap by pressing it between paper towels.

10. Keep your soap product for Lab 2C, where you will test its lather and emulsion abilities.

FIGURE 2.25

Be Safe!

DO NOT use the soap on your skin. It is not safe.

Analysis and Conclusion

1. List two sources of fat, aside from lard, that could be used to make the soap.

2. What was the purpose of
   a) adding the salt?
   b) washing the soap?

Extension and Connection

3. What could you do to your soap to make it more attractive to a consumer?

4. Use another source of fat and repeat the experiment. Create a data table comparing your soaps.

LAB 2C

Emulsions of Hard and Soft Water

Do you know whether the tap water used in your home is soft or hard? How could you test it to find out? Why do you use detergents for many household cleaning jobs? These are some of the questions you will be able to answer after completing this experiment.

Purpose

To investigate the chemical action of soap and detergent in hard and soft water and to determine the quantities of soap and detergent required to form emulsions.

Prediction

Predict how soap and detergent will react in hard and soft water.

Be Safe!

Avoid skin contact with solids and solutions. Dispose of all solutions in the containers provided by your teacher.

Materials

- 100 mL distilled (soft) water
- 2 g magnesium sulfate
- 3 g of commercial soap (bar soap)
- 3 g of commercial detergent
- 3 g of soap made in Lab 2B
- 5 mL commercial hair shampoo
- 2 g sodium carbonate
- 30 mL cooking oil
- safety goggles
- two 250-mL beakers
- stirring rod
- five test tubes
- test tube rack
- rubber stoppers
- 10-mL graduated cylinder
- hot plate
- scoopula
- filter, filter paper, stand
- ring clamp
- lab coat
- electronic balance or scale
- ruler

Procedure

Part One: Testing Soaps and Detergents with Water

1. Create a data table like that in Figure 2.26 to record your observations.

	Test tube 1 (distilled water)		Test tube 2 (tap water)		Test tube 3 (hard water)	
	Height of suds (cm)	Diagram of test tube 1	Height of suds (cm)	Diagram of test tube 2	Height of suds (cm)	Diagram of test tube 3
Commercial soap						
Commercial detergent						
Commercial hair shampoo						
Soap made in Lab 2B						

FIGURE 2.26

2. Add 100 mL of distilled water to a 250-mL beaker. Then add 2 g of magnesium sulfate (Figure 2.27) and stir. Label this beaker "Hard Water." You now have prepared a stock solution of hard water. You will use the hard water for all three parts of this experiment.

3. Arrange three test tubes in a test tube rack and label them "Distilled Water," "Tap Water," and "Hard Water."

4. Half fill the first test tube with distilled water. Half fill the second test tube with tap water. Half fill the third test tube with the hard water you made in step 2.

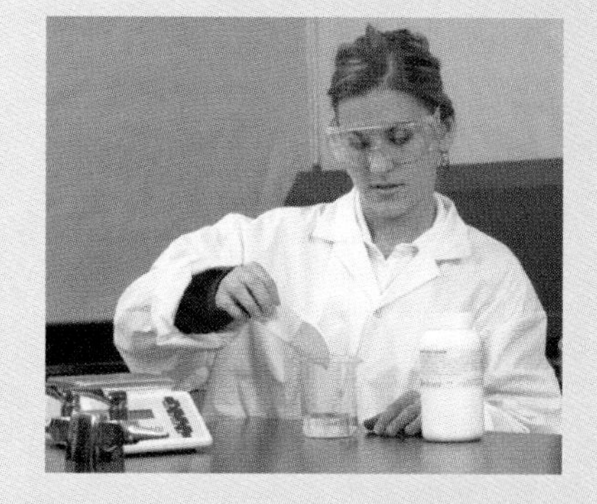

FIGURE 2.27 Remember to use an electronic balance or scale when measuring.

5. Using a metal scoopula, add 1 g from the bar of commercial hand soap to each of the three test tubes. Stopper and shake the first test tube 10 times. Record your observations. Caution: After shaking several times, stop, carefully release the pressure by removing the stopper, re-stopper, and shake.

6. Repeat step 5 for the other two test tubes. Shake each test tube in an identical manner.

7. Dispose of the substances and rinse the test tubes.

8. Repeat steps 4–7 using 1-g samples of a commercial detergent.

CONTINUED

9. Repeat steps 4–7 using 1 mL of a commercial hair shampoo

10. Repeat steps 4–7 using a 1-g sample of the soap you made in Lab 2B.

Part Two: Hard and Soft Water Emulsifying Tests

1. Create two data tables with headings like that in Figure 2.28. Label one table "Hard Water Emulsifying Test" and label the other table "Soft Water Emulsifying Test."

	Observations	Diagram of test tube
Test tube 1		
Test tube 2 Commercial soap		
Test tube 3 Commercial detergent		
Test tube 4 Commercial hair shampoo		
Test tube 5 Soap made in Lab 2B		

FIGURE 2.28

2. Place five test tubes in a test tube rack. Half fill each with the hard water you made in Part One, step 2.

3. Add 3 mL of cooking oil to each test tube.

4. Add nothing to test tube 1. Add a 1-g sample of hand soap to test tube 2; a 1-g sample of detergent to test tube 3; 1 mL of shampoo to test tube 4; and a 1-g sample of soap you made in Lab 2B to test tube 5.

5. Stopper and shake all tubes, one at a time, in an identical manner.

6. Record your observations, then dispose of the substances and wash out the test tubes.

7. Repeat steps 2–6 for the soft (distilled) water.

Part Three: Softening Hard Water

1. Add 2 g of sodium carbonate to the remaining hard water solution in the 250-mL beaker from Part One, step 2.
2. Place the beaker and contents on a hot plate. Heat and stir for 5 minutes. See Figure 2.29.
3. Remove the beaker and allow it to cool.
4. Meanwhile, set up the filtration apparatus as in Figure 2.25 from Lab 2B.
5. When the beaker is cool enough to handle, pour the solution through the funnel to separate the solid (precipitate) from the liquid (filtrate).
6. Half fill a clean test tube with the clear filtrate. Add 1 g of commercial soap to the test tube. Place the stopper on the tube and shake it.
7. Record your observations in a data table with headings like that in Figure 2.28.
8. Dispose of all your materials as directed by your teacher and clean your work area.

FIGURE 2.29

Analysis and Conclusion

1. Which ion is responsible for creating hard water in this lab? Support your answer with observations you recorded throughout the lab.
2. Based on your data, does soap work better in soft or hard water? Use examples from the lab to support your answer.
3. Which works better in hard water—soap or detergent? Would you conclude that shampoo is more like soap or detergent? Explain your answer.

CONTINUED

4. How would you classify your tap water—as hard or soft water? Why?

5. What effect did the addition of sodium carbonate have on the hardness of the water? Explain your answer.

6. Which of the cleansing agents worked best to emulsify the oil in the hard water? Justify your answer.

Extension and Connection

7. Call the 800 telephone numbers provided on some of the commercial products, or write to a few commercial manufacturers of detergent, seeking technical information on the production of the product. Give an oral presentation to the class on what you found out.

8. Collect advertising claims of various cleaning agents. Design experiments to check these claims.

2.4 Chapter Summary

Now You Can...

- Explain the behaviour of emulsifying agents (2.1, Lab 2A, 2.3)
- Identify and collect information on science and technology related careers (2.1)
- Write word equations for condensation and hydrolysis reactions (2.2)
- Explain how the hydrophobic, hydrophilic, or amphiphilic character of organic molecules is related to the presence of O, N, or ions in the molecules (2.3)
- Use lab equipment safely and accurately to investigate emulsions (Lab 2A)
- Research and report on an application of condensation, hydrolysis, or emulsification processes (2.2)
- Safely prepare a common organic product (Lab 2B)
- Investigate and compare the relative quantities of soap and detergent required to form emulsions in hard and soft water (Lab 2C)
- Communicate ideas and results of laboratory experiments (Labs 2A, 2B, 2C)
- Demonstrate an understanding of safe laboratory practices (Labs 2A, 2B, 2C)
- Use data tables to present information (Labs 2A, 2B, 2C)
- Select and use appropriate vocabulary (2.1, 2.2, 2.3, Labs 2A, 2B, 2C)

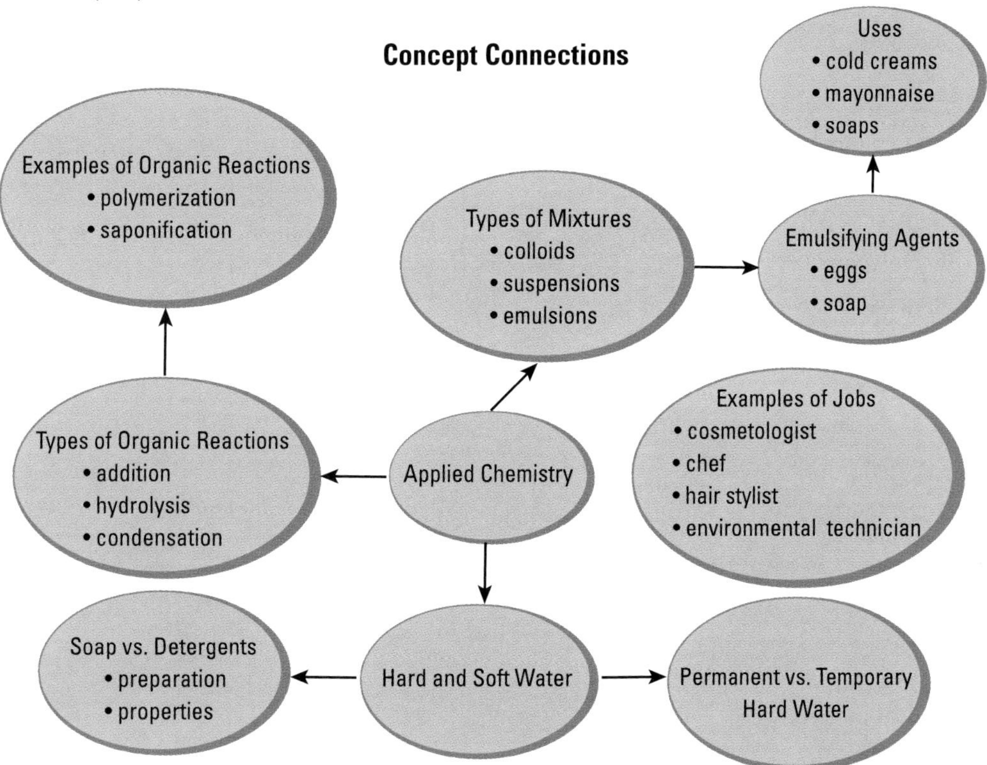

FIGURE 2.30 Compare your completed graphic organizer to the one on this page. How did you do? Can you add any new links to your organizer?

CHAPTER 2 *review*

Knowledge and Understanding

1. For each of the following statements indicate True or False. If False, rewrite the sentence to make it true.
   - **a)** Mayonnaise is a solution.
   - **b)** Emulsifying agents do not allow solutions to mix.
   - **c)** A hydrolysis reaction adds water to a molecule, splitting it in two.
   - **d)** Saponification is the process of adding fats with an acid to make soap.
   - **e)** Soap molecules are only polar.
   - **f)** Hard water contains deposits of calcium and magnesium.
   - **g)** Detergents can be used to help clean up grease spills because they attack the oils.

2. Classify each of the following as a colloid, suspension, or emulsion:
   - **a)** Some bits of solid settle out when the mixture is allowed to stand.
   - **b)** A salad dressing displays the instructions "Shake well before using."
   - **c)** Hand lotion (Figure 2.31).
   - **d)** Tiny bits of solid and water droplets mix.

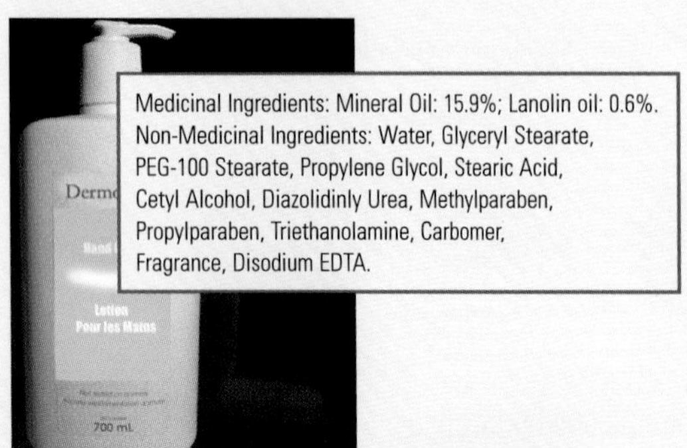

FIGURE 2.31

3. Find three products in your kitchen at home that contain an emulsifying agent such as egg yolk. Which ingredients of the product do you think each emulsifying agent works on?

4. Draw and label a soap molecule. Explain the difference between the hydrophobic and hydrophilic parts.

5. Create a data table giving simple word equations and uses in the workplace, home, or school for the following organic reactions: addition, hydrolysis, and condensation.

6. Use a series of diagrams to explain how agitation of a washing machine helps to lift grease or dirt from fabric.

7. Create a poster or collage that shows everyday examples of emulsions, colloids, and suspensions.

8. Explain the difference between temporary hard water and permanent hard water. How can each type of hard water be softened?

Inquiry

9. Lucas is given a mixture of oil and water. What should he do to make them combine in solution? What should he do to make them form an emulsion?

10. How would you test the tap water in your home to decide whether you would benefit from a water softener?

Making Connections

11. A cosmetologist is a beauty service technician involved in the cosmetic treatment of hair, skin, and nails (Figure 2.32) Name four products that are used by a cosmetologist and identify each as a suspension, colloid, or emulsion. Give reasons for your choices.

12. Enzymes work by breaking the attractive forces that hold a large food molecule together. To do this, enzymes require the presence of water. What type of organic reaction is this? Why would our bodies need enzymes?

13. Pamela spilled maple syrup on her hands, and she easily rinsed it off with water. Patricia spilled butter on her hands, but she could not easily wash it off with water. She needed to use soap. Briefly explain the difference.

14. Gelatin is a common emulsifying agent found in desserts. What other products in the kitchen use gelatin as an emulsifying agent?

Communication

15. Write a letter to the editor of a newspaper, make a web page, or create a poster to explain how water softeners remove the metal ions from water. Explain any health risks related to water softeners.

16. Explain the difference between emulsion-forming and emulsion-breaking agents such as soap, salt, and eggs.

17. You have a summer job teaching sailing at a camp. You are responsible for taking care of the boats when they are not on the water. The director of the waterfront wants you to help her purchase cleaning supplies for the season. What types of cleaners would you recommend? Why?

FIGURE 2.32 A cosmetologist at work

PUTTING IT ALL TOGETHER

Making a Consumer Product

Society has become dependent on chemicals and chemical processes. What would life be like without plastics, glass, medicines, or soap? These products and thousands of others have been developed and produced by industries through many chemistry-based experiments. These industries research constantly in hopes of improving existing products or discovering something new. In this activity you will design, create, and market a new product.

The Project

In groups you will research one of four consumer products and safely prepare it using the processes learned in Chapters 1 and 2. The product may be cold cream, lip gloss, toothpaste, or hair gel. Your job is not only to make the product, but also to ensure that it is environmentally friendly. If your product is fully biodegradable, then you have the best product.

What You May Need

- all materials to safely prepare your consumer product
- access to the Internet and/or library resources
- coloured pens, markers, and/or pencils
- any material needed to package your product

What You Will Do

1. Using classroom and electronic resources, research your chosen chemical product, then design an experiment to make the product.
2. Have your teacher approve your experimental design before you begin.
3. Prepare your product. Record your observations on producing your product. Determine how the method could be improved to prepare a better product.

4. Prepare the product a second time but change the aspects of the production necessary to improve it. Make sure that you have defined beforehand which aspect you will change and explain what you will do to change it.

5. Record your observations after the second trial.

6. Make notes on the following: the raw materials required to make your product, the waste products produced, disposal of waste, and packaging.

7. Develop a name, label, and concept for your product. Labels should include all components of the product listed in order by mass of the component in the product. You may want to look at existing labels of products for ideas.

8. Prepare a product evaluation sheet giving a ranking scale for the various qualities of the product, such as fragrance, texture, colour, "feel," quality, pH, appeal of label, and image of product. (DO NOT taste your product or use it on your skin, teeth, or eyes.)

9. Design an information booklet about your experiment that includes the following:
   - *Background information:* Include the type of reaction, simple word equations, and relevance of your product to society
   - *Materials:* List all materials used and include a diagram of your set-up, if applicable
   - *Procedure:* Numerically list the steps you followed. Be sure to include all safety precautions you took.
   - *Results:* Present all results in the form of data tables, graphs, photographs, or drawings.
   - *Discussion:* Determine, to the best of your ability, the environmentally unfriendly aspects of your product, and describe how you might eliminate them. What health or safety issues are related to your product? How expensive is your product to make? Does this affect the price you would charge for it?

10. Prepare a commercial for a magazine, for television, or for radio that uses the information in your booklet to advertise your product.

CONTINUED

Assessment

1. Present your commercial to the class. Make notes during other groups' presentations.

2. Compare your commercial to those of other groups in your class. What do you think they did well? Did their commercial make you want to buy the product? Explain your answer. What do you like best about your commercial and product? If you were to do this activity again, what changes would you make to your commercial or your product? Why?

Communications– Sounds and Pictures

UNIT 2

CHAPTER 3

Sound Production

From a whisper, to the sound of a bird's wings flapping, to traffic on a busy street, to an airplane taking off, sound is everywhere. The manipulation or production of sound is based on vibrations and waves. Look at Figure 3.1. Why do you think various musical instruments sound different even when they play the same note? Is sound production different in a radio or stereo than it is in a musical instrument? Why or why not? In this chapter, you will learn how sound is produced.

FIGURE 3.1

What You Will Learn

After completing this chapter, you will be able to:

- Describe the properties of a vibrating object and how vibrating objects produce waves (3.1)
- Describe and compare the properties of transverse and longitudinal waves (3.1)
- Explain how frequency, amplitude, and wave shape affect the pitch, intensity, and quality of notes produced by musical instruments (3.2)
- Explain how sound pollution has affected the way we live and work (Case Study)
- Describe what happens when waves interact with one another (3.3, Lab 3C)
- Describe some contributions Canadians have made to the field of communication technology (3.3)
- Identify and collect information on science- and technology-based careers (3.3)

What You Will Do

- Carry out investigations concerning the scientific concepts involved in communications technology (Labs 3A, 3C)
- Estimate the value of some wave-related quantities (Lab 3A)
- Determine the relationships among the major variables for a vibrating object (Lab 3A)
- Conduct an investigation to analyze and explain the production of sound by a vibrating object (Activity 3B)
- Formulate scientific questions about waves (Activity 3B, Lab 3C)
- Select appropriate instruments and use them correctly to collect observations and data (Lab 3A, Activity 3B, Lab 3C)
- Compile, organize, and interpret data (Lab 3A, Activity 3B, Lab 3C)
- Use appropriate word equations to communicate experimental results (Lab 3A)

Words to Know

amplitude
beat
colour
compression
constructive interference
crest
cycle
decibels (dB)
destructive interference
frequency
hertz (Hz)
in phase
interference
longitudinal wave
medium
oscilloscope
out of phase
overtones
period
pitch
rarefaction
sound
timbre
transverse wave
trough
vibrating
wave
wavelength

A puzzle piece indicates knowledge or a skill that you will need for your project, Courier Communication, at the end of Unit 2.

3.1 Sound Wave Production

Most forms of communication involve sound. People have found many ways to transmit sound from one place to another or to vary its properties, or characteristics. We can make it loud, soft, high-pitched, intense, pleasant, or alarming. In this section, we will look at what sound is and what causes its characteristics to change.

Waves

What is sound? How are sounds and waves related? **Sound** is a wave that can be detected by the ear. Waves transmit energy from one place to another, without matter actually moving from where the wave starts to where it ends up. A **wave** is a moving disturbance. There are two main types of waves: transverse and longitudinal.

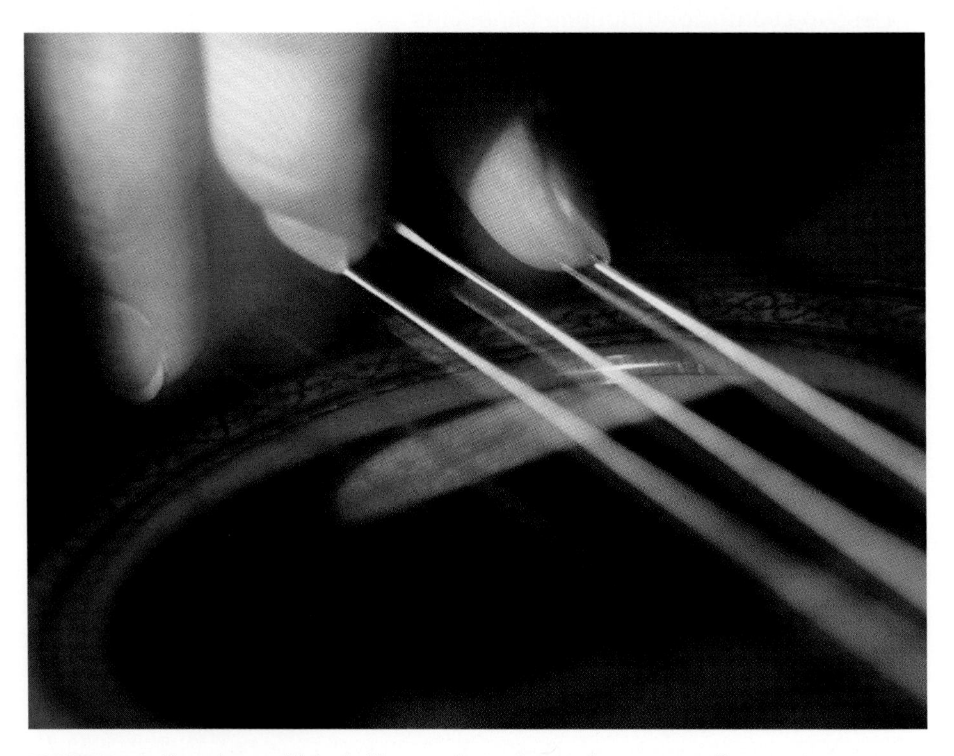

FIGURE 3.2 How do you think plucking a guitar string produces a sound?

Transverse waves move at right angles to the movement of the particles (the substance that is moving). You can make a transverse wave with a coil spring by moving your hand up and down. The wave will travel from your hand to the end of the spring. The spring itself is moving up and down. Give it a try. Refer to Figure 3.3. You can also create a transverse wave by moving the spring from side to side.

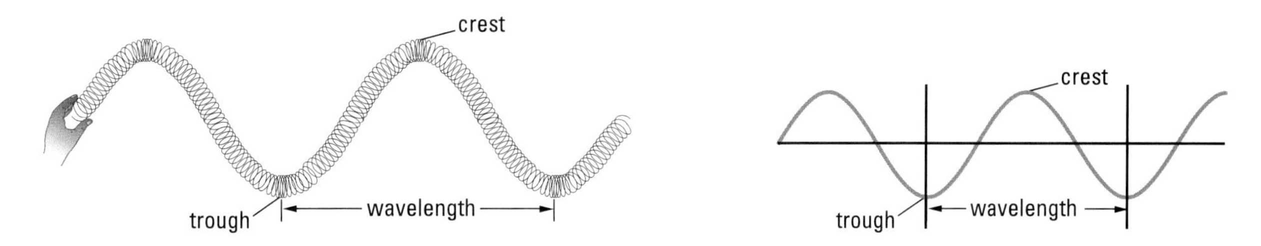

FIGURE 3.3 A transverse wave. Which way are the particles moving? Which way is the wave moving?

In a transverse wave, the high point is called the **crest** and the low point, the **trough**. The distance between successive crests or successive troughs is the **wavelength**. Light is a transverse wave.

Longitudinal waves move parallel to the movement of the particles. You can create longitudinal waves using coil springs (Figure 3.4). Push the spring forward, then pull it back. Keep repeating this action. The disturbance will travel down the spring. Sound is a longitudinal wave.

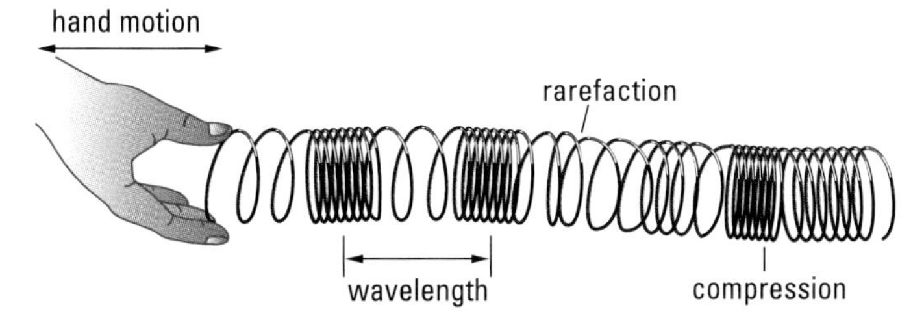

FIGURE 3.4 A longitudinal wave. Which way are the particles moving? Which way is the wave moving?

In a longitudinal wave, the places where the particles are close together are called **compressions**. The places where the particles are stretched apart are called **rarefactions**. The distance between successive rarefactions or successive compressions is the wavelength.

FIGURE 3.5 What is the pendulum's function in this clock?

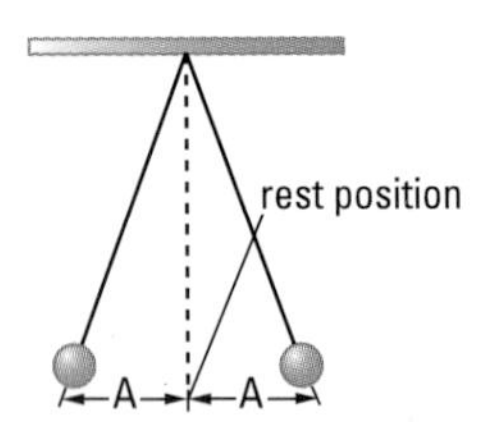

FIGURE 3.6 A simple pendulum. A is amplitude.

What Causes Waves?

A vibrating object produces waves. In science, any object that moves back and forth is called **vibrating**. By this definition, a pendulum in a clock vibrates, even though, in casual conversation, we call it "swinging." See Figure 3.5. Usually, the vibration that produces a wave is so quick that we cannot see it. A simple pendulum vibrates slowly, so it can be used to explain some of the terms we use for waves.

The horizontal distance between the rest position and an extreme is the **amplitude** (Figure 3.6). This is true in both longitudinal and transverse waves.

Two other important terms are frequency and period. The **frequency** of a wave is the number of waves that pass by a certain point every second. Frequency is measured in **hertz (Hz)**. The **period** is the amount of time it takes for each complete oscillation, or **cycle**. In Figure 3.4, a complete back and forth vibration of the spring is one cycle. The cycle is the complete sequence of motion that repeats itself. The period is measured in seconds. Figure 3.7 puts this information together.

Properties of Transverse and Longitudinal Waves

	Type of Wave	
	Transverse	**Longitudinal**
Wave and movement of particles	Right angles to each other	Parallel to each other
Wavelength	Distance between successive crests or successive troughs	Distance between successive rarefactions or successive compressions
Amplitude	Distance between rest position and maximum crest or trough	Distance between rest position and maximum position to or fro
Frequency	Number of crests or troughs that pass a certain point every second	Number of rarefactions or compressions that pass a certain point every second
Period	Amount of time between one crest to the next crest or one trough to the next trough	Amount of time between one compression to the next compression or one rarefaction to the next rarefaction

FIGURE 3.7

Producing Sound

So now you know what the types of waves are, and that sound is a wave. But how does a sound wave produce sound? To produce sound, you need two things: a medium, and a vibrating object.

The **medium** is the substance that a wave travels through. Air is a common medium for sound.

Once you have a medium, all that is necessary to create the sound is a vibrating object. Will swinging a pendulum make a sound? Certainly not one that humans can hear. We need something that vibrates a lot faster than that—something with a higher frequency. How about a stretched elastic band? It will vibrate very quickly if you pluck it.

In the case of the elastic band, the waves are longitudinal. The disturbance happens when the vibrating elastic band pushes air molecules together into a compression, then moves backward creating a rarefaction. The successive compressions and rarefactions move through the air as a sound wave toward the ear. Your brain then turns it into a sound. We will discuss the transmission of sound waves in more detail in Chapter 4.

You will need to know how sound is produced for your project, Courier Communication, at the end of Unit 2.

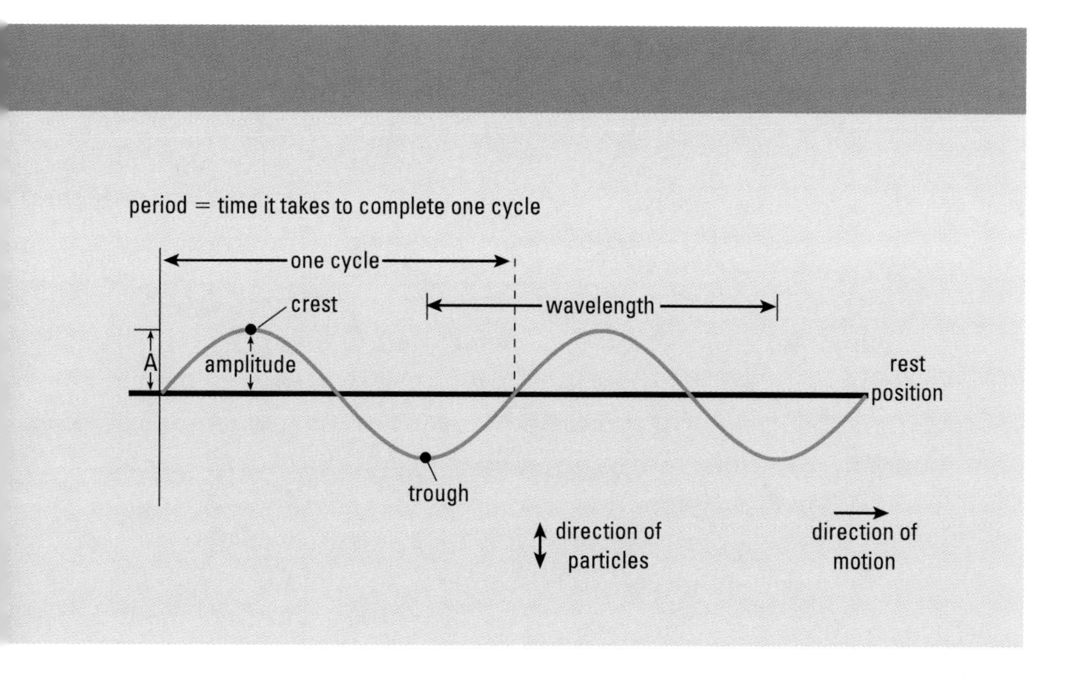

Review and Apply

1. Name or draw five objects you use every day that are specifically designed to produce sounds.

2. To produce sound, you must have a vibrating object. Indicate what is vibrating when each of the following makes a sound:
   a) a guitar
   b) a bell
   c) a squeaky door
   d) your voice

3. Meera works in a pet store. She has been instructed to tell customers not to tap the glass on the aquariums because it bothers the fish. Fish do not have ears. Use your new vocabulary to explain why a tap on glass would bother a fish.

4. Renardo is making a soufflé and insists everyone be very quiet. He claims that if we are too noisy, the puffed-up top of the soufflé will collapse. Explain in your own words how sound might disturb the soufflé top.

5. A stretched elastic band makes a sound when plucked. What could you do to the elastic band to change the sound produced? Try to think of two things you could do.

6. In a graphic organizer, arrange the concepts you have learned about in this section.

7. **Creating a Science Show**

   In a group, develop a segment for a kids' science television program. It should explain the difference between transverse and longitudinal waves. Before you begin, decide on the age of your audience. If you have access to a video camera, you can tape the program. Or, you can create a storyboard showing what each scene of the television program is to look like. A storyboard is made up of small drawings that show what the television camera is to film. It can also include the dialogue the characters are to say. Your segment should be no longer than 10 minutes.

LAB
3A

The Pendulum

A pendulum is a vibrating object. Since sound requires a vibrating object, it makes sense to study one and determine what variables affect its frequency. In Activity 3B, we will use this knowledge to vary the sounds produced by making a musical instrument.

Be Safe!

Properly fasten or hold down the retort stand so that it will not move.

Be careful when swinging the masses not to drop them or to hit anyone nearby.

Purpose

To change different variables of a simple pendulum and analyze the results.

Materials

- 1 m of string
- stopwatch
- retort stand
- utility clamp
- metre stick
- 50-g mass
- 100-g mass

FIGURE 3.8 The set-up for your experiment

CONTINUED

Procedure

1. In this lab, you will perform six trials. Each trial will measure a different variable. Create a table like Figure 3.9 to record your data and calculations.

Trial	Length of string (cm)	Amplitude (cm)	Mass (g)	Number of cycles in 30 s	Frequency (Hz)	Period (s)
1	50	20	50			
2						
3						
...						

FIGURE 3.9

2. Make a loop in one end of the string so that the weights can be changed easily.

FIGURE 3.10

3. Attach the other end of the string to the clamp so that the string is 50 cm long. Keep in mind that you will have to remove and reattach this string several times during the experiment.
4. Attach a 50-g mass to the end of the 50-cm long string.
5. Fasten or hold down the retort stand to prevent it from moving.
6. Pull the weight 20 cm back from the rest position. See Figure 3.10. Start the stopwatch as you let go of the mass.
7. Count how many cycles occur within 30 seconds. Record your results in your table.

In steps 8–12, you will keep all variables the same as above except for the one variable that is changing.

Varying the Amplitude

8. Choose a new amplitude (smaller than 20 cm) and repeat steps 6 and 7.

9. Choose a new amplitude (larger than 20 cm) and repeat steps 6 and 7.

Varying the Length

10. Choose a new length for the string (shorter than 50 cm) and repeat steps 6 and 7.

11. Choose a new length for the string (longer than 50 cm) and repeat steps 6 and 7.

Varying the Mass

12. Change the mass to the 100-g mass. Use an amplitude of 20 cm and a length of 50 cm to repeat steps 6 and 7.

Completing Your Table

13. To calculate the frequency, take the number of cycles and divide it by the 30 s. Your answer is in Hz.

14. To calculate the period, divide one second by the frequency. Your answer is in seconds.

Analysis and Conclusion

1. Does amplitude have any effect on frequency? How do you know? If there is an effect, describe it.

2. Does the length of the string have any effect on frequency? How do you know? If there is an effect, describe it.

3. Does the mass on the end of the string have any effect on frequency? How do you know? If there is an effect, describe it.

4. Write a word equation that expresses the relationship among the variables of the pendulum.

Extension and Connection

5. Make a pendulum that produces a period of one second—that is, every cycle would take one second to complete. To what practical application does this relate?

3.2 Sounds and Music

Almost everyone listens to music. However, not everyone likes the same type of music. People like different kinds of music played on many different musical instruments. What makes an instrument sound the way it does? Why is a flute high-pitched while a tuba is low-pitched? In this section, you will apply your knowledge about sound production from the previous section to explore sound's relationship to music.

Loudness or Intensity of Sound

The loudness of a sound is measured in **decibels (dB)**—the *bel* part of the word named for Alexander Graham *Bell*, the inventor of the telephone. Decibels are a measure of the amount of sound energy entering the ears. How does this relate to the characteristics of a vibrating object? The larger the amplitude of vibration, the more intense the sound.We hear intensity as loudness. Refer to Figure 3.11. Think of it in this way: If you were to hit a large bell using a hammer, you would get an extremely loud sound if you hit it hard compared to hitting it softly.

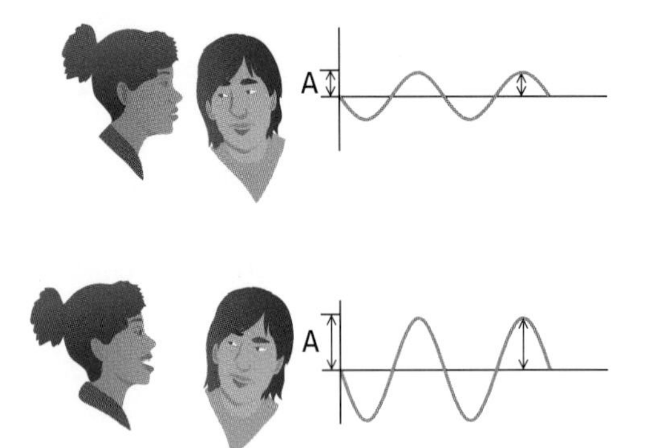

FIGURE 3.11 Whispering produces a smaller amplitude than talking.

Figure 3.13 lists some common sounds and their intensity level in decibels. Sounds higher than 90 dB can cause hearing loss. The degree of hearing loss is a result of both the intensity level of the sound as well as the length of time the person is exposed to the sound. In Section 6.2 you will learn about types of hearing devices for people who suffer from hearing loss.

Depending on the noise, those who are exposed can wear different types of ear protection. Figure 3.12 shows examples of ear protection devices.

FIGURE 3.12 Why is each person wearing ear protection?

Ontario Guidelines for Noise Levels

Ontario occupational exposure limits for workplace noise	Intensity (dB)	Common sounds
	0	The faintest sound that the human ear can hear, threshold of hearing
	20	Average whisper, quiet library
	60	2-person conversation, sewing machine
	70	Busy street traffic
	80	Vacuum cleaner
8 h per day is the maximum exposure.	90	Industrial vacuum cleaner, lawn mower, shop tools, subway trains
4 h per day is the maximum exposure.	100	Chainsaw, circular saw, snowblower, tractor, low-flying jet aircraft
2 h per day is the maximum exposure.	105	Subway platform
	110	Smoke alarm, jackhammer
15 min per day is the maximum exposure.	115	Auto horn, leaf blower
No exposure without ear protection.	120	Loud rock concert, sandblasting
	130	Fireworks; listener will begin to feel pain
	160	Handgun or rifle; a hole is blown in the eardrum

Canadian Centre for Occupational Health and Safety

FIGURE 3.13

Music or Noise?

Noise contains many combinations of frequencies that do not sound pleasant to the ear. What you consider pleasant depends partly on your personality or culture, but there are some combinations that almost everyone finds irritating. These unpleasant combinations are commonly called noise. Musical instruments are designed to produce notes that contain only pleasant combinations of frequencies. Of course, you can make "noise" using a musical instrument, if you do not know how to play it properly!

ScienceWise Fact

Every additional 10 dB means 10 times as much energy, or 10 times more intensity. A circular saw is 100 times louder than a vacuum cleaner.

Pitch

The frequency of a sound wave is called **pitch**. See Figure 3.14. In music, pitch describes how high or low the note is that is produced. Recall that in Lab 3A the amplitude had no effect on the frequency of the pendulum.

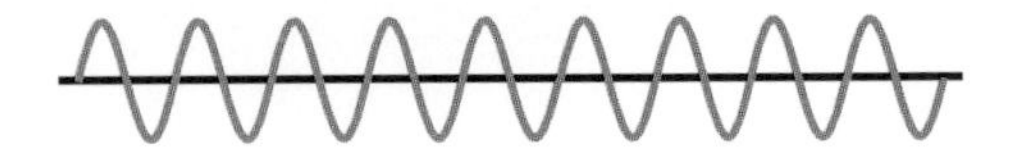

low-frequency sound wave

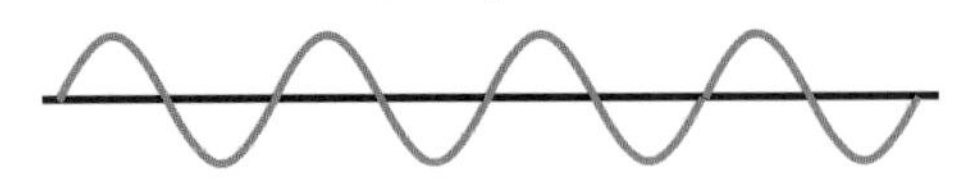

FIGURE 3.14 Which diagram represents a high pitch? A low pitch?

Pitch in a Musical Instrument

In Lab 3A, you found that you could change the frequency of the pendulum by changing the length of the string. This principle is used in musical instruments.

In stringed instruments, we either use strings of different lengths or shorten the strings by placing our fingers on the string (Figure 3.15). Shorter strings have higher frequencies and, as a result, higher pitches.

FIGURE 3.15 Harps have a different string for each pitch. Guitars have only a few strings, and we change the pitch by shortening the string with our fingers.

You could also change the frequency of the string by changing either the density or the diameter of the string. The denser the string, the lower the frequency. Also, the thicker the string, the lower the frequency. Stringed instruments often use denser or thicker strings for lower notes because it would be impractical to use strings that are 2 or 3 m long.

In many wind instruments, we lengthen the air column by closing the holes or by using valves to add an extra length of tube. The longer the air column in the tube, the lower the frequency and the lower the pitch (Figure 3.16).

FIGURE 3.16 Covering the holes or valves makes the tube longer, lowering the pitch. How does the trombone change pitch?

The Timbre of a Musical Note

The **colour** or **timbre** of a musical note relates to the quality of the sound. A guitar has a different timbre than a flute, even if the pitch is the same. The timbre is a function of overtones. **Overtones** are other pitches or tones that overlap the basic, or fundamental, pitch.

There are several ways to change the timbre of an instrument:

- change the material the instrument is made of
- add a sound box
- change what is vibrating (air, string, lips)
- modify the shape of the instrument (long straight tube, curled tube, round sound box, larger or smaller sound box)

Try This at Home

Making Music

Sounds are made by vibrating objects. This activity will help you understand the relationship between frequency and pitch.

What You Will Need

- four identical drinking glasses
- water
- ruler
- metal spoon
- four small pieces of paper

Be Safe!

Be careful when tapping the glasses not to break them.

What You Will Do

1. Write one of the letters C, D, E, and F on each piece of paper. These are the notes you will create using water and your glasses.
2. Place the glasses side by side. Carefully fill the first glass to the top with water. Place the paper with the letter C in front of it.
3. Fill the second glass $\frac{8}{9}$ full. Use your ruler to help you. Place the paper with the letter D in front of it.
4. Fill the third glass $\frac{4}{5}$ full. Use your ruler to help you. Place the paper with the letter E in front of it.
5. Fill your fourth glass $\frac{3}{4}$ full. Use your ruler to help you. Place the paper with the letter F in front of it.
6. During this activity, take time to gently tap the glasses with your metal spoon to make sure you hear differences in pitch.
7. Write a report explaining what happens to the pitch when water is added. Include the following in your report:
   - drawings showing the relationship between the frequency and the pitch
   - what happens when you tap the glass with the spoon

Review and Apply

1. Identify two jobs that require ear protection and explain why it is necessary for each job.
2. If you wanted to make a room quieter, what material should you use? Why?
3. Ben is trying to build a mandolin—a small instrument similar to a banjo. He is concerned because he thinks one of his strings must be 10 m long. What advice do you have for him?
4. Deanna works in a music store. A customer wants to buy a clarinet for his grandfather's birthday. He wants to buy a plastic clarinet because he thinks they sound the same as the wooden ones but are cheaper. What should Deanna tell him? Explain your answer.
5. Add the new concepts from this section to the graphic organizer you started in Section 3.1.

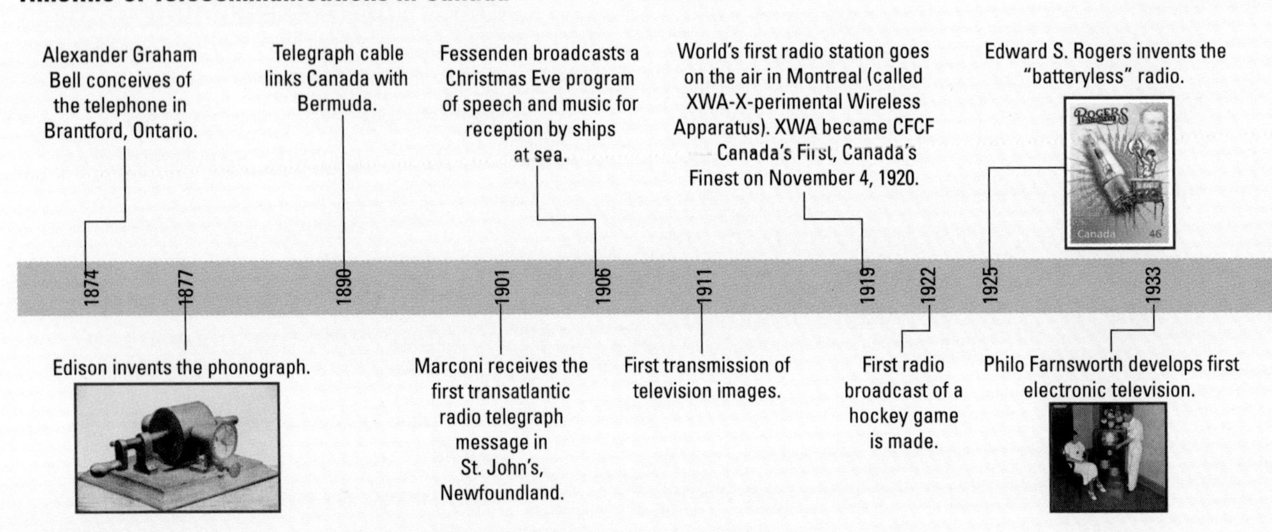

FIGURE 3.17

6 Canadian Inventor: Edward S. Rogers

Edward S. Rogers was born in 1900. By the age of 11, he was already experimenting with radio as one of the first licensed amateur transmitters in Canada. He was the first amateur radio operator in Canada to transmit a signal across the Atlantic.

In 1925, Rogers invented a radio that could be plugged into an AC (alternating current) power source. It was the first radio to operate from household electricity as opposed to batteries. To give people a reason to buy it, he founded Canada's first AC-powered radio station, CFRB (Canadian First Rogers Batteryless).

He was granted the first television licence in Canada in 1931, but died at age 39 before he succeeded in making an impact in the area of television.

Research another Canadian who had an impact on communications technology. Use the library or the Internet. Include the following information:

- how this person made an impact
- what this person invented or made better
- when this occurred, and what else was going on in the country (or world) at that time
- why you chose this person

You may present your findings using a method of communication of your choice (for example, video, voice recording, web page). Be as creative as you like.

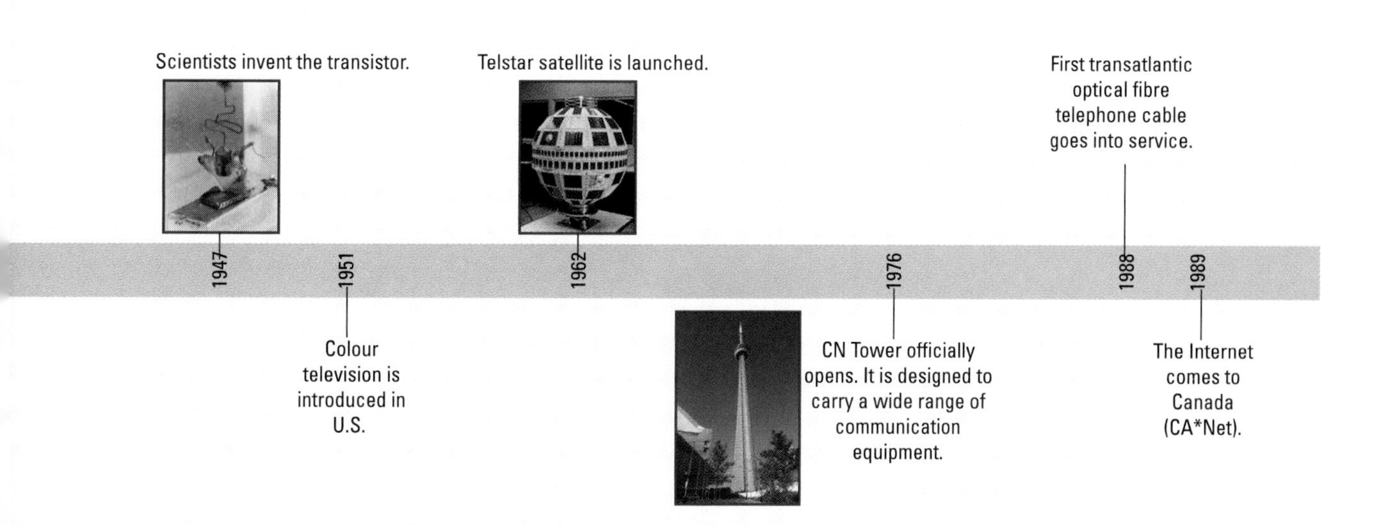

ACTIVITY 3B

Making a Musical Instrument

To design a working musical instrument, you need to know about the principles behind sound production. You now have enough knowledge to design and build a simple box guitar. In this activity, you will find an outline of the basic design, but the details will be up to you.

What You Will Need

- a small- to medium-size box for the sound box
- 1- m piece of wood (about 2 cm thick and 5–10 cm wide) for the neck
- small piece of wood for the bridge
- paint
- two or three screw eyes for tuners
- paintbrushes
- nylon string
- construction paper
- small nails for hitch pins
- tissue paper
- utility knife
- safety goggles
- pencil

Be Safe!

Wear safety goggles when cutting or hammering.

Utility knives are very sharp. Be very careful when cutting.

What You Will Do

1. Decorate the box and neck as desired.
2. Put on your safety goggles when cutting. Measure and cut matching holes through two opposite sides in the box. These holes are for the neck of your guitar. Do not make them too big. The neck needs to fit snugly in the holes.

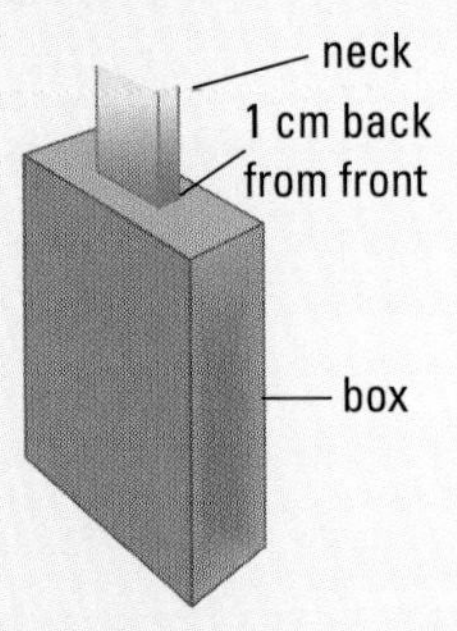

FIGURE 3.18

3. Push the neck through the box so that about 5 cm sticks out from the bottom.

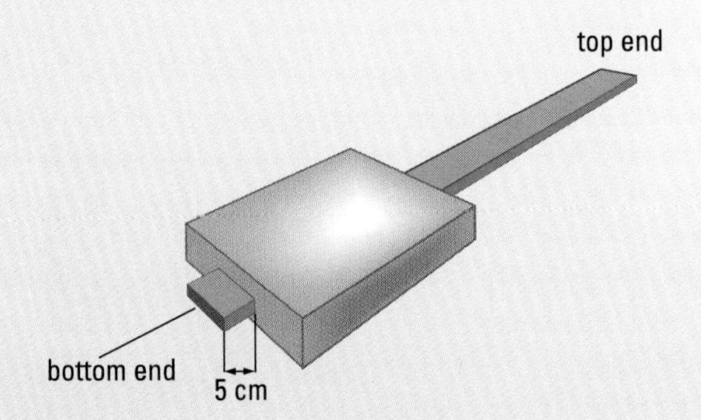

FIGURE 3.19

✓ **Self-Check**

Make sure your guitar looks like Figure 3.19. If not, make any necessary corrections.

4. Decide whether you want two or three strings on your guitar. There needs to be 2 cm between each pair of strings.
5. At each end of the neck, use a pencil to mark where the strings will end. You will want the screw eyes to be at different distances from the nails.
6. Put on your safety goggles. Hammer the nails into the bottom of the neck. The nails will be the "hitch pins" for your strings.
7. Attach a bridge near the end of the neck, just after the nails.

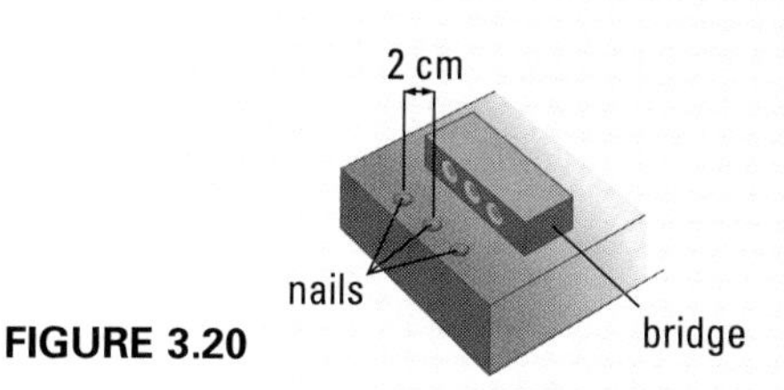

FIGURE 3.20

8. Screw the screw eyes into the top of the neck on a diagonal line.

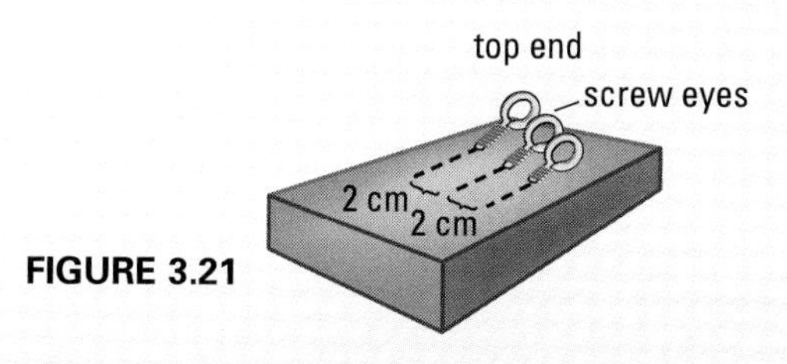

FIGURE 3.21

9. Attach a nylon string from each nail to the corresponding screw eye.

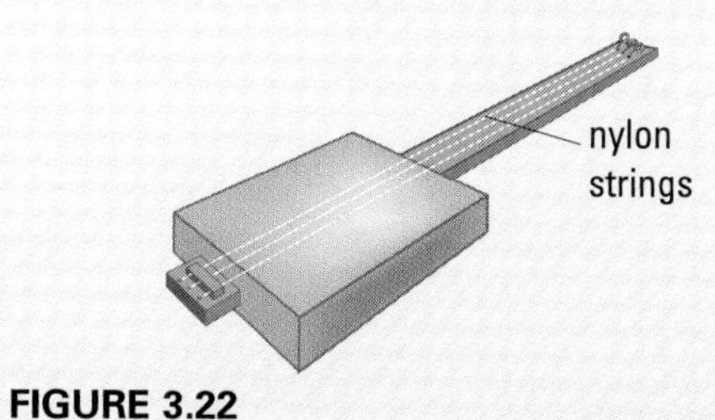

FIGURE 3.22

10. Tune the guitar by turning the screw eyes.
11. Adjust your instrument until you can play "Mary Had a Little Lamb" using at least two strings.

What Did You Find Out?

1. What do you think is the purpose of the sound box? Do you think the shape of the sound box matters? Explain.
2. Stuff your box with a soft material such as tissue paper.
   a) What does that do to the timbre of your instrument?
   b) How does it affect the pitch of the strings?
   c) Explain any differences, using what you know about sound production.
3. How does the sound of your guitar compare to the sound from one of your classmates' guitars? Explain the differences.
4. What do you like best about your guitar? Why?
5. What changes could you make to your guitar to produce a different sound? How could you improve the sound or make it louder? Try two changes and describe what happens for each.

Making Connections

6. Design a home-made wind instrument that children at a day camp might be able to make. Be sure to decide on the age of the children before you start. What other kinds of organizations in your area would be interested in musical instruments?

CASE STUDY

Sound Pollution Around Pearson International Airport

FIGURE 3.23 Jet airplanes can be louder than a rock concert.

Airports are noisy places. A single airplane produces noise with an intensity of 130 dB. In comparison, the average rock concert produces a sound intensity of around 120 dB.

There is no doubt that Pearson International Airport, one of the busiest airports in the world, is important to the local economy. The 28 million people who travelled through the airport in 1998 were responsible for more than $11 billion in revenues for local businesses, personal income, and tax revenue. However, people do live near the airport, and the noise is a very real problem.

a) How do you suggest the nearby municipalities balance the need for comfortable noise levels with the need to earn a living?

The Greater Toronto Airports Authority (GTAA) is responsible for keeping the noise at tolerable levels. It does this in the following ways:

- restricting the operating hours of jet aircraft based on noise certification
- keeping flight paths away from populated areas
- working with surrounding municipalities to reduce the number of homes being built along the flight paths
- working with the community to provide a forum for complaints

FIGURE 3.24 Why do you think new jet airplanes are quieter than older ones?

Noise Certification

Newer planes are quieter. Fewer than 100 older planes visit Pearson International Airport each year. The federal government has adopted rules that require most planes visiting Canada to be the newer, quieter types. Meanwhile, GTAA restricts flights of older aircraft at night-time.

Time-of-Day Restrictions

Though newer planes are quieter, they still have restrictions. Most of these planes are allowed to arrive at or depart from the airport only between 7 A.M. and midnight. Some of the quietest models can arrive at or depart as early as 6 A.M. or as late as 12:30 A.M. Any of these rules may be changed in the event of severe weather, emergencies, air-traffic control problems, or mechanical difficulties.

b) In your opinion, are these rules adequate for the residents living near the airport? Why or why not?

Flight Procedures

The GTAA tries to keep noise to a minimum by keeping the planes as far away from residences as possible. For arrivals, aircraft are required to stay at a certain height until they are lined up with their runway. For departures, they must throttle back—similar to easing up on the gas in a car—once they reach 300 m, so that their engines are not as noisy during takeoff. At night, aircraft use the runways farthest from residences.

Analysis and Communication

The GTAA is always open to suggestions through its Noise Management Committee.

1. Imagine that you live near Pearson International Airport. How could the GTAA improve on these rules? Pretend that you will be attending one of the GTAA's community forums. Choose one GTAA rule and come up with suggestions for improvement. Incorporate your suggestions into a presentation or video.

Making Connections

2. People selling houses that are built close to airports have to warn potential buyers of the noise. When it comes to noise, do you think it is safe to live close to the airport? Why or why not?

3. Research other airports in Ontario. What are the guidelines for airports in Northern Ontario? Are they the same as for Pearson? Explain your answer.

3.3 Interference of Sound Waves

What happens when one wave passes through another? Have you ever leaned over the side of a boat or pier, watching the waves? They seem to break up when they come in contact with an object such as a rock or large structure. Would they do the same if they hit another wave? In this section you will learn what happens when sound waves cross each other's paths as well as some of the practical applications of **interference**, the interaction of two or more waves.

Types of Interference

If two waves are identical in every way, including direction, they are said to be **in phase**. Waves that are not completely identical are said to be **out of phase**. Waves can interfere with each other as they cross paths.

When two crests or rarefactions pass through each other, the new amplitude is the sum of the individual ones. The same is true if two troughs or compressions pass through each other. This type of interference, where the amplitude is increased, is called **constructive interference**. In this case, sound waves get louder. See Figure 3.25.

FIGURE 3.25 Constructive interference

When a crest and trough or a rarefaction and compression pass through each other, the new amplitude is somewhere between that of the crest and the trough or the rarefaction and the compression. In this case, the amplitude will be smaller than either of the original ones. This type of interference is called **destructive interference**. In this case, sound waves get softer. See Figure 3.26. What do you think is meant by "total destructive interference"?

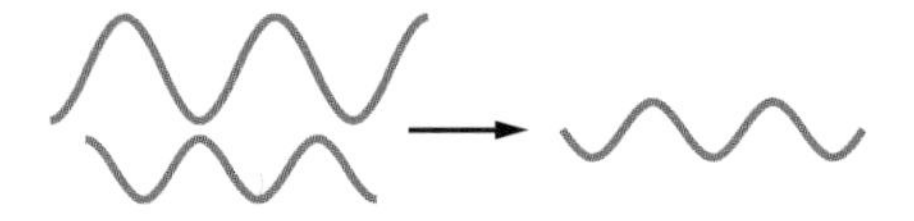

FIGURE 3.26 Destructive interference

An Application of Destructive Interference

People who work in loud environments need to wear earphones to protect their hearing. Active noise reduction (ANR) headsets capture sound from the environment and use computer technology to produce and cycle a second sound wave out of phase. This results in destructive interference inside the headset and helps reduce a worker's exposure to loud noise.

FIGURE 3.27 Pilots use ANR headsets to reduce the noise of the engine. In what other types of jobs would it be useful to wear ANR headsets?

Beats and Interference

Here is something for you to try: Take two crystal glasses. Hit each gently with a spoon and listen carefully (Figure 3.28). Each piece should produce a slightly different tone. When heard together, does the sound seem to warble? Does the sound seem to get louder, then periodically softer? Consider a guitar and a piano. If you play the same note on each, chances are that you will hear the warble again.

FIGURE 3.28 Crystal will produce a tone when tapped. The pitch will depend on the type of crystal and the shape of the piece.

When two sounds have slightly different frequencies, their lines of interference move instead of staying in one place. This is because waves with different frequencies move at different speeds. When a line of constructive interference passes your ear, the sound is louder. When a line of destructive interference passes your ear, the sound is softer. These differences in intensity are called **beats**. Figure 3.29 shows how beats are produced as the sound travels to a listener from two sources of different pitches.

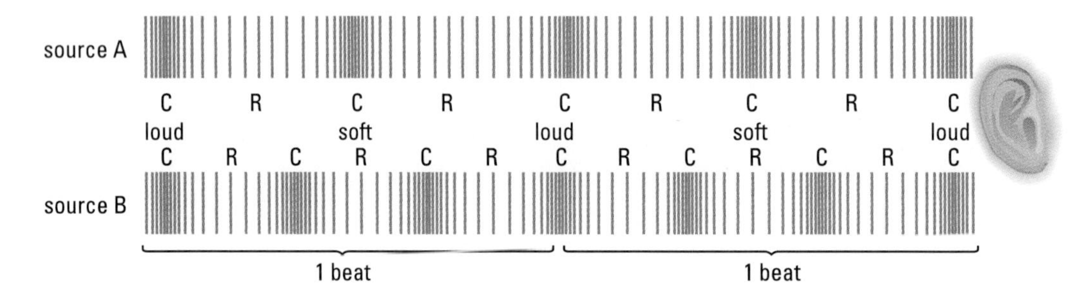

FIGURE 3.29 Production of beats by sources of slightly different frequencies. C is for compression; R is for rarefaction.

FIGURE 3.30 A musician tuning an instrument by ear will listen for beats (or "warble") to see if the instrument is in or out of tune. If you carefully watch a bassist in a rock band, you will see him or her constantly adjusting the strings to stay in tune.

So, why did the two crystal glasses produce the warble or beats? Each glass will vibrate at a slightly different frequency. That means each will produce a sound with a different pitch. Since the frequencies are not the same, the lines of interference will move. You hear those lines as they pass your ear in the form of different intensities of sound—beats.

What about in the case of the piano and guitar? They were playing the same note, or pitch. Why was there a beat there?

In the case of the piano and guitar, the beat is due to the fact that the note played on the guitar will not be identical to the one on the piano. Either the piano or guitar is out of tune. To tune the guitar to the piano, you would play the note on both the piano and guitar and adjust the length of the guitar string until you no longer heard the beat.

Timbre

The timbre of a musical instrument is due to the interference of the overtones. Think back to Section 3.2.

An **oscilloscope** is an instrument that shows the wave forms fed into it. It will work for any type of wave.

Blow a note from a flute into an oscilloscope and you will get a pattern that is different from that of a saxophone, guitar, violin, or bell. This pattern is responsible for the instrument's timbre. Similarly, if you speak into an oscilloscope, you will see a different pattern than that of a classmate. You could call it your voice signature.

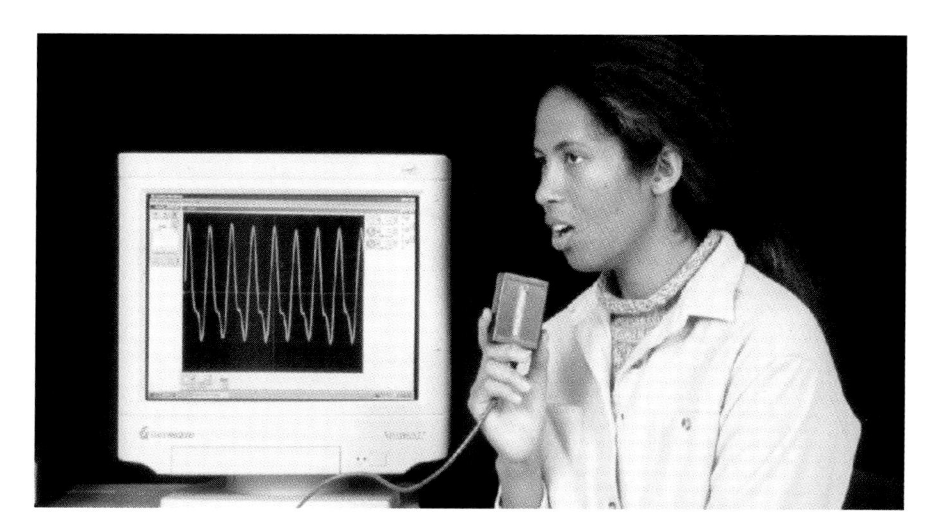

FIGURE 3.31 An oscilloscope displays longitudinal wave patterns as transverse waves.

You will need to know how interference can affect sound for your project, Courier Communication, at the end of Unit 2.

Summary

What you hear depends on many factors. Figure 3.32 summarizes this idea.

How Properties of Waves Affect What You Hear

What you hear	What is responsible
Loud or soft	Amplitude
High or low pitch	Frequency
Timbre	Sum of overtones

FIGURE 3.32

LAB 3C

Interference in Sound Waves

The tines on a tuning fork vibrate out of phase. They move in and out at the same time so that they are always moving in opposite directions. The sound waves moving out from different sides of the tuning fork interfere with each other. In this lab, you will observe the interference pattern that a tuning fork produces.

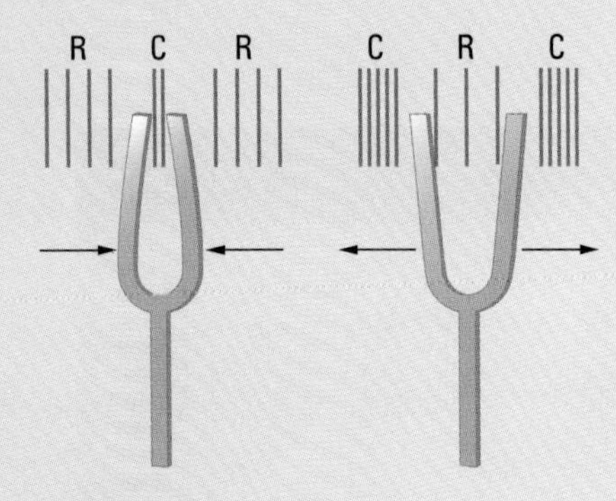

FIGURE 3.33 The sound waves produced by a tuning fork. C is for compression; R is for rarefaction.

Be Safe!

Only hit the tuning fork with a rubber instrument. Hitting it against other surfaces may damage the tuning fork.

Make sure the tuning fork does not touch the ear.

Purpose

To observe the interference pattern around a tuning fork.

Materials

- small rubber mallet
- tuning fork

Procedure

1. Work with a partner. In your notebook, make an observation chart similar to Figure 3.34. The drawings show a bird's-eye view of an observer's head and the tuning fork.

Location of fork	Observations
head fork	

FIGURE 3.34

What You Should Do

❷ Hit the tuning fork with the rubber mallet. You want a tone that lasts for a few moments. If your tone is too short, try again.

❸ Place the humming fork near your partner's ear. Note the angle of the fork.

❹ Turn the fork slowly, allowing your partner to make observations.

What Your Partner Should Do

❶ As the fork turns, note any changes in intensity (louder or softer) in the observations table.

CONTINUED

Analysis and Conclusion

1. Did you or your partner notice any changes in intensity as the tuning fork was turned? If you did not notice any changes, repeat the experiment and listen more carefully.

2. Explain your observations in this lab by referring to Figure 3.35.

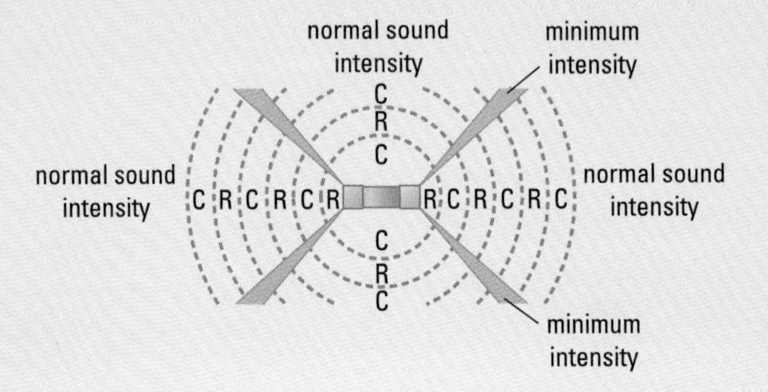

FIGURE 3.35 Overhead view of the interference pattern of the sound waves from a tuning fork. C is for compression and R is for rarefaction.

Extension and Connection

3. Take two identical tuning forks and wrap a rubber band around the tine on one fork. What differences in sound do you notice? Why?

4. Find out how the members of a band tune their instruments before a concert. Make sure to include the link to interference and beats.

Review and Apply

1. Meghan sets up the demonstration room at her electronics store to showcase new, state-of-the-art speakers.
   a) For some reason, the sound is not that great in the armchair in the corner. What advice do you have for her?
   b) Where is the best place to sit to hear the speakers properly? Why?

2. Explain, in your own words, how you could use the beats resulting from interference of sound waves to tune a guitar using a tuning fork.

3. Why are musical instruments generally handmade and expensive? Choose one instrument and create a brochure explaining how it is made and its price. You may have to do research using the library or Internet, or by contacting a music store.

4. Arif and Jason attend a concert but sit in different parts of the hall. After the concert, Jason says he could not hear the bass very well. Arif says he heard the bass quite well. Explain why the bass may not have been as loud for Jason.

5. Add any new concepts you have learned in this section to the graphic organizer you started in Section 3.1.

Job Link

Sound Technician

Bill Wilkinson knows vibrations. He specializes in sound and vibration control in business and the arts.

FIGURE 3.36 Bill Wilkinson

ScienceWise: Where do you work?

Bill Wilkinson: I am a technical representative for Wilrep Ltd. We specialize in sound and vibration control.

ScienceWise: What is that?

Bill Wilkinson: Sometimes a room or space is supposed to be used for concerts or speeches. We engineer the elements in the room so that everything sounds better. We also design mountings for machinery. Machinery will vibrate when it is in use. We design mountings to minimize the vibrations.

ScienceWise: How did you learn all of this?

Bill Wilkinson: I finished high school, then learned on the job. We consult engineers as well, but most of the time it is just to confirm our designs. Sound and vibration control is a very specialized industry, so even if you went to school for it you would still learn most of it on the job.

Surf the Web

Go to **www.science.nelson.com** and follow the links for ScienceWise Grade 12, Chapter 3, Section 3.3. Find out what materials would be required to build a recording studio in the basement of a house.

■

3.4 Chapter Summary

Now You Can...

- Analyze the properties of a vibrating object and how vibrating objects produce waves (3.1)
- Explain the properties of transverse and longitudinal waves (3.1)
- Explain, in qualitative terms, how frequency, amplitude, and wave shape affect the pitch, intensity, and quality of notes produced by musical instruments (3.2)
- Explain what happens when waves interact with one another (3.3)
- Discuss the contributions Canadians have made in the field of communication technology (3.3)
- Explain how sound pollution can have an impact on the way we live and work (Case Study)
- Carry out investigations concerning the scientific concepts involved in communications technology (Lab 3A, Activity 3B)
- Estimate the value of some wave-related quantities (Lab 3A)
- Determine the relationships among the major variables for a vibrating object (Lab 3A)
- Conduct an investigation to analyze and explain the production of sound by a vibrating object (Activity 3B)
- Formulate scientific questions about waves (Activity 3B, Lab 3C)

Concept Connections

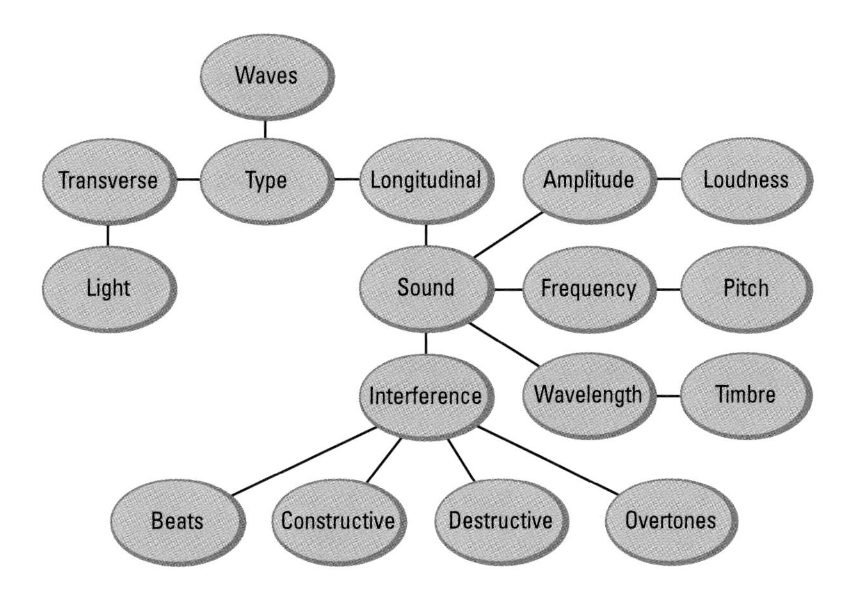

FIGURE 3.37 Compare your completed graphic organizer to the one on this page. How did you do? Can you add any new links to your organizer?

CHAPTER 3 *review*

Knowledge and Understanding

1. In her job as a maintenance worker for the railroad, Manju claims that she can tell if the train is coming, without even looking, just by touching a rail of the tracks. Is this possible? Why or why not?

2. A music store has the following motto: Every musical instrument is unique. Is there any truth to this statement? Explain your answer using what you now know about musical instruments.

3. Doreen is very particular about where she puts the new speakers for her home theatre. Her sister insists that it does not matter where you put the speakers because the sound will be the same. Who is right? Why?

4. Why is a clarinet quieter than a trumpet? List several reasons. If you are unsure of how sounds are produced on these instruments, talk to the music teacher, visit a music store, or do research in the library or on the Internet.

5. LuLing plays her home stereo so she hears a maximum level of 100 dB. Jane likes to turn her personal stereo up high and she hears maximum levels of 110 dB.
   a) How many times louder is the maximum level Jane hears than the maximum level LuLing hears?
   b) How can the level from a small personal stereo be so much louder?
   c) What problems could Jane have when she gets a little older?
   d) What advice would you give Jane?

Inquiry

6. Work with a partner and design an experiment to determine the effect that the type of material used to make a woodwind instrument has on the instrument's pitch. Money is no object. Assume you can use an electronic tuner to determine the pitch produced by the instrument.

7. Using a sound meter, how might you systematically test the sound pollution around a construction site?

8. Design an experiment to test a number of materials to find out which is the best sound insulator. Possibilities include a box, a tuning fork, various materials such as carpet, wood, plastic, fabric (heavy or light), cotton balls, and a sound meter.

Making Connections

9. Investigate an example of noise pollution relevant to your community. Present your information in a poster.

10. Motorboats can be destructive to the environment. Not only do they leak gas and oil into the water, but they can also have a negative impact on surrounding wildlife.
    a) What effect would a motorboat have on a fish in the lake if it passed overhead?
    b) What effect would a motorboat have on a bird standing on the beach?

11. Your community has decided to increase the number of commuter trains running in order to reduce the number of cars on the road and, therefore, reduce pollution. Yet, trains are noisy. Imagine you are on the committee charged with determining ways to make the noise from the trains less intrusive on the surrounding neighbourhoods.
    a) Where should the railroad tracks be located?
    b) What kinds of material should surround the tracks?
    c) When should the trains run?
    d) How can the community provide input on an ongoing basis?

Communication

12. Your teacher has asked you to make a presentation at your school's open house. You are to explain how a guitar produces different sounds. Assume that you will be presenting to people who have no knowledge of how sound is produced. Write down what you would say to the people. Include any relevant diagrams.

CHAPTER 4

Communication Using Waves

We use waves to communicate with one another. These waves are not just sound waves, but also electromagnetic waves (light waves, radio waves, and microwaves). Examine Figure 4.1. How do we communicate with waves? Why are satellite dishes shaped the way they are? How does a telephone work to allow us to communicate? Where do radio waves come from? How can you identify invisible waves? In this chapter, we will explore all of these questions and more.

FIGURE 4.1

What You Will Learn

After completing this chapter, you will be able to:

- Explain, using waves, how energy from communications devices is transmitted, reflected, and absorbed by different kinds of matter (4.1, Lab 4A, 4.2)
- Describe and explain, using examples, the effects produced by the refraction and total internal reflection of visible light waves as they pass through different transparent media (Lab 4B)
- Explain how different forms of energy can be transformed into and transmitted as waves (4.3)
- Examine and describe the operation of transducers that carry out the energy transformations in common communications equipment (4.3)
- Describe the historical development of a significant product of communications technology (4.4)
- Describe, using scientific principles, the functioning of common home and workplace communications technologies (4.4)

What You Will Do

- Select and use appropriate SI units (4.1)
- Communicate using data tables (Labs 4A, 4B)
- Select and use appropriate instruments to collect observations and data (Labs 4A, 4B, 4C)
- Demonstrate the skills to plan and carry out investigations (Labs 4A, 4B, 4C)
- Construct and test a prototype of a communications device and resolve problems as they arise (Lab 4C)
- Describe the impact of developments in communications technology on the way we work and on our social environment (Activity 4D)

Words to Know

amplifier
analog system
anode
carbon microphone
cathode
cathode ray tube
composite video signal
dynamic microphone
electromagnetic waves
focal point
input transducer
output transducer
phosphor
photoelectric effect
photosites
pixels
radar
receiver
reflection
refraction
total internal reflection
transducer
transmitter

A puzzle piece indicates knowledge or a skill that you will need for your project, Courier Communication, at the end of Unit 2.

4.1 Transmission and Reflection of Waves

Knowing how waves are transmitted and reflected will be useful for your project, Courier Communication, at the end of Unit 2.

How do waves travel from one place to another? Recall from Chapter 3 that waves travel differently in different media or if their surroundings are changed. In Activity 3B, you noticed that you changed the sounds your musical instrument produced when you changed aspects of your instrument.

In this section, we will discuss sound waves and electromagnetic waves. **Electromagnetic waves** are transverse waves (Figure 4.2). They include X-rays, microwaves, radio waves, and visible light waves. These waves do not need a medium to travel through. They are able to travel through vacuums. Sound waves are longitudinal waves, and they do need a medium to travel through.

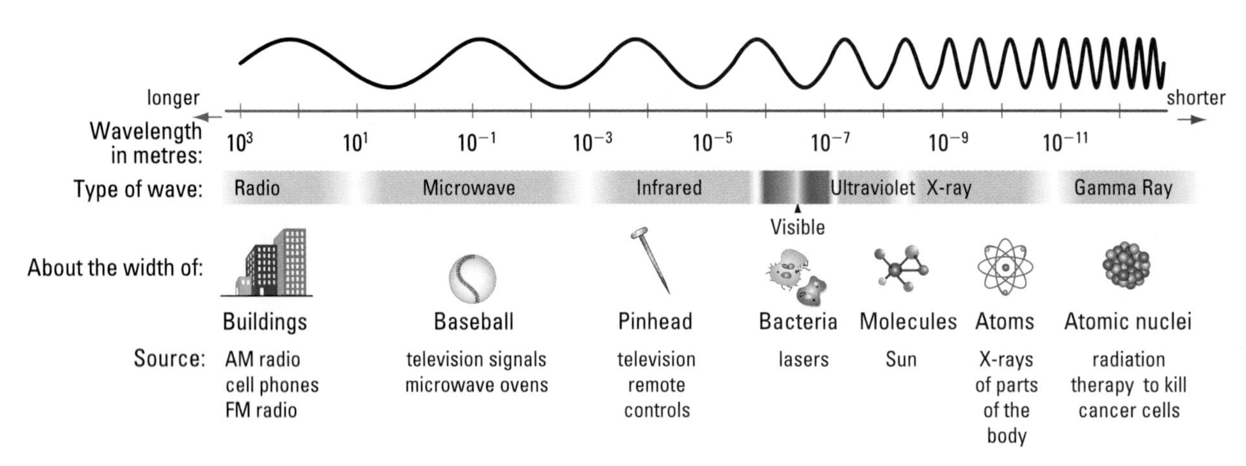

FIGURE 4.2 The electromagnetic spectrum

Transmission

When designing ways to transmit and receive waves, whether they are sound or electromagnetic, there are four variables to consider:

- amplitude
- frequency
- speed
- wavelength

Amplitude

The amplitude of a wave changes over time. You know that the farther away from the source a sound wave travels, the quieter it gets. You also know that the loudness of a sound depends on the amplitude. What is happening?

A wave requires energy to travel from point A to point B in a given medium, such as air or water. As it travels along, it loses energy. The farther it is from the source, the more energy it has lost. The amplitude is the part of the wave that is affected by this loss of energy. Refer to Figure 4.3.

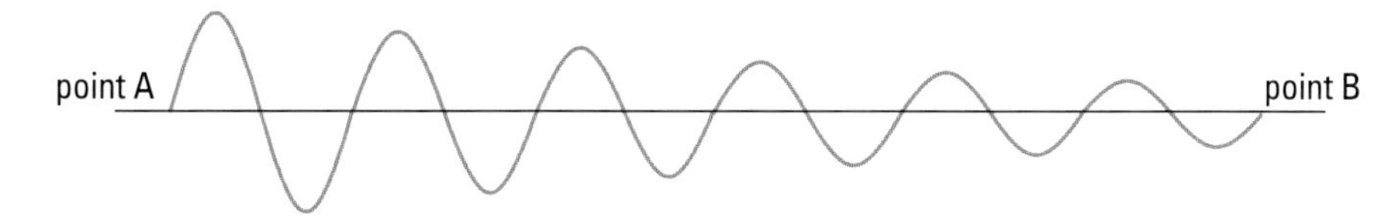

FIGURE 4.3 Waves lose energy as they travel away from the source. This energy loss causes the amplitude to decrease.

Frequency

As you learned in Lab 3A, amplitude has no effect on frequency. In fact, neither wavelength nor speed affects frequency. A vibrating object has a frequency that depends on the nature of the object (for example, the length and thickness of a violin string). That frequency is transferred to the wave. For example, if a violin string has a frequency of 10 000 Hz when it is plucked, the sound wave produced will also have a frequency of 10 000 Hz. That frequency will not change unless the medium changes.

Speed

The speed of a wave depends on the nature of the medium. Sound waves travel faster through water than air. Light waves travel faster through air than glass. Speed is defined as the distance travelled per unit time. It is usually measured in metres per second (m/s).

To find the speed at which you walk, divide the distance you travelled by the time it took to get from the start to the end of your trip. See Figure 4.4.

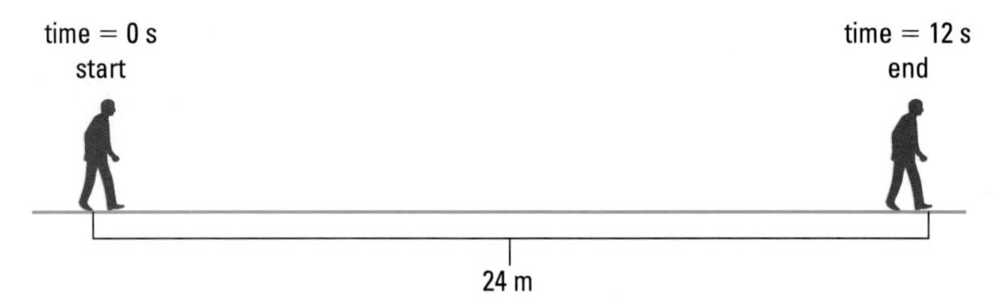

FIGURE 4.4 How fast is this person walking? Be sure to express your answer using the appropriate SI units.

ScienceWise Fact

Waves travel at different speeds, depending on the medium. A fast medium is one in which a wave can travel relatively quickly. For example, the speed of sound in air is 332 m/s, but in fresh water it is 1500 m/s. Water would be considered a fast medium for sound. Notice how well sound travels in water and how loud sounds are when you are hearing them under water.

If you want to know the speed of a wave, divide the distance the wave travelled in one cycle (wavelength) by the time it took to travel that distance (period of the source). See Figure 4.5.

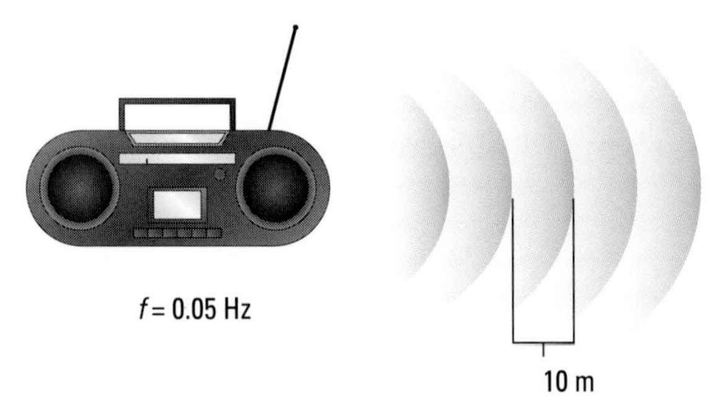

FIGURE 4.5 To calculate the speed of a wave, divide the wavelength by the period of the source. This calculation has two steps. Calculate the period first. (Remember that the period is 1 divided by the frequency!) Then calculate the speed. Compare your answer with those of two classmates.

Wavelength

If the speed of a wave can be calculated from the wavelength and frequency, what does the wavelength of a wave depend on? The wavelength depends on frequency and the speed of the wave.

If you increase the frequency of a wave, the wavelength will decrease. If you decrease the frequency, the wavelength will increase.

If the wave travels through a faster medium, the wavelength will increase. If the wave travels through a slower medium, the wavelength will decrease.

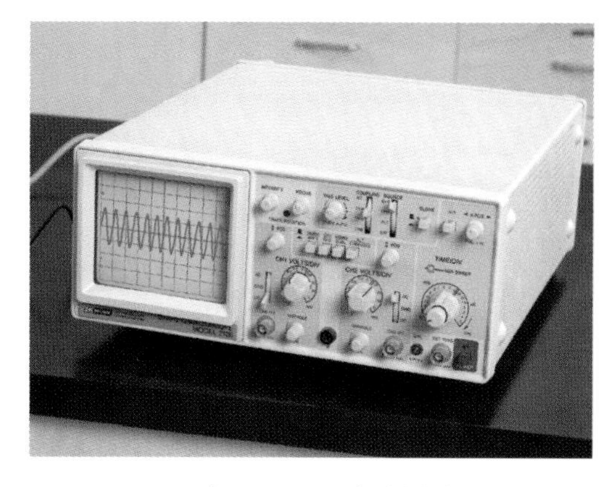

FIGURE 4.6A Appearance of a high-frequency wave on an oscilloscope. Compare its wavelength to the one in Figure 4.6B.

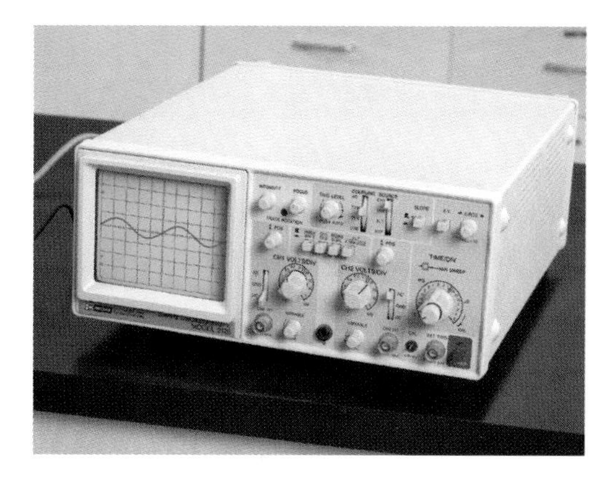

FIGURE 4.6B Appearance of a low-frequency wave on an oscilloscope. Compare its wavelength to the one in Figure 4.6A.

Reflection

What happens to these four variables—amplitude, frequency, speed, and wavelength—when a wave meets an obstacle? An echo is the result of reflected sound waves. The image in a mirror is the result of reflected light waves. In fact, a wave will travel in a straight line unless something gets in its way. **Reflection** occurs when a wave hits a medium different from the one in which it has been travelling. Soft and rough surfaces do not reflect well, but hard and smooth surfaces do. For example, if you look at a smooth piece of aluminium foil, you can see your reflection. However, if you crumple the same piece of foil, you will have a hard time seeing your reflection.

What is actually happening to the amplitude, frequency, speed, and wavelength of the wave as it is reflected? You will investigate this behaviour in Lab 4A.

Absorption

If some of the energy of a wave is absorbed, the amplitude is affected. In school gymnasiums sounds are very loud, because the sound waves bounce off the smooth surfaces and very little of their energy is absorbed. In rooms with carpets and curtains the sounds are more muted, because the carpets and curtains absorb some of the energy of the sounds.

Waves that encounter a barrier will be reflected and absorbed. The amount of reflection or absorption depends on the material the barrier is made of.

LAB 4A

Transmission and Reflection of Transverse Waves

Light is a transverse wave. What variables affect the way it is transmitted and reflected?

Purpose

To investigate the way a transverse wave travels and how it is reflected.

Materials

- long spiral spring
- metre stick
- stopwatch

Procedure

1. Read both parts of the Procedure and create a data table to record all your observations.

Part One: Transmission and Reflection in a Spring with a Fixed End

2. You will create a pulse at your end of the spring by moving your hand quickly from the rest position to one side and then back to the rest position (Figure 4.7). Practice a few times before doing it for real.

3. Stretch the spring to a length of approximately 4 m on a smooth clean floor. Place a metre stick at a right angle to the spring, in front of your partner.
4. Your partner should hold the far end of the spring so that it does not move during step 5. Remember to record your observations for each step.
5. Create a pulse.

FIGURE 4.7 Moving your hand by half a cycle will produce a pulse. Practice before working with your partner. Move your hand from the start position to the left, then back to the start position.

6. When you create a pulse, you and your partner will each be responsible for making different observations. You will use a stopwatch to count how many seconds it takes for the wave to travel from your end to your partner's end. Your partner will try to estimate the amplitude of the pulse. Does the amplitude stay the same from the start of the spring to the end? Explain your answer.
7. Create a pulse like the one in step 5. Does it reflect back to you? Has the reflection changed it in any way? If so, how?
8. Repeat step 5 using a pulse with a larger amplitude than that in step 7.
9. Repeat step 5 with a smaller amplitude than that in step 7.
10. Create two pulses, one after the other. Does the second pulse catch up to the first? Describe or draw a diagram of what happens.

Part Two: Transmission and Reflection in a Spring with a Free End

1. Dangle the spring from a high point in the classroom so that one end does not quite reach the floor. (You may have to bunch up the spring at the other end.) You want the spring to be able to move freely at the bottom.
2. Repeat steps 5–9 from Part One.

FIGURE 4.8

Part Three: Observing Interference in a Ripple Tank

1. Your teacher will set up a ripple tank to show different wave patterns. She or he will produce waves using a wave generator.
2. Observe the speed of the wave in each of the following situations:
   - when the wave hits three different barriers (wood, plastic, barrier wrapped in fabric)
   - when a wave travels through different depths of water
3. Compare these results to the ones you had with the spring. Describe any similarities and differences.

CONTINUED

Analysis and Conclusion

1. Does amplitude affect the speed of the pulse? Explain using observations from your investigation.
2. Does it matter if the end of the spring is fixed (as in Part One) or free (as in Part Two)? Explain using observations from your investigation.
3. What happens to the speed of a wave as it travels through a medium?
4. In Part Three, what effect did each barrier have on the reflection of the wave? Explain your answer.

Extension and Connection

5. Why do water waves seem to be farther apart and faster out in a lake but seem to be closer together and slower as they get close to the beach?
6. Describe how you could measure the speed of sound waves using the reflection of sound from an outside high wall of the school.
7. Farhanna tells Cheryl that when she hears the school bell from under the water in the pool, the pitch is lower. Is she correct? Explain. (Hint: Which property of sound do we sense as pitch?)

Review and Apply

1. Television cable companies transmit electromagnetic waves along wires from a central location to your home. They have to amplify the signal every few kilometres. Why?
2. You work for a radio company that needs to transmit a 750 000 Hz signal with a wavelength of 400 m. How fast will the signal travel? (Hint: This is a two-step question).
3. Radio signals are electromagnetic waves. Use your knowledge of waves to explain why severe weather (such as a heavy rain or a blizzard) affects the transmission of radio signals.
4. Dominic stands in a canyon and yells "Echo!" as loud as he can. He yells only once but hears the word repeated several times before it fades out. Explain why that happened.
5. In a graphic organizer, organize the concepts you have learned in this section.

Electromagnetic Interference Can Be a Problem

You already know from Section 3.3 that waves can interfere with each other. The most common wave interference that we experience every day is from radio waves.

Just think of electrical devices that produce radio waves:

- radios
- cordless phones
- computers
- cellular phones
- televisions

In fact, you can even hear the interference. Try talking on a cordless phone while sitting next to a computer screen, or placing a cell phone next to an FM radio. You will probably hear static. That is the interference from radio waves.

Usually, electromagnetic interference is a minor bother. However, it is the reason why people are not allowed to have cell phones turned on in airplanes or hospitals. Imagine if interference caused the navigation system in an airplane to stop working, or caused a heart monitor in a hospital to give the wrong readings. In some instances interference can be dangerous.

FIGURE 4.9 Find out why we are advised to turn off cell phones at the gas pump. Do you think this is a good idea? Explain your answer.

4.2 Wave Applications

Waves can be used to receive and send information. What types of devices do we use to do this? And how do they work? In this section, you will learn about some of the devices we use.

Radar

Radar is used for many different purposes, but usually to identify the presence or speed of something. For example, airplanes use radar to detect the presence of other objects in the sky. Meteorologists use radar to detect the presence of clouds and storms. Police use radar to detect the speed of approaching cars.

Radar uses radio waves that cannot be seen or heard by people. They are very easy to produce and can travel very long distances. They are extremely easy to detect with electronic devices, even when they are faint (small amplitude). Look at Figure 4.10 to see how radar is used.

A radar set includes a transmitter and a receiver. **Transmitters** are objects that produce waves. **Receivers** are objects that detect waves. When a radar set is switched on, it produces a high-intensity radio pulse. A pulse is the result of a single, rather than continuous, disturbance. This pulse then travels until it hits a solid object. When it hits a solid object, it bounces off, as all waves do. Eventually, it reaches the radar set again. The radar set detects how long it took for the pulse to come back and uses this information to calculate the distance of the object. Sophisticated radar can detect the shape and speed of an object—due to the fact that radio waves can be reflected.

FIGURE 4.10 An air traffic controller using radar to monitor and control the movement of aircraft around the airport.

Satellite Dishes

Why are satellite dishes shaped like shallow bowls? Satellite dishes are built to detect radio waves sent from satellites that are hundreds of kilometres away. By the time the waves reach the satellite dish, they are quite weak. Satellite dishes are shaped as they are because when waves bounce off them, the waves all go to one point (if it is a parabolic dish) called a **focal point**. With so many waves being reflected to one central spot, the signal becomes strong enough to be detected. See Figure 4.11.

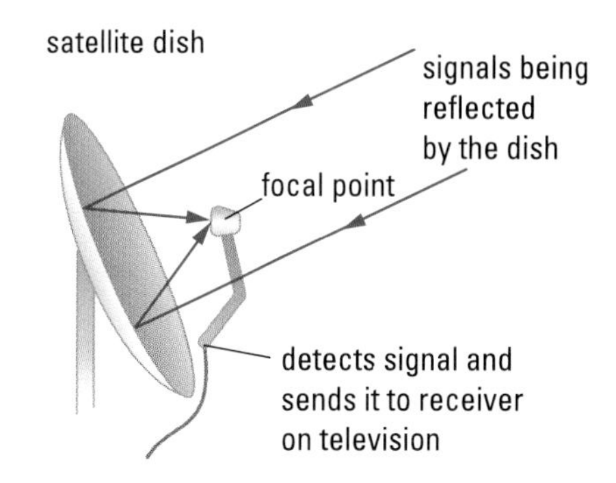

FIGURE 4.11 The parabolic shape of a satellite dish allows the received waves, or incident waves, to be reflected to one point.

Transmission of Radio Waves

How are radio waves transmitted from one place to another? Two components are needed: a transmitter and a receiver.

Both the transmitter and the receiver have wire antennas that send or collect radio waves. A transmitter works by sending an electric current through the antenna. Whenever an electric current travels through a conductor, a magnetic field is produced. This magnetic field will travel outward in all directions and create an electric current when it hits the antenna in the receiver.

Consider this example. Suppose we connect a wire to the two terminals of an energy source (for instance, a battery). A magnetic field will be produced surrounding the wire. We can change the current in this wire by opening and closing a switch in the circuit and that will cause the magnetic field to change as well. This is a form of transmitter. If we connect a second wire to a voltmeter and place it near the transmitter wire, then the changing magnetic field of the transmitter will induce an electric current in the second wire. This current will vary in the exact same way as it does in the transmitter (Figure 4.12). This second wire is the receiver.

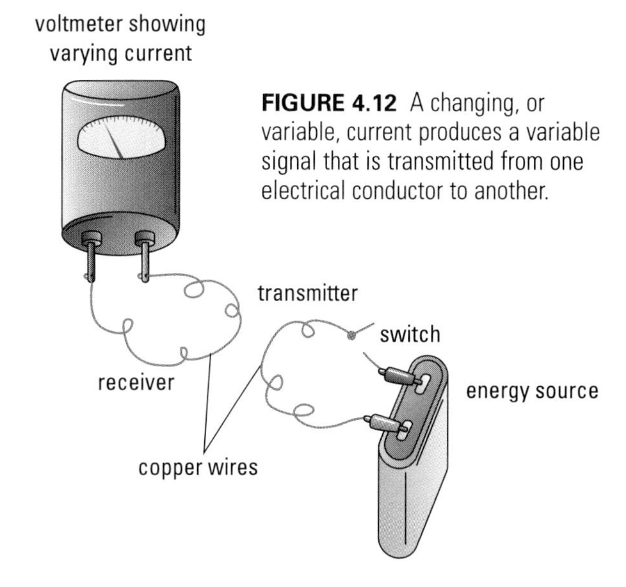

FIGURE 4.12 A changing, or variable, current produces a variable signal that is transmitted from one electrical conductor to another.

Transmitters and receivers can be located a great distance apart from each other. How do radio waves travel this distance?

We already know electrons moving up and down an antenna will induce a magnetic field around that antenna. This magnetic field will in turn induce an electric field beside itself. This electric field will induce another magnetic field that induces another electric field. This continues on and on. (See Figure 4.13.) As the current changes in the transmitter's antenna, the magnetic field it generates changes, too. This induces a different electric field that induces a different magnetic field and so on. Eventually, a magnetic field is generated near a receiver's antenna and an electric current is induced in the antenna that varies in exactly in the same way as the original one did.

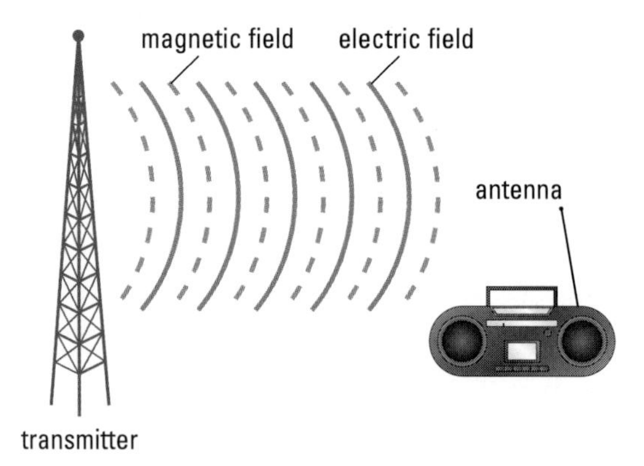

FIGURE 4.13 Radio transmission is a chain reaction. The electric current produces a changing magnetic field that induces a changing electric field that induces a changing magnetic field that induces a changing electric field, and so on.

There are thousands of radio waves surrounding you right now. That is why you can turn on a radio just about anywhere and still get a signal.

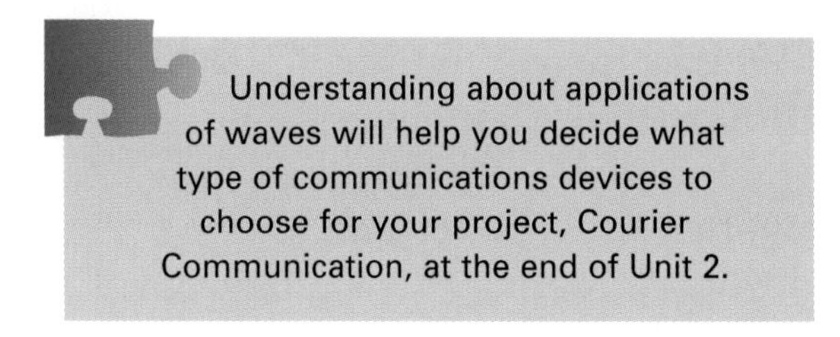

Transmission of Light Waves: Fibre Optics

You have learned how radio waves can be used to transmit and receive information. Radio waves are a type of electromagnetic wave. We also use other types of electromagnetic waves for sending and receiving information, such as light waves. Light waves containing information are often sent from one place to another using glass fibres. Fibre optics makes use of a property of light waves called **refraction**. When a knife is placed in a glass of water, it appears bent. Refraction causes this illusion. Waves refract when they travel from one medium, such as water, to another, such as air, because of the difference in optical density. If the angle at which the wave hits the boundary is large enough, all of the wave will reflect, rather than refract, off the boundary, and none will pass through into the second medium. This is called **total internal reflection**. (See Figure 4.14.)

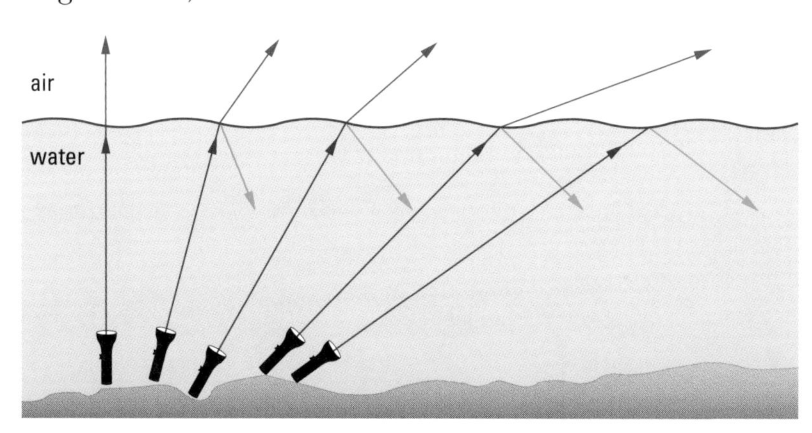

FIGURE 4.14 If the angle at which the wave hits the boundary is large enough, the entire wave reflects off the boundary between the two media.

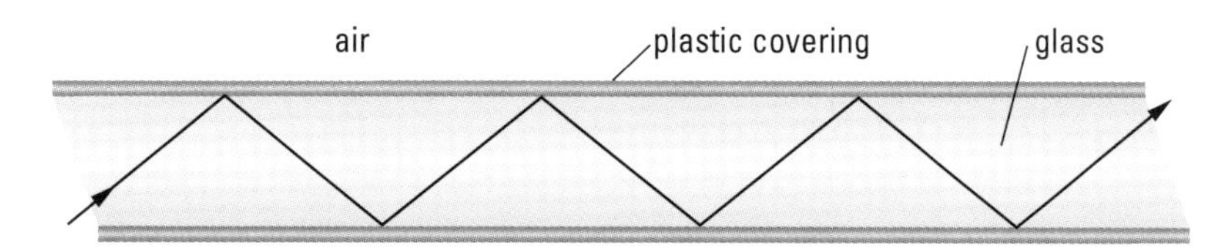

FIGURE 4.15 In a fibre optic cable, light waves bounce off the inside walls.

Fibre optic cables use light waves for carrying information. By starting the transmission at just the correct angle, the light waves bounce off the insides of the cable (Figure 4.15). The light waves can travel very long distances within the cable itself. They are not affected by weather or nearby electric wires or magnetic fields.

FIGURE 4.16 Fibre optic cable is a very efficient medium for transmitting waves.

LAB 4B

Refraction of Light

Waves travel at different speeds in different media. When a wave travels from one medium to another, this change in speed results in the wave changing direction slightly, or bending. This bending is called refraction.

Be Safe!

Be careful when using ray boxes; they get very hot and can cause burns.

Purpose

To investigate what happens to waves when they enter different media and to determine how light refracts as it passes from a less dense to a more dense medium and vice versa.

Materials

- ray box (single slit)
- semicircular glass block
- protractor
- blank piece of paper

Procedure

1. Draw a data table like the one in Figure 4.17 in your notebook. Label it "Air to Glass."

Angle of incidence	Angle of refraction	Observations
0°		
10°		
20°		
60°		

FIGURE 4.17

2. Draw x- and y-axes on the blank paper.

3. Place the glass block on the paper so that the straight side is along the x-axis. This line is now the horizontal. The other line should go through the middle. This middle line is the normal, 90°to the horizontal. See Figure 4.18.

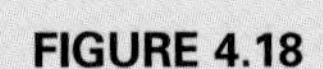

FIGURE 4.18

Part One: Air to Glass

4. Shine the light through the flat edge of the block at the first angle listed in your data table for the angle of incidence. Measure the angle of the light inside the block. This is your angle of refraction. Record it in your data table.
5. Record any other observations.
6. Repeat steps 4 and 5 for all of the angles in your data table.

Part Two: Glass to Air

7. Make another data table like that in step 1. Give this one the title "Glass to Air."
8. Shine the light through the round edge of the block toward the middle of the flat side, at the angle of incidence shown in your chart (Figure 4.20). Measure the angle of refraction and record it.
9. Record any other observations.
10. Repeat steps 8 and 9 for all of the angles in your data table.

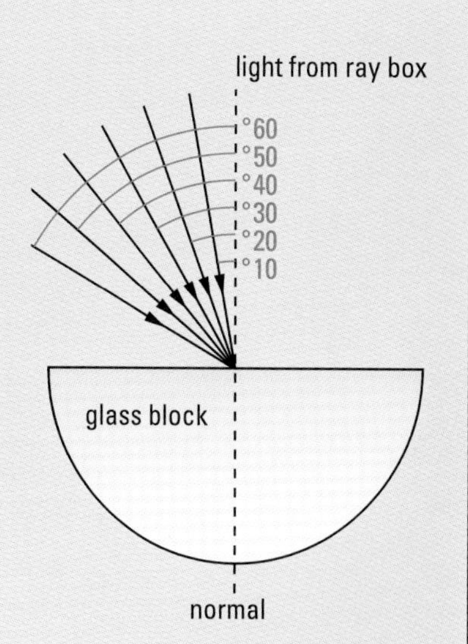

FIGURE 4.19
Measure all angles from the normal.

FIGURE 4.20

Analysis and Conclusion

1. What happens to light when it travels from air to glass or glass to air at an angle of 0°?
2. When light travels from air to glass, does it refract toward or away from the normal? Does any of the light reflect back through the air? Explain your answers.
3. When light travels from glass to air, does it refract toward or away from the normal? Does any of the light reflect back through the glass? Explain your answers.

Extension and Connection

4. Describe what happened to the light when it had angles of incidence of 60° and 50° going from glass to air.
5. Total internal reflection makes fibre optics possible. Why do the fibres have to be kept fairly straight without sharp corners?

Review and Apply

1. You have a job as an air traffic controller. Your former high school science teacher has asked you to speak to the class about how your job uses radar. Create a poster or video to present your information. Research any additional information needed.

2. Explain why parabolic microphones are shaped the way they are.

FIGURE 4.21 Parabolic microphones are used when the sound to be detected is very faint.

3. All electrical appliances produce faint radio waves. Why? How do you know?

4. Find out where fibre optics are being used in communication. What are the advantages of fibre optics over electrical wire?

5. Add the new concepts from this section to the graphic organizer you started in Section 4.1.

6. **Radio Waves versus Light Waves**

   With a partner, design an experiment to test if radio waves interfere with visible light waves. Use a radio and a light as your materials. (Hint: Look at the set-up for Lab 3B.)

4.3 Energy and Communication

You have seen how we can use waves to send or receive information. It is only a small step to see how we can use waves in communications devices. How does the sound of your voice enter your telephone, then travel to someone else's phone? In this section, you will learn how different types of energy (sound, electrical, and mechanical) are changed back and forth in order for us to communicate.

Transducers

Sound is a form of energy. To get a sound to travel along a copper wire, we must convert it to electrical energy and then back to sound again. We use transducers to do this. **Transducers** convert energy from one form to another. This conversion, from sound to electrical energy, is done through an input transducer. **Input transducers** convert non-electrical energy—like a person's voice—to electrical energy.

Look at Figure 4.22 to see how an input transducer works. A permanent magnet has a coiled wire within its field. This coil is attached to a thin diaphragm. When sound waves hit the diaphragm, the diaphragm moves. Since the coil is attached to this diaphragm, it moves too. The coil will move with exactly the same frequency and amplitude as the sound. When the coil moves, it moves relative to the magnetic field around the permanent magnet.

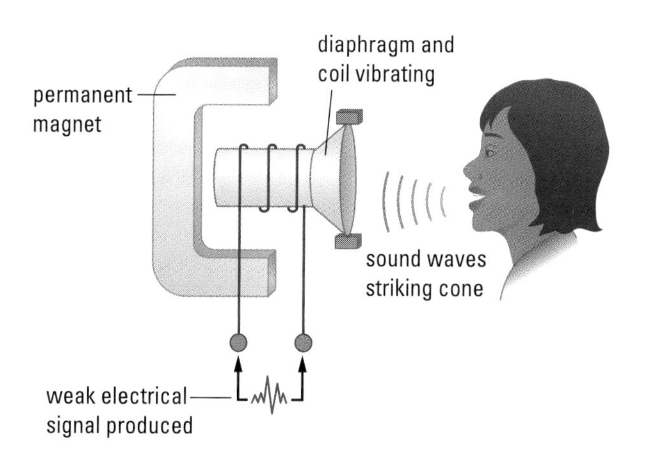

FIGURE 4.22 This is how an input transducer works. See if you can explain to a partner how it works.

Recall when you studied electricity. When a coiled conductor moves relative to a magnetic field, electricity is produced. The current will be an alternating current, moving back and forth in the wire. As the diaphragm moves forward, so will the current. As the diaphragm moves back, so will the current. The movement of the current will be comparable, or analogous, to that of the sound because it moves in a similar way. That is why we call this an **analog system**.

How do we hear the sound that has been converted to electrical energy? This is done using an output transducer. See Figure 4.23. An **output transducer** converts electrical energy into non-electrical energy. To do this, it simply reverses the process of the input transducer.

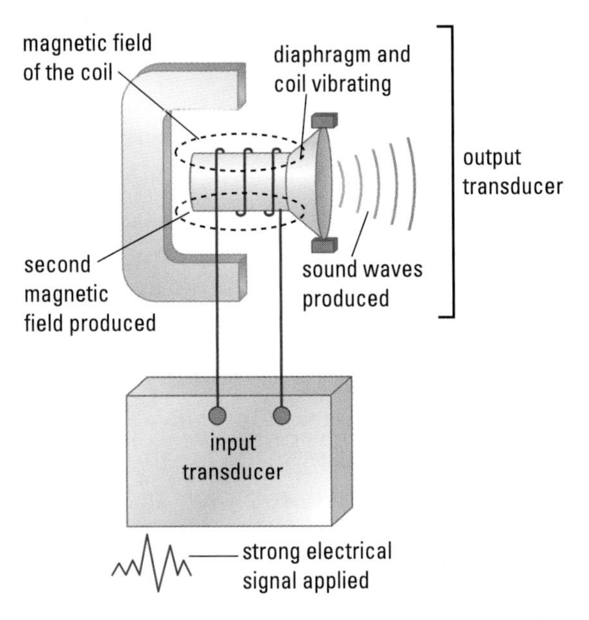

FIGURE 4.23 Here is how an output transducer works. How is it similar to an input transducer? How is it different?

From Input to Output

Remember that the current produced in the coil is analogous to the sound that produced the current. The magnetic field around the coil will interfere with the magnetic field around the magnet. This causes the diaphragm to move back and forth. This vibration of the diaphragm produces a sound identical to the one that entered the system at the input end. See Figure 4.24.

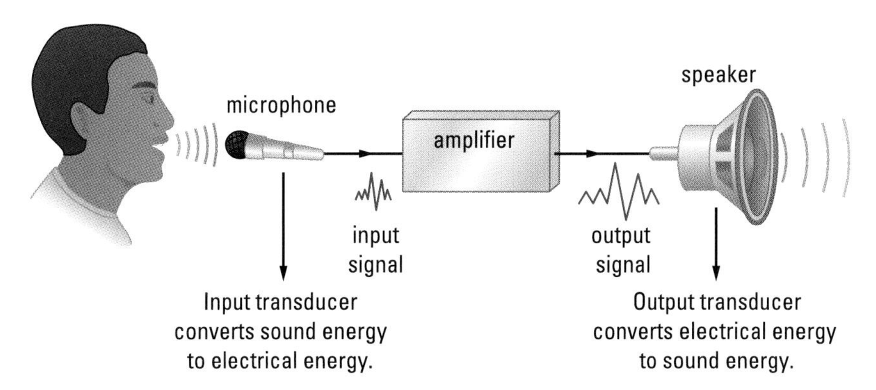

FIGURE 4.24 How input and output transducers relate to each other

There are many applications of this system. In fact, with all that you now know about waves and transducers, it is easy to understand how many different communications devices work.

Types of Communications Devices

Microphones

A microphone contains an input transducer. A microphone modifies a current, analogous to the sound input, through the wire at the other end of the microphone. Wireless microphones send radio waves to a receiver.

There are different types of microphones. The input transducer described in this section is called a dynamic microphone. **Dynamic microphones** contain a permanent magnet. See Figure 4.25.

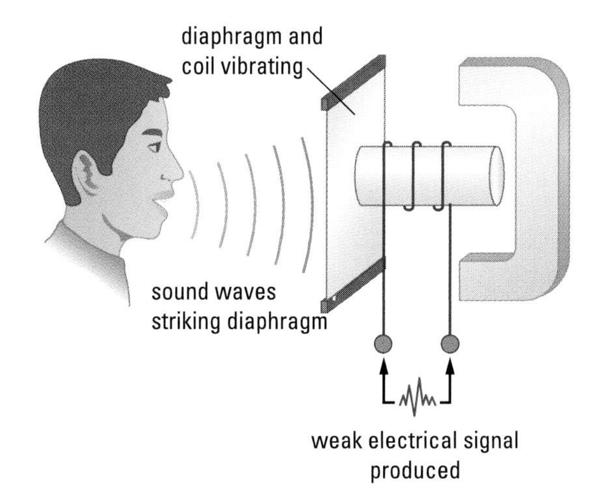

FIGURE 4.25 A dynamic microphone

Another common type of microphone is the **carbon microphone**. A carbon microphone is an input transducer that uses carbon instead of a magnet to generate an analog signal.

Did you know that carbon, the material in a pencil, conducts electricity? However, it does not conduct as well as copper. That is why some resistors contain carbon. In the case of the microphone, it consists of a metallic container filled with carbon granules. A diaphragm is placed over the open end of the container. Wires attached to the container are connected to an electrical circuit such that a current flows through the carbon granules. Sound waves cause the diaphragm to vibrate. This changes the electrical resistance of the carbon granules as the diaphragm compresses them (Figure 4.26). As the resistance changes, so does the current going through the conductor. The current will again be analogous to the sound.

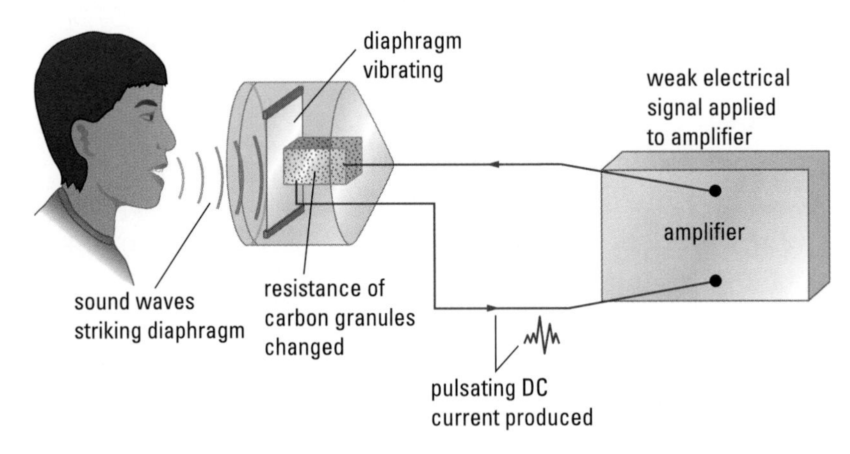

FIGURE 4.26 A carbon microphone. How is it similar to a dynamic microphone? How is it different?

Amplifiers

The signal produced in the case of many input transducers is not strong enough to travel very far. It does not have enough energy. **Amplifiers** are used to strengthen the signal produced.

Most analog signals sent along copper wires need a boost every so often, depending on the type of signal (television, telephone). Amplifiers are used to do this.

Telephones

The telephone was the first application of transducers converting sound to electricity and back to sound again.

When you pick up a phone handset, a circuit is opened between your home and the central office of the phone company. The central office usually generates the dial tone that you hear when you pick up a phone (which tells you that you are connected). When you dial the number, the central office connects you to the line of the person you are calling.

You can probably guess how your voice travels down the line to the person at the other end. The handset consists of a microphone and a speaker. In older telephones, you would have spoken into a carbon microphone. Modern telephones use electronic microphones. In either case, an analog signal is produced. It travels along the copper wire connecting you to the central office. This signal is sent to whomever you called. At that end, it is converted back to sound.

To save energy, telephones only transmit a narrow frequency—just enough to understand a human voice. This explains why people's voices sound different over the phone than they do in person.

ScienceWise Fact

Telephones filter out all frequencies except those between 400 and 3400 Hz. The average person can hear sounds between 20 Hz and 20 000 Hz.

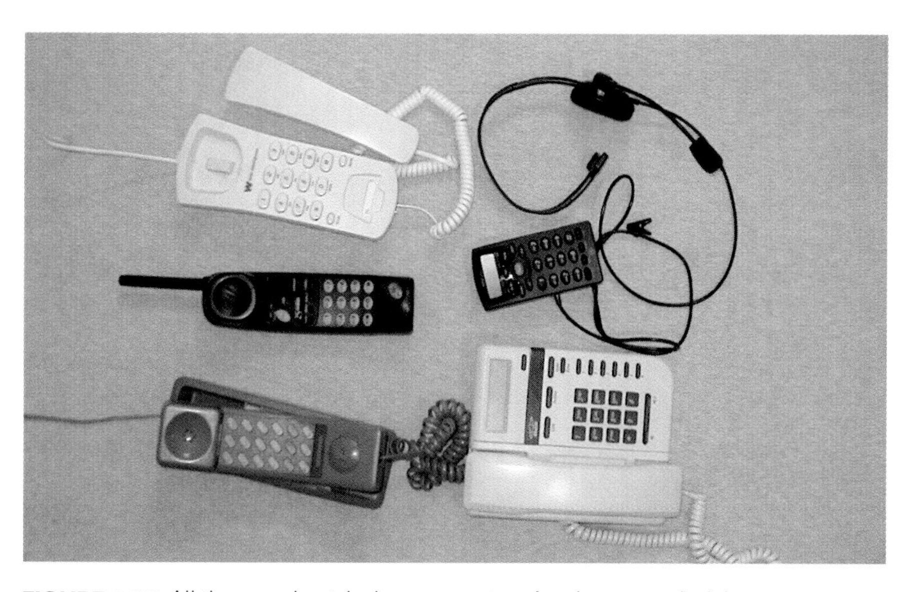

FIGURE 4.27 All these analog telephones operate using the same principles.

AM and FM Radios

You already know how radio waves are produced and transmitted. How do we send information using radio waves?

To send an analog signal, something about the radio wave must be changed. We have a choice of changing either the frequency or the amplitude of the wave.

Let us examine what a radio wave looks like. If we graph the size of the current over time, it will look like Figure 4.28. The mathematical name for this characteristic shape is a sine wave. This wave is the carrier radio wave. The waves that represent the information (for example, voices and music) are placed on top of the carrier wave.

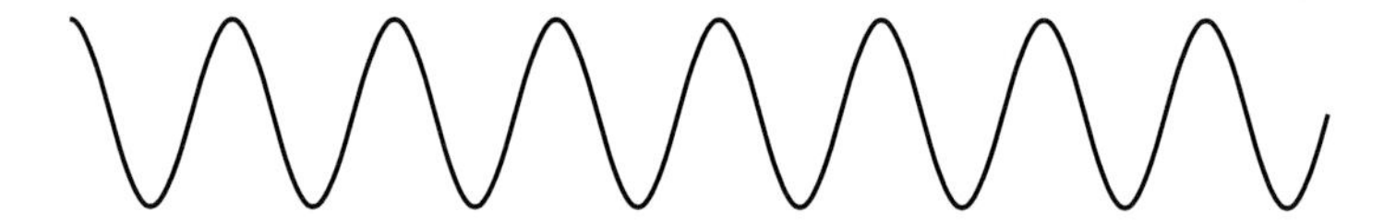

FIGURE 4.28 The sine wave has a typical shape, as shown here.

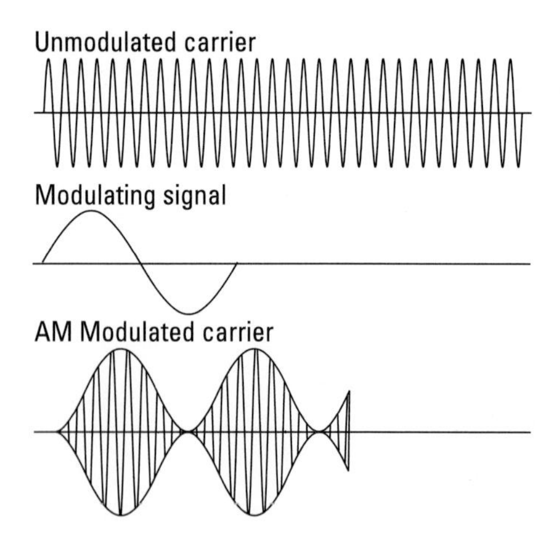

FIGURE 4.29A In AM (amplitude modulation) transmission, the input transducer translates the sound into a wave whose *amplitude* changes in the same way as the sound producing it. This wave is placed on top of the carrier wave. This produces a new wave made by adding the signal produced by the sound to the carrier wave. This new wave is transmitted to the receiver and changed back to a sound at that end.

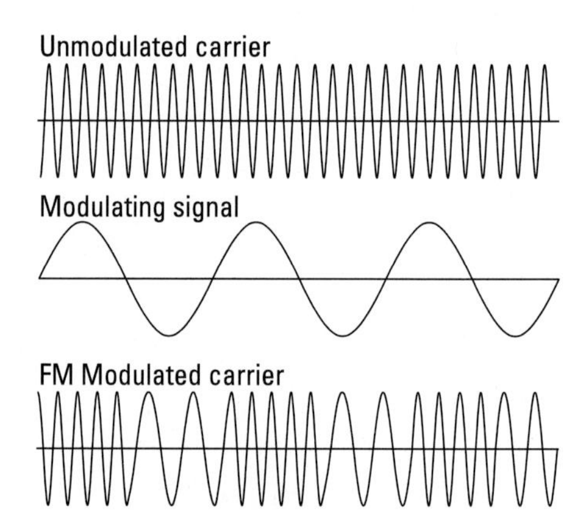

FIGURE 4.29B In FM (frequency modulation) the input transducer translates the sound into a wave whose *frequency* changes in the same way as the sound producing it. This wave is placed on top of the carrier wave to produce a new wave. FM frequencies are measured in megahertz, MHz. (1 MHz = 1 million Hz.)

When these radio waves reach the antenna of a receiver, the electrons in the antenna will move back and forth in a way analogous to the radio waves. An output transducer, or speaker, then converts this varying current to a sound.

Your radio tuner allows you to choose which radio waves to listen to. Its job is to filter out unwanted frequencies. If you wish to listen to 640 AM, you adjust the tuner so that it picks up only waves with a frequency of 640 kHz. Your antenna increases the amount of metal the radio waves can come in contact with. An amplifier in your radio increases the strength of the signal so that you can hear it clearly through the speaker.

Technology: Analog or Digital?

We have spent a great deal of time on analog signals in this chapter, even though digital systems are becoming more common. Why? Because the waves that enter and exit digital communications devices are very similar to the analog signals that travel through wires or various media.

A digital signal is simply a signal that turns analog signals into numbers. It does this by sampling a wave (such as a sound wave) at various intervals. It assigns a number to the amplitudes it records.

The advantage of digital transmission is that it is easy to reproduce a set of numbers. Remember that analog systems are subject to fading as they travel. Copying analog signals is also a problem because the copy is never as intense as the original. Since the analog signal is a combination of varying intensities, the signal becomes distorted. Digital systems do not have this same problem.

Investigate digital devices found in your home or workplace to determine if an analog equivalent existed in the past.

Review and Apply

1. For each of the following devices, indicate if it contains an input transducer, an output transducer, or both. Explain your reasoning.
   a) headphones
   b) telephone
   c) radio
   d) microphone

2. Children often play a game by tying two paper cups together with a long string and using them like two telephones. Why does this work? Explain what would happen in the following situations. What if you:
   a) held the string taut?
   b) let the string go slack (loose)?
   c) wet the string?

3. Describe the similarities and differences in the construction and operation of a microphone and a loudspeaker.

4. Radio stations must have a licence from the government in order to operate. Why is it important for the government to control what frequency a radio station uses?

5. Identify the wave in Figure 4.30 as either FM or AM and explain your choice.

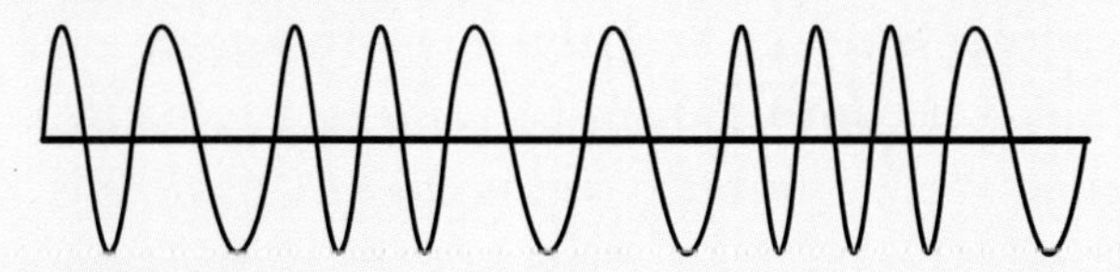

FIGURE 4.30

6. Add the new concepts from this section to the graphic organizer you started in Section 4.1.

7. **Design a Speaker**

   In a group, design a speaker made out of copper wire, a magnet, and household items. Compare your design with those of other groups. Identify ways you could modify your design. Make the changes.

Job Link

System Administrator, Telecommunications

The telephone system in hospitals is complex. It requires an entire department of people just to take care of it. Pierre Charbonneau is the System Administrator in the Telecommunications department of the Ottawa Hospital. He is responsible for the day-to-day operation and maintenance of the telephone system at a large regional hospital.

FIGURE 4.31 Pierre Charbonneau

ScienceWise: What do you do?

Pierre Charbonneau: I manage the telephone system at the hospital.

ScienceWise: How did you get to your current position?

Pierre Charbonneau: I started at another company in the Information Systems department. That company had a policy that everyone should know something about another department, just in case we were needed. I chose telecom (telecommunications).

ScienceWise: And you just learned the rest on your own?

Pierre Charbonneau: I received on-the-job training from the telecom people. This allowed me not only to learn new skills, but to practise them as well. At the Ottawa Hospital, I started on the computer help desk, but continued to update my skills through telecom seminars and peer mentorship. Eventually, I moved to the telecom department permanently.

Surf the Web

Wireless technology has become increasingly common. What is the difference between a cell phone and a satellite phone? Go to **www.science.nelson.com** and follow the links for ScienceWise Grade 12, Chapter 4, Section 4.3. Create a data table comparing the two types of phones. Include any relevant diagrams or pictures.

LAB 4C

Building a Telephone Network

Your school probably does not have the equipment for you to make a telephone, but, given a few telephones, you can easily set up a telephone network within your own classroom.

Purpose

To build a simple telephone network, similar to the ones that Alexander Graham Bell's company developed in the early years of the telephone, and to analyze the advantages and disadvantages of the arrangement and suggest improvements.

Materials

- three household telephones (the simpler the better)
- copper wire or alligator clips
- one 9-V battery or other energy source
- one 300-ohm resistor
- three phone jacks
- switches (optional)
- screwdriver
- wire strippers

FIGURE 4.32A The inside of a phone jack

Procedure

1. Set up your telephones as shown in Figure 4.32B. The copper wires should be attached to the screw terminals for the green and red wires on the phone jacks.

   ✓ **Self-Check**

   Make sure your telephones look similar to this before continuing to the next step.

2. Try to talk to each other on the phones. What happens? Explain your answer.

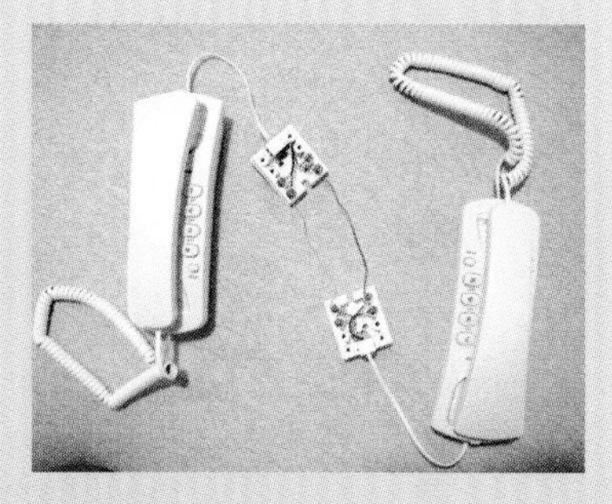

FIGURE 4.32B

Be Safe!

If you use bare copper wire, do not allow the wires to cross once they are connected to the energy source.

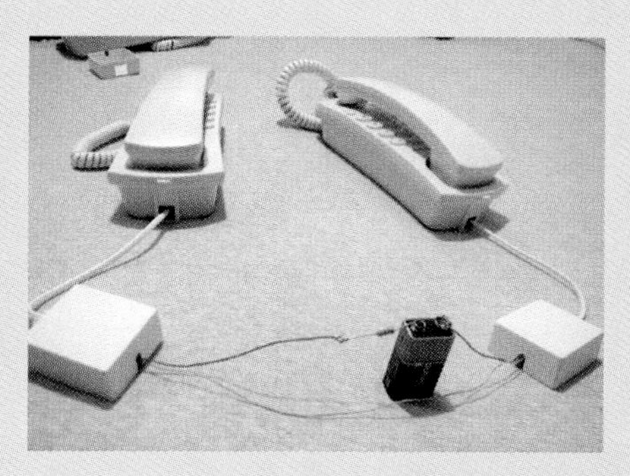

FIGURE 4.33

❸ Add the resistor and energy source to the circuit as shown in Figure 4.33.

✓ **Self-Check**

Make sure your telephones look similar to this before continuing to the next step.

❹ Try to talk to each other on the phones. What happens? Explain your answer.

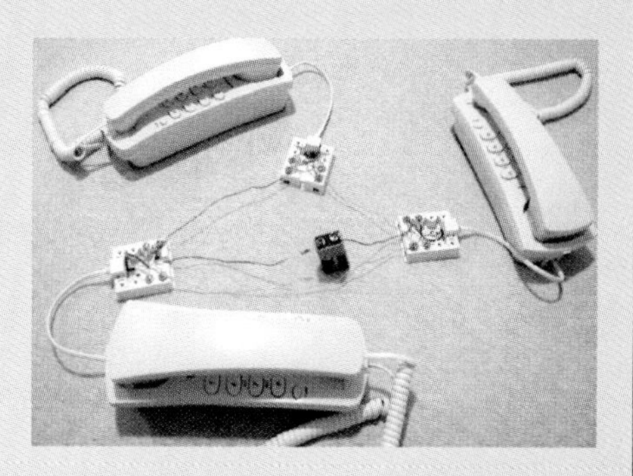

FIGURE 4.34

❺ Add a third phone to the circuit as shown in Figure 4.34.

✓ **Self-Check**

Make sure your telephones look similar to this before continuing to the next step.

❻ Try to talk to each other on the phones. What happens? Explain your answer.

Analysis and Conclusion

❶ What is necessary in order for the telephone system to work?

❷ Imagine that the network you built is the phone network in a small town.

a) List one advantage of your system.

b) List two disadvantages of your system.

c) Make suggestions as to how you could improve this system.

Extension and Connection

❸ How could you incorporate switches into your system so that people could have private conversations? Draw a diagram to show your design. Build it if you have time.

ScienceWise Fact

During a power outage, or blackout, phones still work. That is because all telephone companies have a backup energy source. However, if the blackout went on long enough, the phones would eventually stop working.

4.4 Communications Technology

In the previous section, you learned how different types of energy were changed back and forth to allow us to communicate using telephones and radios. Television, another form of communication, is an application of waves, where light energy is converted to electrical energy then back again. But how is it done?

Television: Light to Electricity

All metals conduct electricity, but some metals are better at conducting electricity than others. If you shine a light on certain metals, electrons are emitted. This is called the **photoelectric effect**. When light is shone on a metal such as selenium, selenium's resistance to the flow of electricity changes. The more intense the light, the more current can flow. A video camera uses this effect.

The picture that a video camera captures is made up of thousands of individual bits of the original scene. Look closely at a photograph in a newspaper. See Figure 4.35. You will notice that it is actually made up of a series of dots. A video camera produces a similar picture. It sends each dot of the picture, one at a time, to the television studio, but does it so quickly that it looks as if it were sent to your television all at once.

The television picture is read in lines. A typical North American television screen is made up of 525 lines of dots.

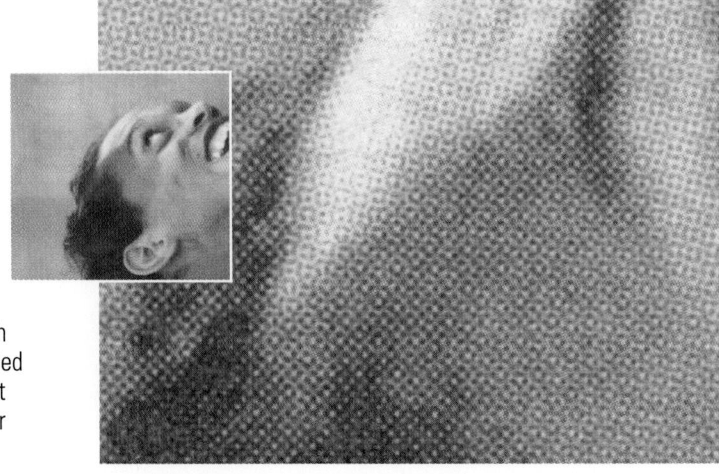

FIGURE 4.35 This is a photograph of a newspaper photograph magnified several times. What is similar about a television image and a newspaper photograph?

A video camera uses its lens to focus light on a screen. The screen is a mosaic of photosites. **Photosites** are made of materials that are sensitive to light. Their resistance changes based on the amount of light absorbed. The photosite becomes capable of absorbing a large amount of electric charge when intense light hits it. When dim light hits it, it becomes capable of absorbing only a small amount of electric charge.

FIGURE 4.36 Red, green, and blue are the additive primary colours. Combining them in varying degrees produces all the colours that humans can see.

The photosites create a video picture, 30 times per second, by recording light intensity. We usually call these video pictures "frames." When you pause a videotape while watching it on television, you are looking at one video picture or frame.

How do we create or show colour? The cheapest method is to have individual filters, which are red, green, or blue, on each of the photosites. The human eye senses all colours by means of only three pigments in the cells of the retina. These pigments are sensitive to red, green, and blue light. We can create any colour we want the viewer to see by combining different amounts of red, green, and blue light. Refer to Figure 4.36.

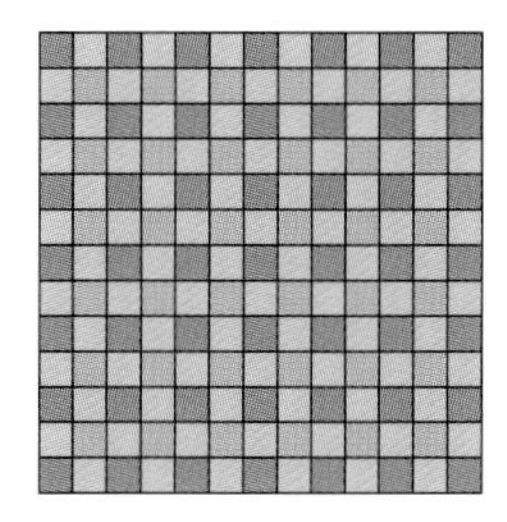

FIGURE 4.37 The Bayer filter

The Bayer filter (Figure 4.37) is the most common arrangement for colour filters. In Figure 4.38, we can see the red, green, and blue photosites that make up the image. The photosites are close enough together to be seen by the eye as a single colour.

A video camera will rapidly record every other line of dots or **pixels**, then go back and record the remaining lines. This signal is sent through a cable or on radio waves. Along with this signal is information to help your television know what part of the image it is receiving.

The sound is recorded separately from the image using a microphone, and transmitted as described in Section 4.3.

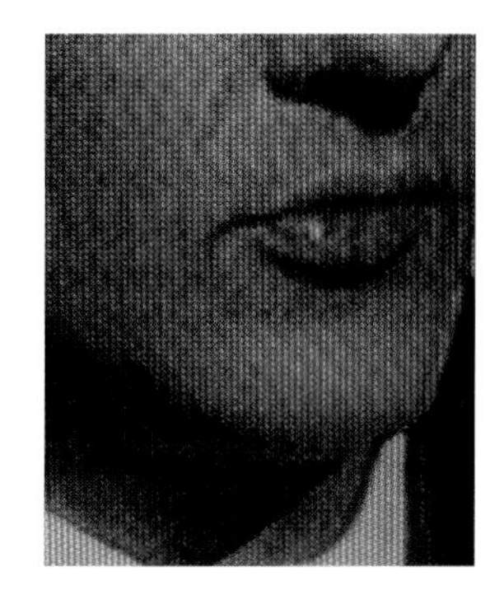

FIGURE 4.38 A close-up of a TV image. The colours are spaced in a way that allows us to see a complete colour picture.

Television: Electricity to Light

A television screen converts electricity to light. The picture you see on a television screen is made up of many pixels that, put together, form an image.

The picture tube has a **cathode** (negative terminal) at one end that produces electrons. These electrons are attracted to the **anode** (positive terminal) nearby. An accelerator anode increases the speed of the electrons. When they hit the phosphors on the television screen, the screen glows wherever the electrons hit. **Phosphors** are chemicals that glow when hit with electrons. Another name for the picture tube is the **cathode ray tube**.

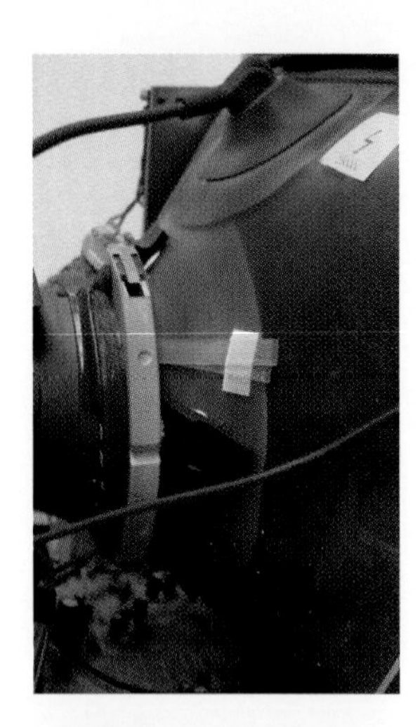

FIGURE 4.39 A television tube consists of a cathode ray tube surrounded by coiled wires that are able to move the beam of electrons.

In a television, coiled conductors surround the cathode ray tube. See Figure 4.39. When electricity flows through the coiled conductors, a magnetic field is produced. This field interferes with the electrons, making them change direction. One coiled wire moves the electron beam up and down and the other, left and right. The size of the current going through the coil determines how much the electron beam is drawn away, or deflected.

The point is to make the electron beam scan from left to right and top to bottom in an orderly way. The beam of electrons will hit one point on the screen for a moment, then move on to another spot. The smaller the area affected at each stop, the clearer the picture. The more times the entire screen gets scanned per second, the more natural the movement on the screen appears. The standard in North America is 525 lines, 30 times per second. Televisions with more lines produce sharper pictures.

Colour Television

Colour television adds another feature. That is, it uses three electron beams: one for each of the three colours: red, green, and blue. The phosphors on the colour television screen are arranged in the same way as the photosites in the video camera. Some are red, some are green, and some are blue.

What Is a Television Signal?

In order to get a colour image on a television screen, the television just needs to be told which colour beam to shoot at the screen and when to shoot it.

Television signals are broadcast on radio waves or sent through cables. A colour television signal contains five pieces of information for your television set:

- intensity information for the beam as it paints each line
- horizontal retrace signals to tell the television when to move the beam back to the beginning after the end of each line
- vertical retrace signals to tell the television when to move the beam from bottom right to top left (This signal is sent 60 times per second.)
- a signal indicating colours
- sound

A **composite video signal** consists of the first three pieces of information. This is enough information for a silent, black-and-white video. It is amplitude modulated. Look back at Section 4.3 about AM radio if you are unsure of what that means. The sound signal is frequency modulated.

Traffic Jam on the Airwaves: HDTV to the Rescue?

There are billions of cell phones, pagers, television channels, and radio stations. North America is running out of space (bandwidth) for broadcasting information. One way to fix this problem is to send the information digitally. Numbers do not require as much bandwidth as an analog signal. At present, cell phones and pagers are digital, but this is not enough. Most TV stations in North America use an analog signal. However, they are now in the process of changing to digital broadcasting.

High Definition Television (HDTV) is the adopted standard for digital television in Canada and the United States. It provides a sharper image than analog television. HDTV has over 700 lines of resolution, compared to 525 lines of resolution supplied by most analog televisions. The dimensions of the picture are different too. The ratio for HDTV is 16:9 instead of the 4:3 on an analog television.

The CRTC (Canadian Radio-television and Telecommunications Commission) is encouraging all television broadcasters to ensure that "by 31 December 2007, two thirds of each broadcaster's schedule is available in the HDTV format." An analog television cannot recognize this digital signal.

Does this mean you have to throw away your television? Not necessarily. An external digital converter can be used to convert the signal from digital to analog.

Whether you choose to use a converter or buy a digital television, HDTV is the wave of the future.

FIGURE 4.40 Which is the HDTV image? How do you know?

ACTIVITY 4D

Effect of Telecommunications Innovations on Society

What are we able to do now that we could not do before we had telephones, radio, or televisions?

In 1980, the Internet as we know it did not exist, cordless and cellular phones were too expensive for most people, and personal computers (known as microcomputers) had just been invented. In this activity, you will design a poster demonstrating how telecommunications innovations have affected society.

What You Will Need

- pen
- paper
- markers
- construction paper
- Bristol board

What You Will Do

1. With your group, brainstorm things involving telecommunications that we can do now that could not be done in 1980. You may want to ask people who were born before 1980, or do research on the Internet or in the library.
2. Relate the list you made in step 1 to the workplace. What kinds of things involving telecommunication are we able to do in the workplace now that could not be done in 1980?
3. Relate the list you made in step 1 to society. What kinds of things are we able to do in society now that we could not do in 1980? Think locally (your school area, city, or province) and then globally (your country, the world).
4. Relate the list you made in step 1 to the environment. What kinds of things are we able to do that we could not do in 1980? Think of things we can do now to find out more about the planet and the environment. Think of ways telecommunications innovations can help the environment. Also think of ways telecommunications innovations may damage the environment.

5. Design a poster that includes your answers to steps 2–4. Use as much paper as necessary. While designing your poster, answer the following questions:
   a) What message would you like to get across?
   b) Who is the target audience?
   c) Where will your poster be displayed?
   d) Depending on your answer to (c), will people have a great deal of time to read the poster?
   e) How will you catch the attention of your target audience?

6. Make your poster.

What Did You Find Out?

Compare the ideas in your group's poster to those in your classmates' posters.

1. Were your ideas similar? If so, which one(s)?
2. Were there any ideas you had not thought of? If so, which one(s)?
3. What changes would you make to your poster? Why?

Making Connections

4. List four careers that may involve designing this kind of poster.
5. Think of the telecommunications devices you use every day. What would happen if you did without one of them for a day? Try it. Summarize how this did or did not affect your typical daily routine.

Review and Apply

1. Abboud works at an advertising agency. He has decided to take a frame from the videotape of a commercial he wants to use in a campaign. When he blows it up on the photocopier, the picture in the frame is unclear. Explain what happened.

2. Bjonka runs the electronics department at the local department store. This morning, she noticed a customer place a magnet on the screen of one of the televisions. Now, that television screen shows all the colour in the centre of the screen but nowhere else. What happened? Do you think it can be fixed? Explain your answer.

3. Home movie theatres have become quite popular. High-quality speakers usually use powerful magnets. The speakers that are placed next to the television must be lined with metal shielding. Why?

4. Right now, there are more televisions on Earth than there are people. Television screens and computer monitors contain materials that are hazardous to the environment.
   a) Think of ways that we can dispose of or recycle television sets to minimize damage to the environment. How could the government encourage or enforce your ideas?
   b) Using the computer, design a flyer that a municipality might produce, explaining how people should dispose of old television sets and computer monitors.

5. Add the new concepts from this section to the graphic organizer you started in Section 4.1.

6. **Create a Timeline**

The three most important communications devices in our society—telephone, television, and radio—were developed around the same time.

Use the Internet or the library to research the development of the telephone, television, or radio. Create a timeline. Your timeline should include names of scientists, entrepreneurs, and engineers, as well as dates and illustrations. Don't forget to include recent developments of the device you have chosen.

4.5 Chapter Summary

Now you can...

- Explain, using waves, how energy from communications devices is transmitted, reflected, and absorbed by different kinds of matter (4.1, Lab 4A, 4.2)
- Describe and explain, using examples, the effects produced by the refraction and total internal reflection of visible light waves as they pass through different transparent media (Lab 4B)
- Explain how different forms of energy can be transformed into and transmitted as waves (4.3)
- Examine and describe the operation of transducers that carry out the energy transformations in common communications equipment (4.3)
- Describe the historical development of a significant product of communications technology (4.4)
- Describe, using scientific principles, the functioning of common home and workplace communications technologies (4.4)
- Select and use appropriate instruments to collect observations and data (Labs 4A, 4B, 4C)
- Construct and test a prototype of a communications device and resolve problems as they arise (Lab 4C)
- Describe the impact of developments in communications technology on the way we work and on our social environment (Activity 4D)

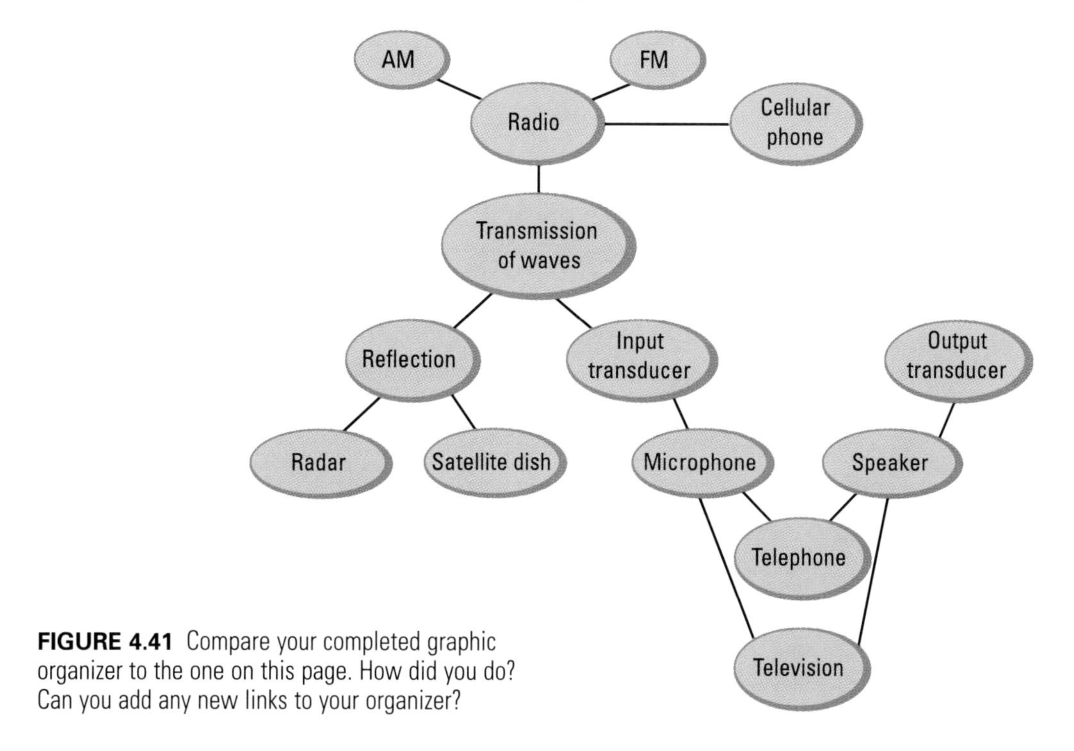

FIGURE 4.41 Compare your completed graphic organizer to the one on this page. How did you do? Can you add any new links to your organizer?

CHAPTER 4 *review*

Knowledge and Understanding

1. What are the frequencies of your favourite AM and FM radio stations? If radio waves travel at 300 000 000 m/s, calculate the wavelength of each of your favourite two radio stations. (Remember that speed = frequency × wavelength.)
2. Explain, in your own words, why transmitters and antennas are always made of uninsulated metal.
3. Distinguish between each of the following terms:
   a) input transducer, output transducer
   b) dynamic microphone, carbon microphone
   c) FM radio, AM radio
   d) reflection and refraction of waves
4. What property of metals allows us to convert light energy into electrical energy and vice versa?

Inquiry

5. Work with a partner to decide how you could determine the speed of the water waves coming into shore at the beach. What equipment would you need?
6. Keisha has built a radar set that can produce and detect radio waves. If the radar set sends a signal with a speed of 350 m/s and detects the echo 5 s later, how far away is the object?
7. Jim has found a microphone in his basement. He needs to determine if it is a dynamic microphone or carbon microphone before he can decide if he will use it with his computer. How could you figure out which type of microphone it is?
8. Our environment is filled more and more with radio waves and electromagnetic fields. Do these kinds of radiation pose an environmental or health risk? Use various sources such as the Internet or library to find out current views on this issue.
9. One theory about birds that migrate is that they are able to find their way through a "sixth sense" that detects magnetic lines from the north to south pole of the Earth.
   a) Might our communications technologies interfere with this "sixth sense"? Why or why not?
   b) Design an experiment to determine if radio transmissions affect a bird's sense of direction.
   c) Would it be a good idea to perform the experiment you designed in b)? Why or why not?

Making Connections

10. Devon has leased some of his farmland to a local radio station. The radio station has built a transmitter there. Ever since the transmitter was in use, Devon has heard static coming out of his stainless steel toaster. What do you think is happening?

11 Suppose you owned an electronics store that sells many televisions. You know that digital televisions are the newest thing, but they are expensive and many people cannot afford them.

a) How would you determine whether people might invest in a digital television before buying the inventory?

b) How would you determine how much floor space should be devoted to analog televisions and how much floor space should be devoted to digital televisions?

c) How might you convince someone who already owns an analog television to buy a digital one? (You may have to do research on the Internet or at the library to answer this question.)

12 Our communications technologies require specialized equipment (at both the consumer's end—television sets—and the producer's end—television studios, cables) and communications towers. Do these requirements have any impact on the environment? For example, when a telephone is manufactured, electricity is used. What impact does that have on the environment? When we throw away a television, we need somewhere to dispose of it. What does that do to the environment? List environmental impacts in a data table like that in Figure 4.42, which you will create in your notebook. You may need to do research on the Internet or at the library.

Communication

13 Your eight-year-old neighbour, Ravinder, wants to know why echoes happen. How would you explain it to her?

14 Some residents in a small community are convinced that the new radio tower that has been set up just outside town produces radioactive radiation. As a member of the municipal council, it is your job to explain to them (a) that no radioactive radiation is being produced, and (b) how radio waves are transmitted. Design a flyer to explain how radio waves are transmitted.

Communications device	Environmental impacts when communications device is... Manufactured	Used	Disposed
Telephone (including central office, telephone wire, and telephone poles)			
Radio (including transmitters)			
Television (including studios, transmitters, satellites, and so on)			

FIGURE 4.42

PUTTING IT ALL TOGETHER

Courier Communication

Courier companies move packages from place to place for businesses. The courier companies must be fast and efficient in order to stay competitive and make a profit. The central office alerts couriers when calls come in for a pickup or delivery. This requires an efficient and reliable communications network.

FIGURE 1B Couriers must be able to get from one place to another very quickly. Why are bike couriers often used in urban areas? Identify some of the dangers of this job.

The Project

You are a communications consultant—a person that companies hire to examine their communications needs and to design a solution that will work. A local courier company has hired you to do a cost–benefit analysis of the various options available for communicating with its group of couriers. A cost–benefit analysis is a comparison of the advantages and disadvantages, including the price, of similar products. You will decide on the two options you feel are best, and do a presentation of the information.

What You May Need

The materials needed will vary from group to group, depending on your design. You may need the following materials:

- paper, pencil
- Bristol board
- calculator
- computer
- spreadsheet program
- presentation software
- overhead projector

What You Will Do

1. The courier company has 10 drivers, and on an average day delivers 100 packages. Assume the courier company operates in your local municipality or region. When someone calls the company for a pickup, the dispatcher gives the driver the pickup address and any other important information. The drivers never come into the central office, but must advise the dispatcher if they are delayed or if they have to take an unusual route. Dispatchers need to have an idea where each driver is in order to know which one is closest to each pickup point.

You will choose two of the following possible communications solutions to investigate based on availability, cost, and efficiency:

- pay phones
- pagers
- cellular phones
- CB radio
- Global Positioning System (GPS)
- Mobile Data Terminals (MDT)
- two-way radios (point-to-point or with a repeater)

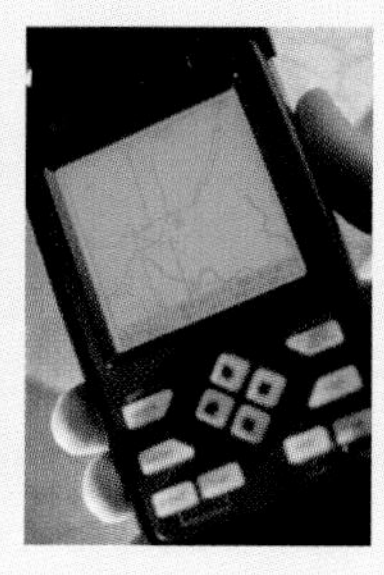

FIGURE 2B The type of communications device you use depends on the information you need to transmit and the area in which the signals will be transmitted.

2. Keep in mind the following:
   - You can mix and match any of the solutions in step 1.
   - There are many types of pagers and cellular phones, each with different costs.
   - There are usually several suppliers for the equipment.
   - You need to consider the cost of start-up as well as the ongoing costs.
   - The number of transmitters in your area will have an effect on some solutions.
   - The weather in your area may affect some solutions.
   - The terrain in your area may affect some solutions.

FIGURE 3B If there are few transmitters in your area, cellular phones may not be a reliable solution.

3. Design a chart to keep track of your information. Research the two communication solutions you chose in step 1 for cost of equipment and monthly operating costs. In some instances, different companies may provide the service. Be sure to investigate each one to figure out which is cheapest.

CONTINUED

4. Decide which of your two choices is the best solution for the courier company. Support your decision based on availability, cost, and efficiency. Report on what you found when researching the other option so that the company has a choice. Remember that the company may prefer to spend a lot of money now and very little in monthly charges or it may prefer to spend very little now and more on a monthly basis.

   List three advantages and three disadvantages for each of the two solutions you have chosen.

5. Once you have chosen your solution, describe a typical pickup scenario.
   - How will the drivers be contacted?
   - How will they get the necessary information?
   - How can they contact the dispatcher once they are on their way to the pickup destination?

6. Organize your findings in a 20-minute presentation. If possible, use audio-visual equipment. Your presentation must include the following:
   - description of how each solution works and the cost
   - your recommendation for the company with three supporting reasons
   - a handout that you can leave with the company so that decision-makers can refer to it

7. Answer the following questions:
   a) What turned out to be the most important consideration when you were deciding what was best for the courier company?
   b) Hiring a consultant costs money. Why do you think a company would choose to hire a consultant?
   c) Telecommunications has changed dramatically in the past 20 years. Could your solution have been used in 1980? Why or why not?

Assessment

8. Make notes during other groups' presentations. What did you think they did well? What did you like best about your presentation? If you were to do this activity again, what changes would you make to your presentation? Why?

Medical Technology

UNIT 3

CHAPTER 5: Medical Technology and Genetic Disorders
CHAPTER 6: Medical Advances
PUTTING IT ALL TOGETHER: The Future of Medical Technology

CHAPTER

5

Medical Technology and Genetic Disorders

Genetic disorders (or **genetic diseases**) are medical conditions that can be passed on from parent to child. Since they cause permanent changes in how the human body functions, genetic disorders can be difficult to treat. In this chapter, you will learn about the important role of medical technology in detecting and treating this type of human illness. You will also learn about some other ways that medical technology related to genetic disorders is used. Look at the photographs in Figure 5.1. How does knowledge of genetic material help to protect the health of the mother and her unborn child? What role does medical technology play in the foods in the grocery store? How is knowledge of genetic material used to solve crimes?

FIGURE 5.1

What You Will Learn

After completing this chapter, you will be able to:

- Differentiate between DNA, chromosomes, and genes (5.1)
- Describe the role of chromosomes in inheritance (5.1)
- Explain how genetic disorders may occur (5.2)
- Describe how karyotypes and pedigrees are used to determine inheritance of genetic disorders (5.2)
- Understand some terms used in medical and reproductive technology (5.1, 5.2, 5.3)
- Describe scientific and technological principles involved in genetic engineering (5.3)
- Provide examples of how science and technology have influenced the diagnosis and treatment of human illness (5.2, 5.3)

What You Will Do

- Follow the transmission of traits using models of chromosomes (Activity 5A)
- Analyze the inheritance of a genetic disorder using a pedigree (Activity 5B)
- Work as a member of a team to research and present information on an issue related to genetic technology (Activity 5C)
- Compile, organize, and interpret data, using appropriate formats and treatments (Activities 5A, 5B, 5C)
- Select and use appropriate ways to communicate scientific ideas and results (Activities 5A, 5B, 5C)

Words to Know

allele
base pair
biotechnology
chromosome
cloning
deoxyribonucleic acid (DNA)
dominant allele
gametes
gene
genetic disorder
genetic trait
genetic engineering
genome
karyotype
meiosis
mitosis
mutation
pedigree
recessive allele
technology
trait

A puzzle piece indicates knowledge or a skill that you will need for your project, The Future of Medical Technology, at the end of Unit 3.

5.1 Inheritance of Traits

We each have a combination of features that makes us unique. Perhaps you are tall with dark hair and skin. Maybe you are of average height with green eyes and curly eyelashes. Many of your unique features will have been inherited, or passed on, from your biological parents.

FIGURE 5.2 Different dog breeds have specific genetic traits that are passed on to the next generation. What are some of the genetic traits of these two examples?

DNA: The Genetic Material

Traits are characteristics of an organism. For example, the traits of a cat include a long tail, pointed ears, and soft hair. **Genetic traits** are traits that are inherited and can be passed on to the next generation (Figure 5.2). Not all the traits that organisms have are inherited. For example, if a particular cat has lost part of its tail in an accident, the short tail is not a genetic trait of that cat and would not be passed on to any of its kittens.

Genetic traits are passed from generation to generation through our genetic material, a molecule called **deoxyribonucleic acid**, or **DNA**. DNA contains all the instructions, written in a chemical code, for a particular individual. DNA is like a blueprint for a particular living thing.

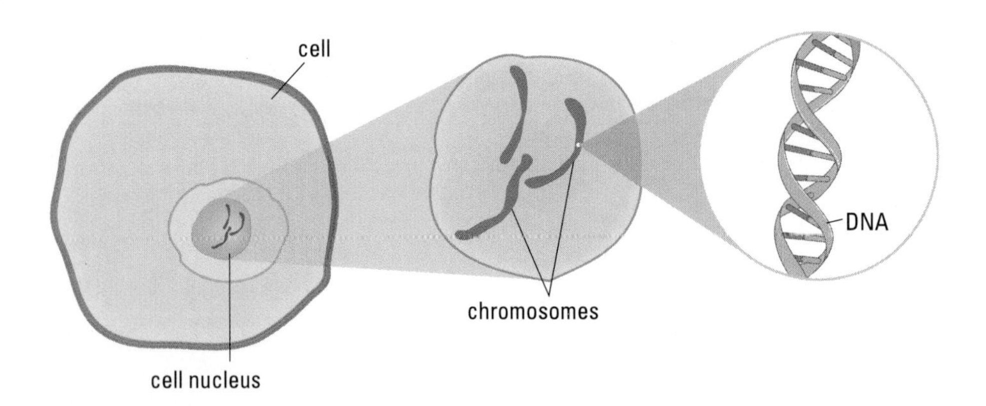

FIGURE 5.3 The genetic material, DNA, is packaged into chromosomes that are found in the cell nucleus.

DNA is a large molecule arranged in a double helix, a shape that is similar to a spiral staircase (Figure 5.3). The double helix is made up of chemicals called **base pairs**, which can be thought of as letters in an alphabet. Just as letters are arranged into words, these base pairs are arranged into different genes. A **gene** is a piece of DNA that contains the code for a particular genetic trait. Genes are usually hundreds of base pairs in length. In humans, DNA is found in the nucleus of every cell of the body, in structures called chromosomes. A **chromosome** is a piece of DNA containing many different genes. Each chromosome can be thought of as a paragraph composed of many different words. Most organisms have more than one chromosome. Humans, for example, have 46 individual chromosomes. The **genome** of an organism is made up of all the genes on all the organism's chromosomes. The genome is like a complete story made of many paragraphs (chromosomes), which are made up of many words (genes), composed of different letters (base pairs).

Chromosomes and Mitosis

Every cell in your body contains all your genetic information, packaged into 46 different chromosomes. During your lifetime, you will make many new cells, all of which must contain a complete set of these 46 chromosomes. New cells (daughter cells) are made by the process of mitosis. **Mitosis** is a form of cell division that gives rise to daughter cells with the same amount of DNA as the parent, or starting, cells. Figure 5.4 shows the products of mitosis for an organism that contains eight chromosomes.

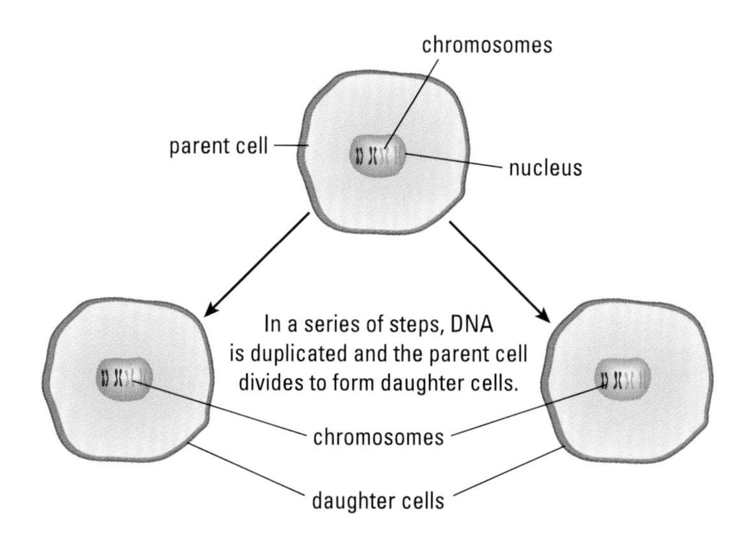

FIGURE 5.4 Products of mitosis. After mitosis, both daughter cells have the same number of chromosomes as the parent cell.

Chromosomes and Meiosis

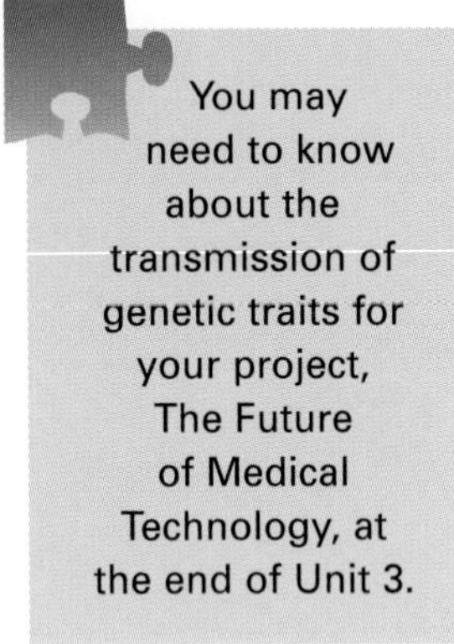

You may need to know about the transmission of genetic traits for your project, The Future of Medical Technology, at the end of Unit 3.

You have probably noticed that the offspring of organisms that undergo sexual reproduction have some traits from the male parent and some from the female parent. This is because half of the DNA of the new individual is donated from the male parent and half from the female parent, through the process of meiosis. **Meiosis** is a form of cell division that gives rise to daughter cells with half the amount of DNA as the parent cell. **Gametes**, or egg and sperm cells, are produced by meiosis.

Recall that the cells of the human body contain 46 chromosomes. These chromosomes are actually a set of 23 pairs of chromosomes. Through meiosis, each gamete receives one member of each of these chromosome pairs, or 23 chromosomes. Figure 5.5 shows the products of meiosis for an organism with two chromosome pairs.

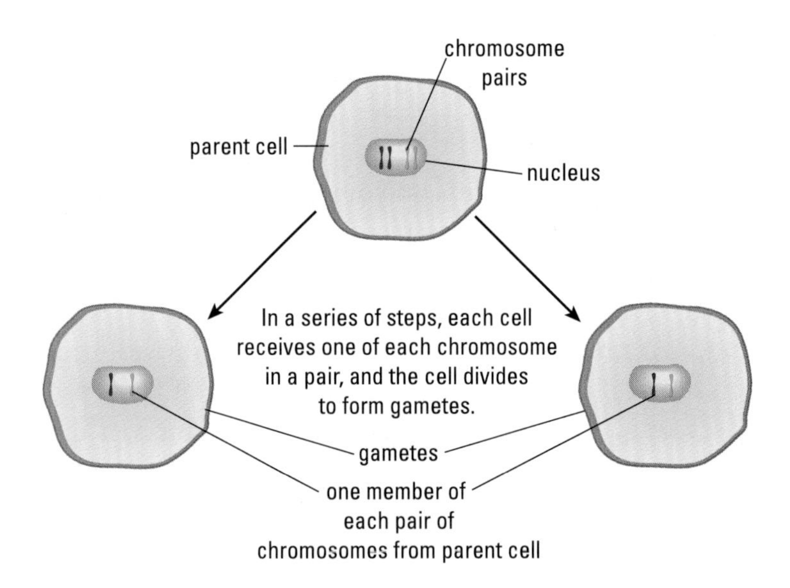

FIGURE 5.5 Products of meiosis. During meiosis, one member of each pair of chromosomes is distributed to each gamete. In humans, each gamete has a total of 23 chromosomes.

Females produce only egg cells, and males produce only sperm cells. During sexual reproduction, an egg and sperm cell unite to form a new individual. All of the chromosomes from the egg and sperm cell are combined, so the new individual again has two copies of each chromosome. One of these copies comes from the male parent and one from the female parent, so the new individual has a mixture of genes from both its parents. For example, a new human individual would again have 46 chromosomes, or 23 chromosome pairs.

Chromosomes and Transmission of Traits

Although they are very similar, the two chromosomes in a chromosome pair are not identical. Recall that each chromosome carries a set of particular genes. Both chromosomes in a pair carry the same set of genes. For example, both chromosomes in a chromosome pair may carry a set of genes that includes the genes for the traits of eye colour and height. However, each chromosome in the chromosome pair could carry different forms of these genes. Each different form of a specific gene is called an **allele**. For example, one chromosome in the pair may carry the alleles for brown eyes and short height. The other chromosome in the pair may carry the alleles for blue eyes and average height (Figure 5.6).

Each new individual therefore has two copies of each gene, and each copy is carried on one chromosome of a chromosome pair. Each gene can have more than one form, or allele. The two gene copies in a chromosome pair can be the same allele, or different alleles. However, usually only one of these alleles appears as a physical trait in the individual. This is determined by whether the allele is dominant or recessive. When an individual has two different alleles for a gene, the allele that appears physically is called a **dominant allele**. The allele that does not appear physically is called a **recessive allele**.

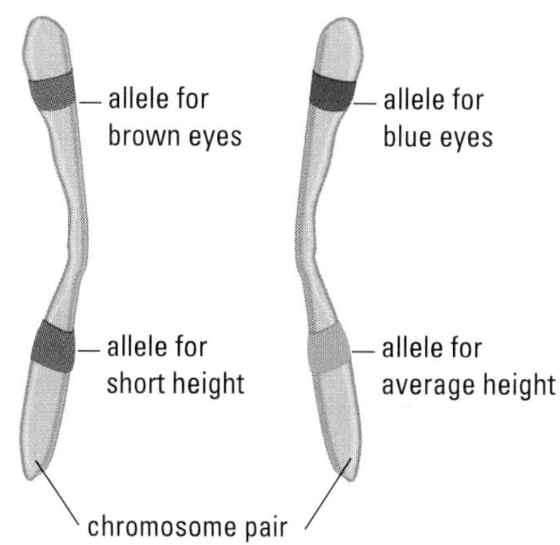

FIGURE 5.6 An allele is a particular form of a gene.

Recessive and Dominant Inheritance

To understand dominant and recessive alleles, consider the case of purebred dogs. When a dog is a purebred, it has certain recognizable traits that always occur in that dog breed. For example, a beagle always has long, floppy ears. The parents of a purebred beagle also had long, floppy ears, as did their parents, and their parents before them, down a long line of parents. The gene for ear shape is carried on one chromosome pair in the dog's DNA. One chromosome in this pair came from the female parent and one from the male parent. Since the parents, grandparents, and great-grandparents all had floppy ears, we know that both copies of the gene for ear shape are the allele for floppy ears. In other purebred dogs, both copies of the ear-shape gene may be an allele for other ear shapes.

When the parents are different breeds, however, all the puppies will have a mix of the traits of both breeds (Figure 5.7). What traits show up in the puppies is determined by whether an allele is dominant or recessive. In the example in Figure 5.7, a floppy-eared male was crossed with a pointed-eared female. Since the puppies get one chromosome from the male parent and one from the female, all the puppies have one allele for floppy ears and one allele for pointed ears. The puppies all have the floppy ears of their male parent, so this allele must be the dominant allele. The allele for pointed ears from the female parent must then be the recessive allele. The trait that arises from the dominant allele is called a dominant trait. The trait that arises from the recessive allele is called a recessive trait.

Whether a trait is dominant or recessive determines how likely it is that offspring will have a particular trait. This information is important to understanding genetic disorders in humans and other organisms, and can be used to diagnose genetic disorders. You will learn more about genetic disorders in Section 5.2.

FIGURE 5.7 When a purebred dog with two alleles for pointed ears (pointed/pointed) is crossed with a purebred with two alleles for floppy ears (floppy/floppy), the puppies all receive one allele for pointed ears and one allele for floppy ears (pointed/floppy). Why do the puppies all have floppy ears?

Review and Apply

1. Describe the function of DNA.
2. Put the following terms in order from most complex to least complex: DNA, cell, chromosomes, and genes.
3. Draw a diagram that shows what happens to chromosomes during the formation of egg and sperm cells. Label your diagram, and include the name of this process.
4. Describe what happens to chromosomes when egg and sperm cells unite. Explain how this process relates to the inheritance of traits.
5. You run a business breeding purebred dogs. One of the male dogs has long hair, and a female has short hair. All the puppies from these dogs have short hair. State which of these traits is dominant and which is recessive.
6. Organize the concepts you have learned in this section in a graphic organizer.

Try This at Home

Can You Do This?

Can you roll your tongue? Tongue rolling is a dominant trait, so if you can roll your tongue, you have at least one allele for tongue-rolling. Answer the following questions to determine which of these other human traits you have:

a) Do the bottoms of your earlobes hang freely?
b) Do you have dimples in your cheeks?
c) Do you have a cleft chin?
d) Does your hairline dip down in the centre of your forehead?
e) Do your thumbs bend back at the end?
f) Do you have blue eyes?

1. In this list, a), d), e) and f) are recessive traits. How many recessive traits did you have?

2. For each of the recessive traits you have, state whether you have no, one, or two allele(s) for the trait.

ACTIVITY 5A

Transmission of Traits

Figure 5.7 showed the physical traits of puppies from a cross of two different breeds of dogs. The parent dogs carry either two alleles for floppy ears or two alleles for pointed ears.

In this activity, you will use models of chromosomes to follow the transmission of these alleles when two purebred dogs are crossed. You will follow the transmission of these alleles from the parents to the offspring shown in Figure 5.7, and relate this to dominant and recessive inheritance.

What You Will Need

- materials to make models of chromosomes, such as modelling clay, construction paper, or felt
- paper and pencil

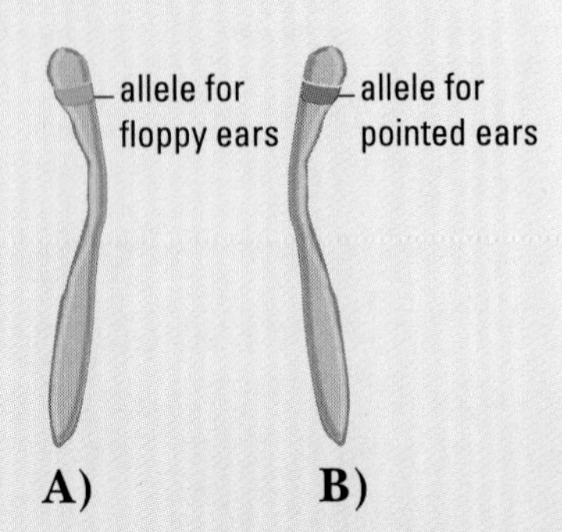

FIGURE 5.8 The gene for ear shape has two possible alleles. The left chromosome A) carries the allele for floppy ears, and the right chromosome B) carries the allele for pointed ears.

What You Will Do

1. The chromosome that carries the gene for ear shape is represented in Figure 5.8. The chromosome in Figure 5.8A has the allele for floppy ears, and the chromosome in Figure 5.8B has the allele for pointed ears.

   Make models to represent the chromosome pair that carries the gene for ear shape for each of the following dogs:

   a) one male purebred dog and one female purebred dog with pointed ears

   b) one male purebred dog and one female purebred dog with floppy ears

2. Sketch each of your models in your notebook. Beside each sketch, write down the appropriate physical traits of the dog that is represented.

3. Gametes are formed by the process of meiosis, which produces daughter cells containing one member of each pair of chromosomes from the parent cell. Using your chromosome models, create models of the chromosome that carries the gene for ear shape, for all the possible products of meiosis for the male and female dog in step 1a).

4. Sketch each of the models in your notebook, and state which allele for ear shape each gamete contains.

5. During sexual reproduction, two gametes combine to create a zygote. Each gamete receives one member of a chromosome pair from each of its parents. Using your chromosome models, create models of all possible chromosome pairs that could be produced by crossing the male and female dog in step 1a).

6. Sketch each of the models in your notebook, and state the ear shape of the puppies that would result from each combination of chromosomes.

7. Predict the two alleles for ear shape that would occur in zygotes obtained by crossing the two dogs in step 1b). Include all possible combinations in your answer. State the ear shape of the puppies that would result from each combination of chromosomes. You may want to repeat steps 2–6 to help you make your prediction.

8. Repeat steps 2–6 for the case of a female purebred pointed-eared dog crossed with a male purebred floppy-eared dog.

What Did You Find Out?

1. The trait of floppy ears is dominant over that of pointed ears. Give all the possible combinations of alleles for ear shape for the puppies with floppy ears.

2. Two of the crosses that you modelled produced puppies with floppy ears. Which were they? Do these puppies have the same alleles for ear shape? Explain your answer.

Extension

3. Repeat steps 1–6 for a cross of a male and female dog, both of which have the chromosome pair shown in Figure 5.9.

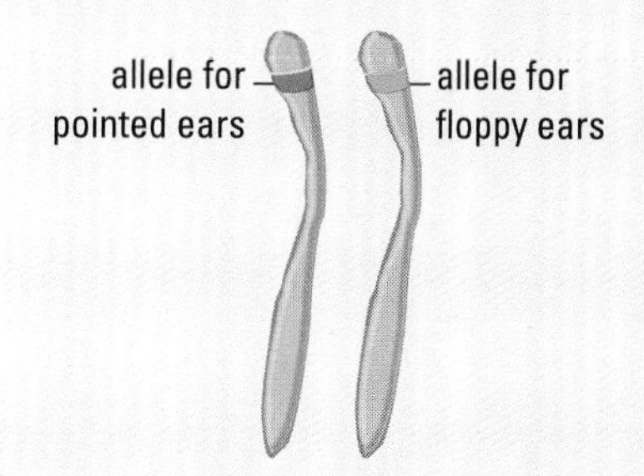

FIGURE 5.9 What ear shapes might be found in the puppies from a cross of animals that both carry this chromosome pair?

Making Connections

4. Based on your observations with the model chromosomes, why do dog breeders value purebred animals more than mixed-breed dogs?

5.2 Genetics and Human Health

You may need to know about genetic disorders for your project, The Future of Medical Technology, at the end of Unit 3.

A **genetic disorder** is an inherited problem with a function of the body. Some genetic disorders are quite common, such as myopia, a condition that limits a person's ability to see objects in the distance. Other genetic disorders are rare, such as hemophilia, a disease that affects a person's ability to form blood clots.

Genetic disorders are caused by **mutations**, or changes in the DNA of the affected individuals. Mutations can be very obvious, such as chromosomes that are missing parts or are rearranged, or a change in the total number of chromosomes. Other mutations are just small changes in the DNA code. However, even these small changes can cause serious disorders, so it is important to be able to detect any mutations.

Karyotypes

A **karyotype** is a photograph of all the chromosomes in the nucleus of a single cell. To make a karyotype, a cell is first treated with a chemical that colours, or stains, the chromosomes, so that they can be seen. The chromosomes are then photographed and the photograph organized so that both members of a pair are beside one another. Karyotypes are used to determine the number, size, and shape of all the chromosomes in an organism. Major changes in the chromosomes can be detected on a karyotype.

Recall that humans have a total of 46 chromosomes per cell, or 23 pairs of chromosomes. Two of these chromosomes are sex chromosomes, which are the chromosomes that determine if a person is male or female. Figure 5.10 shows the karyotypes of a male and a female without any genetic diseases. Males have one X and one Y chromosome, and females have two X chromosomes. The sex chromosomes of a male are the only chromosomes in which each member of a pair looks significantly different from the other. A karyotype is labelled first with the number of chromosomes and second with the type of sex chromosomes. For example, a male would be 46 XY, and a female would be 46 XX.

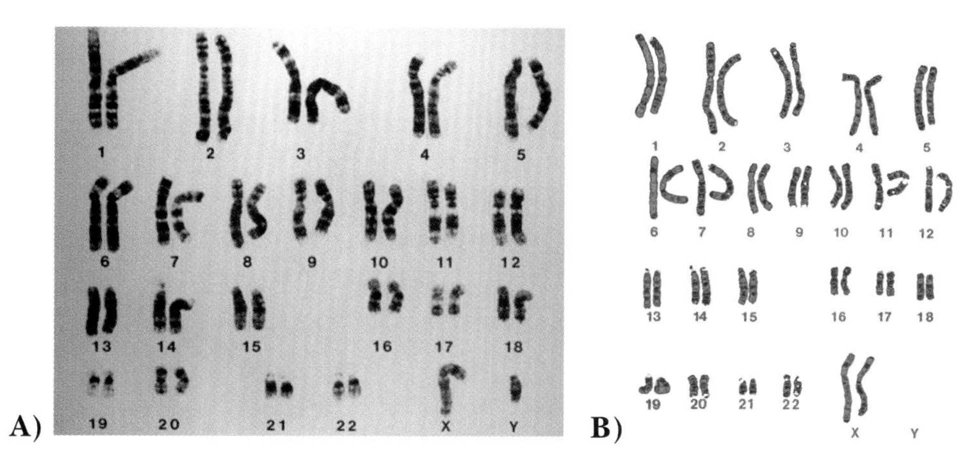

FIGURE 5.10 The karyotype of A) a male and B) a female. There are 23 pairs of chromosomes in each karyotype, arranged according to their lengths. Find the X and Y chromosomes of the male. How do these compare to the two X chromosomes of the female?

Some genetic disorders are caused by changes in the chromosomes that can be detected in a karyotype. For example, people with Down syndrome have three copies of chromosome 21, rather than just two. People with Down syndrome therefore have a total of 47 chromosomes (Figure 5.11).

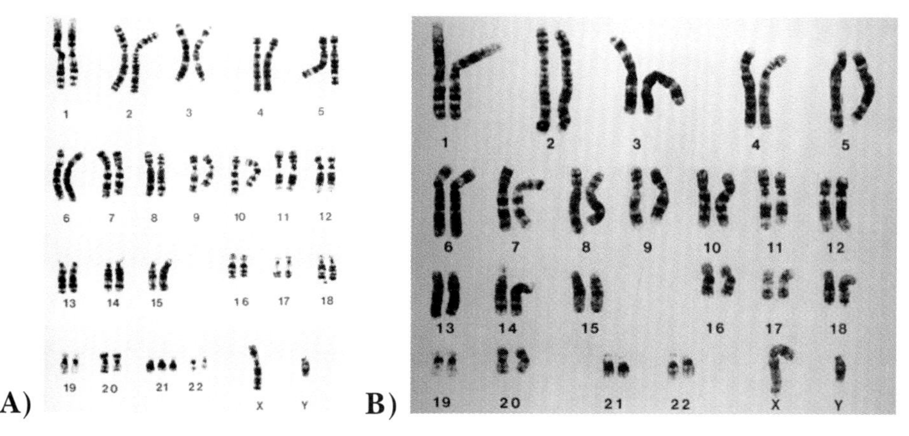

FIGURE 5.11 Karyotypes of males A) with and B) without Down syndrome. Look at the chromosomes labelled "21." Describe the change in the chromosomes of the person with Down syndrome.

The extra chromosome 21 causes many differences in the physical traits of affected people. People with Down syndrome usually have short, stocky bodies and a thick neck. Many have a large tongue that can make speech more difficult. Down syndrome can also make a person more susceptible to infections. They may also have problems with their heart and other organs. Although people with Down syndrome usually have below-average intelligence, there is a wide range of mental capacity among affected individuals.

Pedigrees

Only a few genetic diseases are caused by changes in the DNA severe enough to be detected in a karyotype. Another way of deciding if a disease is genetic is to use a pedigree. A **pedigree** is a graphical representation that shows to whom a person is related and how. Pedigrees can also be called family trees. Pedigrees trace the presence and absence of a particular trait in each individual within a group of related people.

There are a number of different ways that the individuals with a particular trait can be identified. In the example in Figure 5.12, males are represented by squares and females by circles. Individuals with the trait of tallness are denoted by the colour green, and individuals without the trait, by the colour red. As you can see, pedigrees are useful for analyzing the inheritance of any trait, not just genetic diseases. For example, pedigrees can be used to help establish whether or not someone is likely to be a member of a particular family. This can be important in some legal matters, such as establishing parental rights.

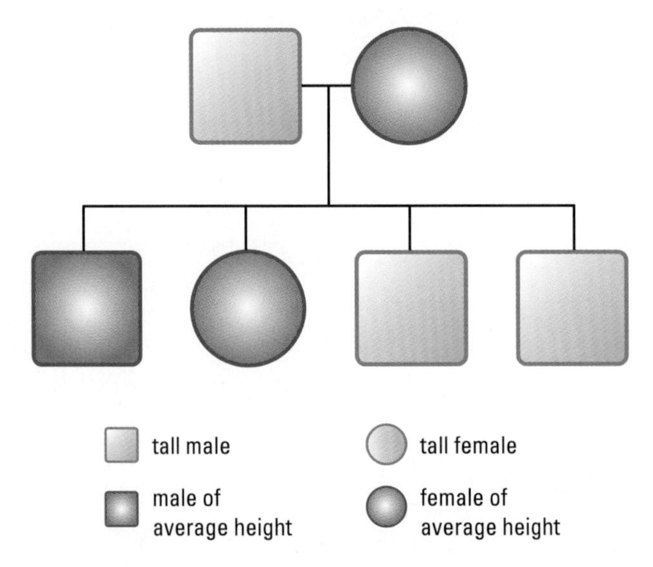

FIGURE 5.12 A pedigree shows which particular individuals, in a group of biologically related people, have a particular trait, in this case tallness.

Drawing a Pedigree

ACTIVITY 5B

Albinism is a genetic disorder in which a person produces much less pigment than normal, or no pigment. People with this condition are described as albino. In this activity, you will create a pedigree for a family affected by this genetic disorder.

FIGURE 5.13 People with albinism produce little or no pigment in their skin, hair, and irises.

What You Will Need

- pencil and paper

What You Will Do

1. The pedigree must indicate whether each individual in the group is male or female, and which individuals are affected by albinism. Decide on the symbols you will use.
2. The grandmother and grandfather of the family were not albino. Start the pedigree with these two individuals, and then add the information in steps 3–7.
3. These grandparents had three children—two boys and one girl. One of the boys was albino.
4. Each of the three children married a person of the opposite sex who was not albino.
5. The girl of the family had two children, a boy and a girl. Neither of these children were albino.
6. The boy with normal pigmentation had three children, two girls and one boy. None of these children were albino.
7. The boy who was albino had four children, two girls and two boys. One boy and one girl had normal pigmentation. The other boy and the other girl were albino.

What Did You Find Out?

1. Explain how your pedigree shows that albinism is a genetic disorder.
2. Look at the number of people in this family who inherited albinism. Based on this information, do you think albinism is a dominant or a recessive allele? Explain your answer.

Making Connections

3. Pedigrees are also used in animal breeding, to track traits that are wanted in a particular animal. Having the pedigree of an animal can increase its worth. Suggest reasons why a known pedigree can be helpful to animal breeders.

Review and Apply

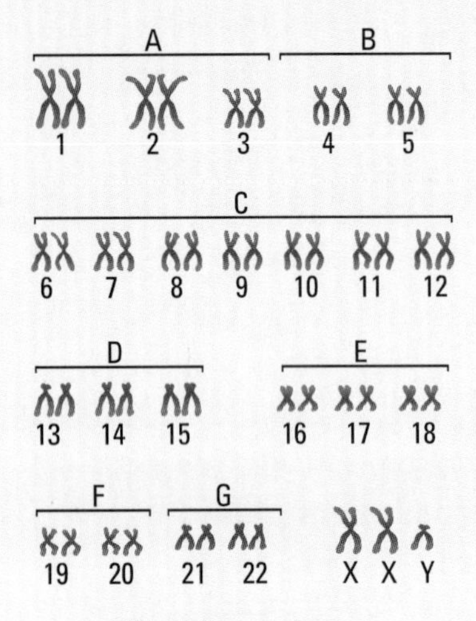

FIGURE 5.14 Karyotype of a person with Klinefelter syndrome.

1. Review the definition of the term "genetic disorder," and explain what causes a genetic disorder.
2. In paragraph form, describe how a karyotype is made.
3. Can all genetic disorders be detected on a karyotype? Explain your answer.
4. Describe the information that is included in a pedigree.
5. Figure 5.14 shows the karyotype of a person with a genetic disorder called Klinefelter syndrome. Describe any differences between the chromosomes of this person and a person without Klinefelter syndrome. You may wish to review the karyotype in Figure 5.10A, which was prepared from a person with no genetic disorders.
6. When you visit the doctor, you may be asked to fill out a questionnaire that includes questions about the health of your biological relatives. Explain why this information can assist the doctor in monitoring your health.
7. Add the new concepts from this section to the graphic organizer you started in Section 5.1.
8. **Animal Tracks**

You and two classmates have started a business breeding purebred Siamese cats. Siamese cats can have different colours on their "points" — their ears, paws, noses, and tails. You have two males, Scooch and Taz. Scooch is a chocolate point, but he has crossed eyes. Taz is a seal point, but his eyes are straight. Jasmine is your female cat. She has chocolate points and uncrossed eyes. Jasmine has just given birth to a litter of kittens, all of which have chocolate points and crossed eyes. If chocolate points is a recessive trait and crossed eyes is dominant, which male is the father of the kittens? Draw a pedigree to present your answer.

Surf the Web

Find out more about Klinefelter syndrome. What are the symptoms of this disorder? How is it detected? Is there a treatment? Start your research by visiting **www.science.nelson.com** and following the links for ScienceWise Grade 12, Chapter 5, Section 5.2.

■

Job Link

Teacher Assistant

Teacher assistants help elementary and secondary school teachers and counsellors in their duties.

FIGURE 5.15 A career as a teacher assistant can be challenging and rewarding.

Responsibilities of a Teacher Assistant

Depending on where they work, teacher assistants may perform any or all of the following responsibilities:

- assist students with lessons under the direct supervision of a classroom teacher
- assist in preparation of learning materials and environment
- accompany and supervise students during activities in school gymnasiums, laboratories, libraries, and resource centres, and on field trips
- assist students with special needs—such as those with mental or physical disabilities—with their mobility, communication, and personal hygiene
- carry out behaviour modification, personal development, and other therapeutic programs, under supervision of professionals such as special education instructors, psychologists, or speech-language pathologists
- monitor students during recess or lunch

Where Do They Work?

- public and private elementary and secondary schools

Skills for the Job

- ability to work with children
- good verbal and written communication skills
- outgoing, friendly personality

Education

- most jobs require completion of secondary school
- some positions may require college courses or specialized training and experience

5.3 Biotechnology

You may need to know about biotechnology for your project, The Future of Medical Technology, at the end of Unit 3.

What do you think of when you hear the word "technology?" You probably think of a piece of machinery, such as a computer or a vehicle. **Technology** is the application of scientific facts to solve human problems. **Biotechnology** is the application of biological science facts to solve human problems. Biotechnology can offer new ways to treat and cure diseases, among other benefits, but it also causes concerns among some people.

Cloning

The oldest biotechnology is cloning. **Cloning** is the process of making a genetic duplicate of an organism. Cloning is used in agriculture to increase the number of plants. Many plants can be cloned simply by taking a cutting and placing the cutting in water until it develops roots (Figure 5.16). The value of clones is that the agricultural producer knows exactly what the characteristics of the new individual will be. For plant producers, this allows them to know exactly the shape and colour of flowers that a plant will produce, for example. By cloning the most valuable plants, growers can increase their profits. You will learn more about cloning plants in Section 7.2.

FIGURE 5.16 Many houseplants can be cloned by taking cuttings. The cutting will grow its own roots, producing a clone of the parent plant.

In 1997, scientists in Scotland successfully produced a living clone of a sheep. This sheep, named Dolly, was the genetic duplicate of her mother (Figure 5.17). Scientists have since successfully cloned other animals. By cloning animals, they hope to be able to produce many copies of animals with the traits that humans value the most. For example, cloning could be used to produce herds of identical pigs that all produce leaner meat, or herds of identical cows that all give more milk. Other examples of benefits

FIGURE 5.17 Dolly, the first cloned agricultural animal

of cloning are to increase the number of endangered species, especially those that are difficult to breed in captivity. Cloning could also be used to produce identical animals for medical research that might save human lives.

Surf the Web

Find out more about how cloning is being used to help bring back the numbers of endangered species. Start your research by visiting **www.science.nelson.com** and following the links for ScienceWise Grade 12, Chapter 5, Section 5.3.

Concerns about Cloning

Cloning of plants and animals has the potential to produce hundreds of individuals all of the same genetic makeup. This situation could change the genetic diversity of some species. Many scientists are concerned that cloning could cause unforeseen consequences. For example, cloned plants or animals would all have exactly the same ability to fight off disease. If one clone could not fight off a particular disease, then none of them could. As a result, all the clones could be wiped out at once.

Other people worry that human clones will be produced, and argue that this is not morally right. Many governments have written laws to limit or ban cloning involving humans, but nevertheless, some scientists are working toward this goal.

Genetic Engineering

Genetic engineering is biotechnology that involves adding or changing specific genes in an organism. Genetic engineering is used in medicine and agriculture to produce particular traits in an organism.

For example, genetic engineering is used to produce large quantities of human insulin. Insulin is a chemical that is normally produced in the body, which regulates the metabolism of sugar. Diabetes is a genetic disorder in which a person does not produce enough insulin. Many diabetics must take daily injections of insulin in order to survive.

Before the 1980s, insulin for treatment of diabetes was obtained from pig tissues. Insulin from pigs is not exactly the same as human insulin, and it is expensive to produce. Today, large quantities of human insulin are available at a lower cost because of genetic engineering.

ScienceWise Fact

Insulin was discovered by two Canadian scientists, Frederick G. Banting and Charles Best. In 1923, Banting was awarded the Nobel Prize for this discovery, but not Best. Banting so strongly believed that the discovery was made by shared work with his fellow scientist that he gave half of the substantial cash award to Best.

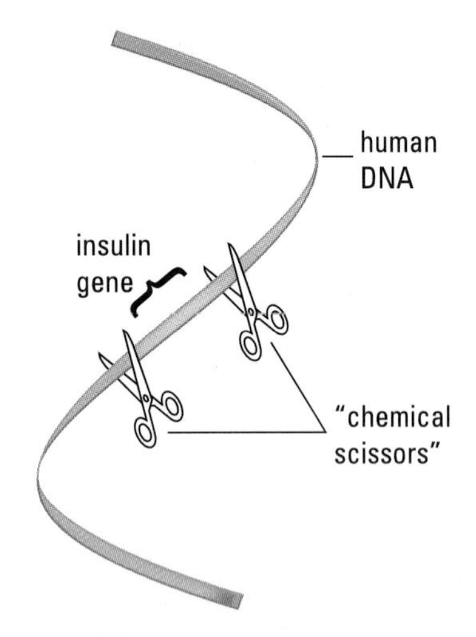

FIGURE 5.18 The gene for insulin is removed using special "chemical scissors" that cut only in particular places along the DNA. Only part of the DNA is shown in the diagram.

The first step in the genetic engineering of human insulin is to prepare a sample of human DNA. The gene for insulin is then removed by treating the DNA with a special chemical that cuts DNA at particular places. These chemicals can be thought of as chemical scissors (Figure 5.18).

The next step is to prepare DNA from a bacterial cell. This DNA is in the form of a circle, which is first cut open with "chemical scissors." The piece of DNA containing the gene is then inserted into the bacterial DNA, and the circle is sealed up again. The bacterial DNA with the human gene is then transferred into another bacterial cell (Figure 5.19).

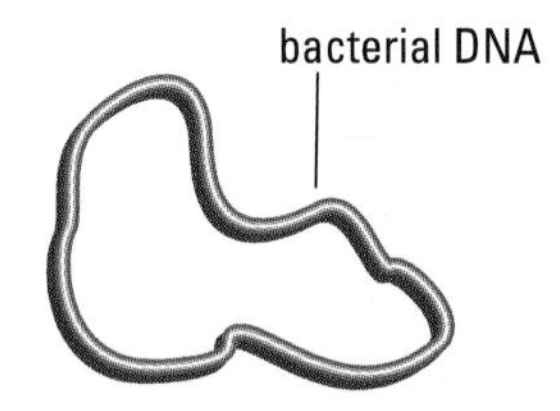

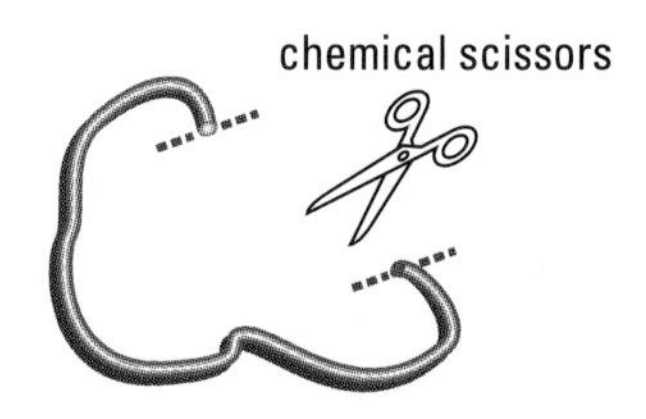

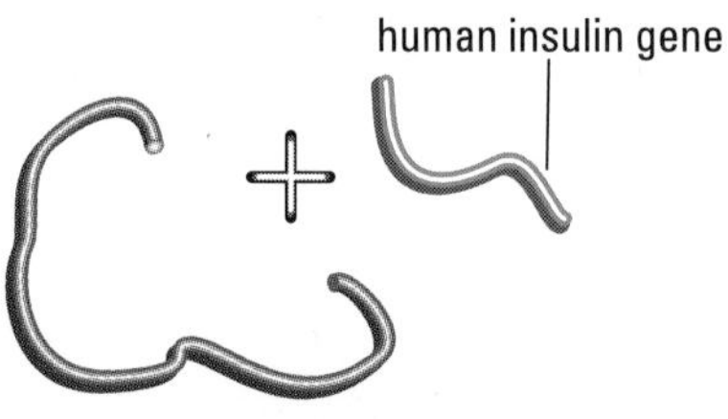

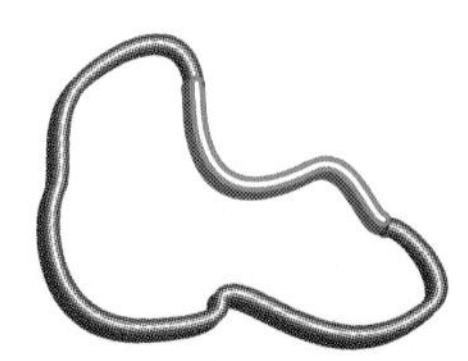

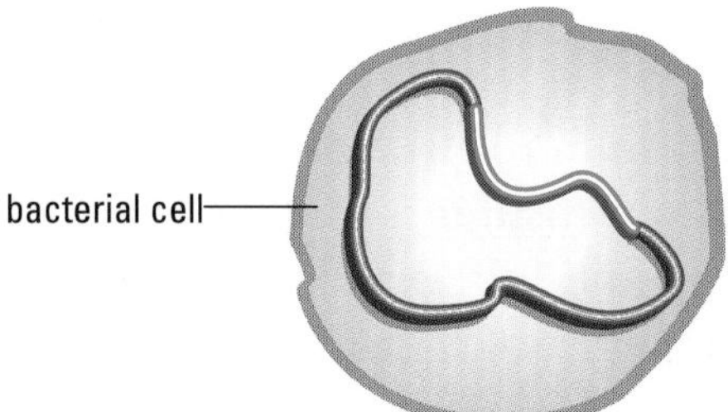

FIGURE 5.19 The human insulin gene is placed into bacterial DNA, then transferred to a bacterial cell (steps 1–5).

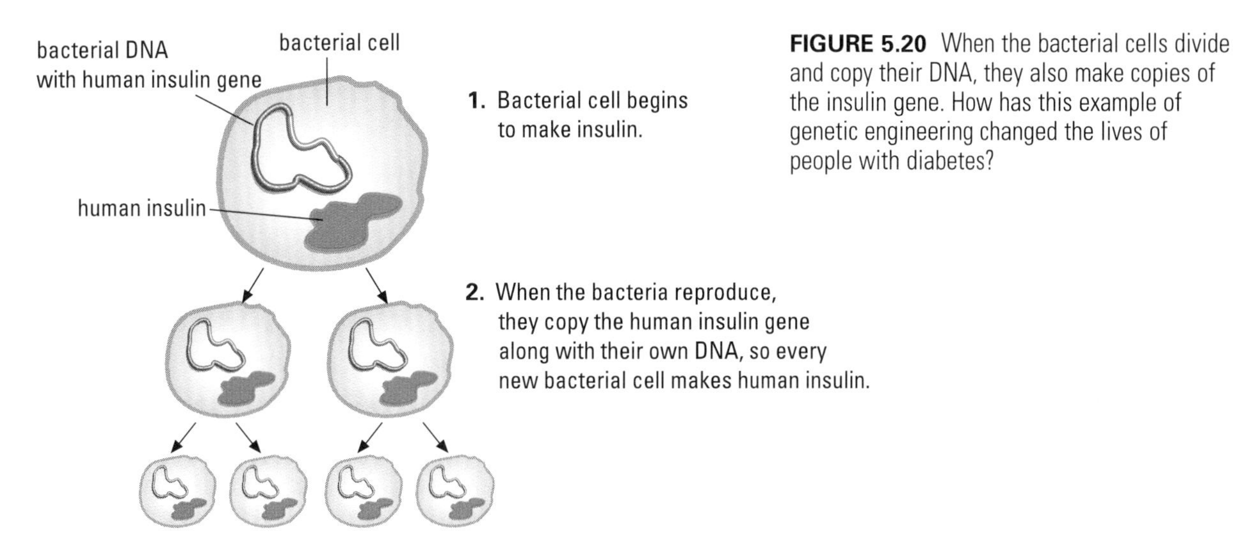

FIGURE 5.20 When the bacterial cells divide and copy their DNA, they also make copies of the insulin gene. How has this example of genetic engineering changed the lives of people with diabetes?

Because the bacterial cell cannot tell the DNA of the insulin gene from its own DNA, the bacterial cell begins to make human insulin (Figure 5.20). Bacterial cells reproduce very quickly. Like all cells, they make copies of all their DNA when they reproduce, including, in this case, the inserted human insulin gene. As a result, there are soon millions of bacterial cells making insulin. This is much more insulin than can be produced in an animal. Since bacteria are single cells, it is also a lot easier to harvest the insulin from bacterial cells than from animals.

Concerns about Genetic Engineering

The use of genetic engineering to produce medicines is quite common and has been accepted as beneficial by most people. However, some scientists use genetic engineering in ways that may cause concern. For example, genetic engineering can be used to place the genes of other organisms into crop plants. This may be done to make a plant resistant to a disease or to a chemical spray, or to change the taste or nutritional value of the food. Some people are very uncomfortable with this process. For example, a person who is a vegetarian may not wish to eat fruits and vegetables that contain animal genes. There is also some evidence that the introduced genes can cross into other plants. No one can predict the effects this may have on the environment.

FIGURE 5.21 Biotechnology is used to produce some of our food. For example, tomatoes have been genetically engineered to improve their flavour.

ACTIVITY 5C

DNA Evidence and Crime

In this activity, you will work as a member of a team to research and present information on the use of DNA evidence in court. You will find out how the evidence is collected, and what the rules are for presenting the evidence in court.

Most DNA evidence is produced by a biotechnology technique called DNA fingerprinting. DNA fingerprinting uses repeated patterns in an individual's DNA as a tool for identification. By comparing the patterns in two DNA samples, scientists can determine if both samples are from the same person, from two related people, or from two unrelated people. DNA fingerprinting is used to compare the DNA (from blood, hair, skin cells, or other biological evidence) left at a crime scene to the DNA of a criminal suspect. Figure 5.22 shows the basic steps in DNA fingerprinting.

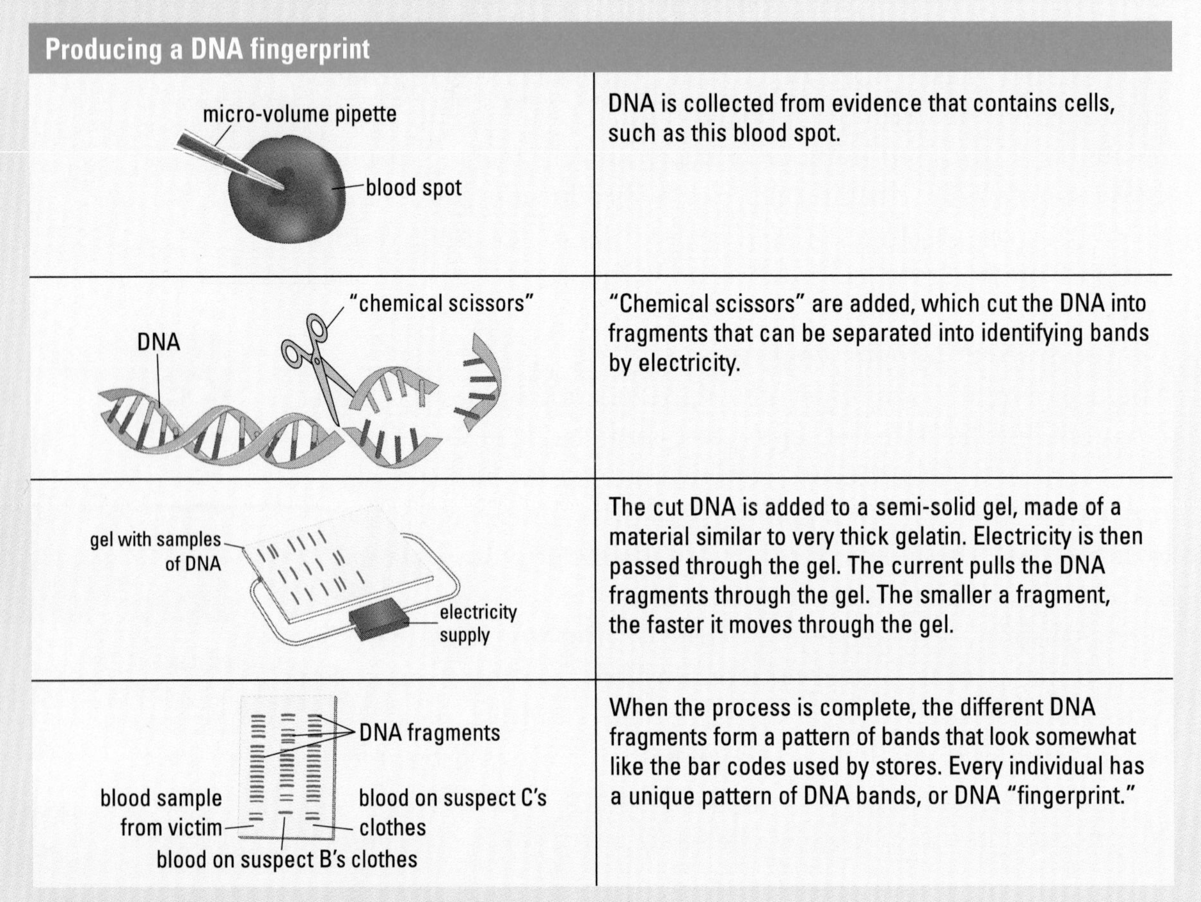

FIGURE 5.22 Each person has a unique DNA fingerprint, which can be obtained from even microscopic pieces of biological material.

What You Will Need

- pen and paper
- access to library or Internet resources to obtain recent news reports

What You Will Do

1. Working in a group, research recent news reports that concern the use of DNA fingerprinting in criminal investigations. Choose one of the cases to research more closely.
2. Using library and Internet resources, collect as much information as you can on how DNA fingerprinting was used in the case. Find out what kind of sample was used to produce the DNA fingerprint, whether other types of evidence were also important, who produced the DNA fingerprint, and how the DNA fingerprint was matched to a specific person.

What Did You Find Out?

1. How important was evidence from DNA fingerprinting to this case?
2. How was the DNA evidence collected?
3. What role did other evidence play in this case?
4. Prepare a poster or portfolio summarizing the case and how DNA evidence was used.

Making Connections

5. In June, 2000, the government passed the DNA Identification Act. This law gave the Royal Canadian Mounted Police the right to create and maintain a database of DNA fingerprints. Judges may order criminals convicted of violent crimes (such as sexual assault) to provide samples containing DNA, which is used to create a DNA fingerprint. This DNA fingerprint is kept in the national DNA databank. These DNA fingerprints can be used as evidence in future crimes, and also to investigate unsolved crimes.

   Prepare a convincing argument for or against the requirement that anyone accused of a serious crime must supply police with a sample of their DNA. Present your argument as an oral presentation, a video, or a written report.

Review and Apply

1. What is a clone?
2. In a paragraph, describe at least one advantage of, and one concern about, cloning.
3. You are the foreman of a large dairy farm. Describe how biotechnology could affect this workplace.
4. In point form, outline the steps of genetic engineering. Use diagrams in your answer.
5. Genetic engineering can produce plants with particular traits very quickly. In a paragraph, outline at least two advantages of this to an agricultural worker.
6. State at least two advantages of producing insulin by genetic engineering over collecting it from animals.
7. You own a bakery that specializes in baked goods that contain no animal products. Explain how your business might be negatively affected by genetic engineering of agricultural crops.
8. Add the new concepts from this section to the graphic organizer you started in Section 5.1.

Surf the Web

DNA evidence has been used to free some Canadians who were wrongfully convicted of murder. Who were these individuals? Find out how DNA fingerprinting helps the courts to convict the right person. Start your search by visiting **www.science.nelson.com** and following the links for ScienceWise Grade 12, Chapter 5, Section 5.3. Prepare a point-form summary of your findings.

■

5.4 Chapter Summary

Now You Can...

- Describe hereditary material (5.1)
- Explain the role of chromosomes in the transmission of genetic traits (5.1)
- Describe how karyotypes can be used to help identify abnormalities in chromosomes of an individual (5.2)
- Use a pedigree to form a hypothesis about the nature of a genetic disorder (5.2)
- Describe the process of genetic engineering (5.3)
- Discuss some advantages of and concerns about biotechnology (5.3)

Concept Connections

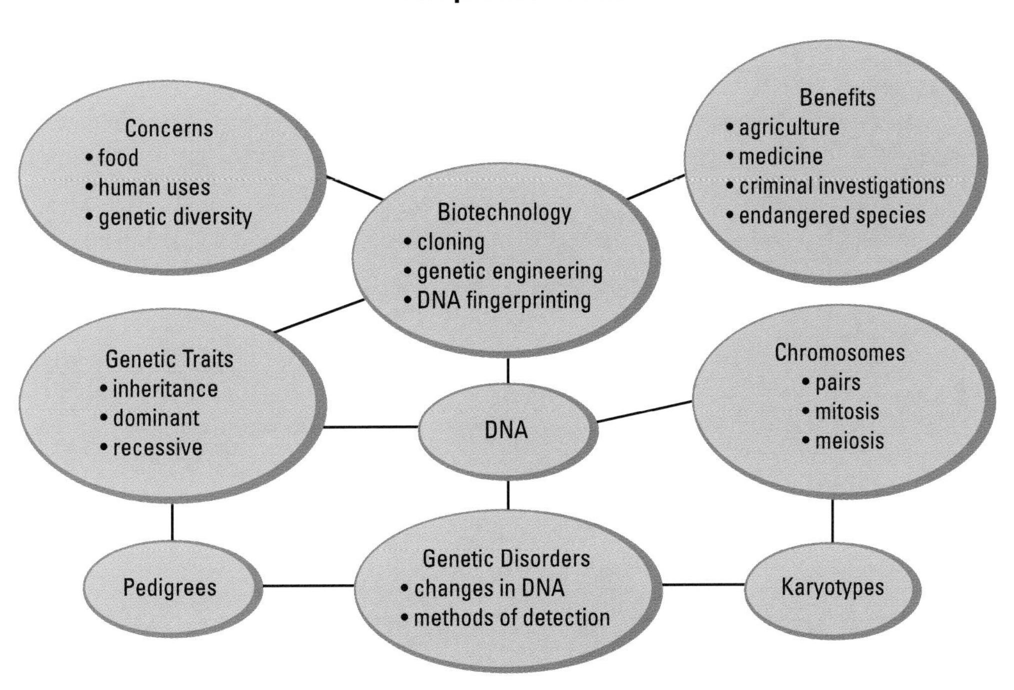

FIGURE 5.23 Compare your organizer with the one on this page. How did you do? What new links can you add to your organizer?

CHAPTER 5 *review*

Knowledge and Understanding

1. In your notebook, match the concept in the left-hand column with its description in the right-hand column of Figure 5.24.

2. In point form, describe how traits are passed down from biological parent to child.

3. Explain the difference between a karyotype and a pedigree.

4. Compare the number of chromosomes in a cell that resulted from mitosis, to the number of chromosomes in a cell that resulted from meiosis.

5. Describe at least three types of change in DNA that can cause genetic disorders.

6. Summarize how cloning is used by the plant nursery industry.

7. Explain how genetic engineering has improved the life of people who use insulin.

1. DNA	**A.** inherited problem with the functions of the body
2. Chromosomes	**B.** photograph of chromosomes that has been sorted into pairs
3. Karyotype	**C.** packages of DNA
4. Genetic disorder	**D.** graphical representation that shows how and to whom a person is related
5. Gene	**E.** the application of biological science facts to solve human problems
6. Pedigree	**F.** the genetic material
7. Dominant allele	**G.** an allele that appears physically in the offspring of two individuals with different alleles for a gene
8. Recessive allele	**H.** involves adding or changing specific genes in an organism
9. Cloning	**I.** an allele that does not appear physically in the offspring of two individuals with different alleles for a gene
10. Genetic engineering	**J.** a piece of DNA that determines a particular genetic trait
11. Biotechnology	**K.** process of making a genetic duplicate of an organism

FIGURE 5.24

Inquiry

8. Figure 5.25 shows the pedigree for a family that is affected by a genetic disorder, which is shown in blue. Identify the family members that are affected. Deduce whether this disorder is a recessive or dominant trait.

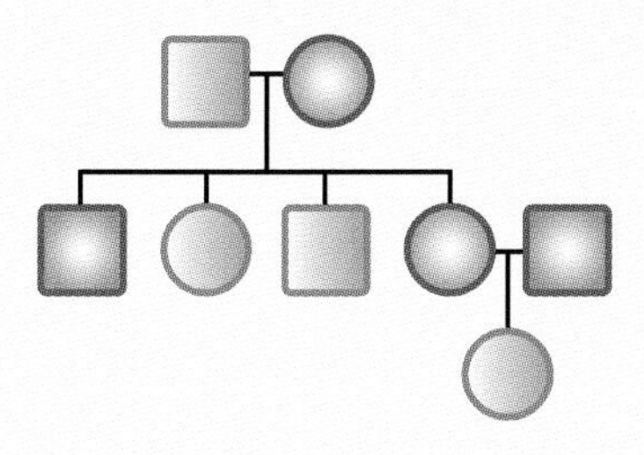

FIGURE 5.25

9. A newborn child shows symptoms of a genetic disorder caused by the presence of an additional chromosome. Which would be most useful in determining if the child has the disorder—a karyotype or a pedigree? Explain your answer.

10. In a certain type of aquarium fish, gold colour is a dominant trait and white is recessive. A gold-coloured fish is crossed with a white fish. Half of the offspring are white and half are gold. Produce a pedigree for these fish. Using the pedigree, deduce the alleles that were carried by each of the parent fish.

Making Connections

11. Read the employment section of a current copy of your local newspaper. Identify a job where some knowledge of genetics, genetic disorders, or biotechnology would help you to carry out that job. Write a paragraph about the job and its connections to these areas.

12. Genetic engineering is used to make large quantities of many medicines. In point form, outline the effects that this use of genetic engineering has had on people's health and on the economy.

Communication

13. With a partner, make a collage out of newspaper and magazine clippings to show the ways that biotechnology and genetics affect daily life at home, at school, and in the workplace.

14. Write a letter to the editor of your local newspaper stating whether you agree with the use of biotechnology to produce foods. Use facts to justify your ideas.

CHAPTER 6

Medical Advances

If you have ever used a thermometer to see if you had a fever, you have used a medical technology. Medical technology helps all of us maintain or improve our health. Some medical technology helps people with particular medical conditions to carry out their daily tasks more easily. Consider the medical technologies shown in Figure 6.1. How has X-ray technology helped in the diagnosis of medical conditions? What kinds of artificial limbs are available? What is ultrasound technology used for in medicine?

FIGURE 6.1

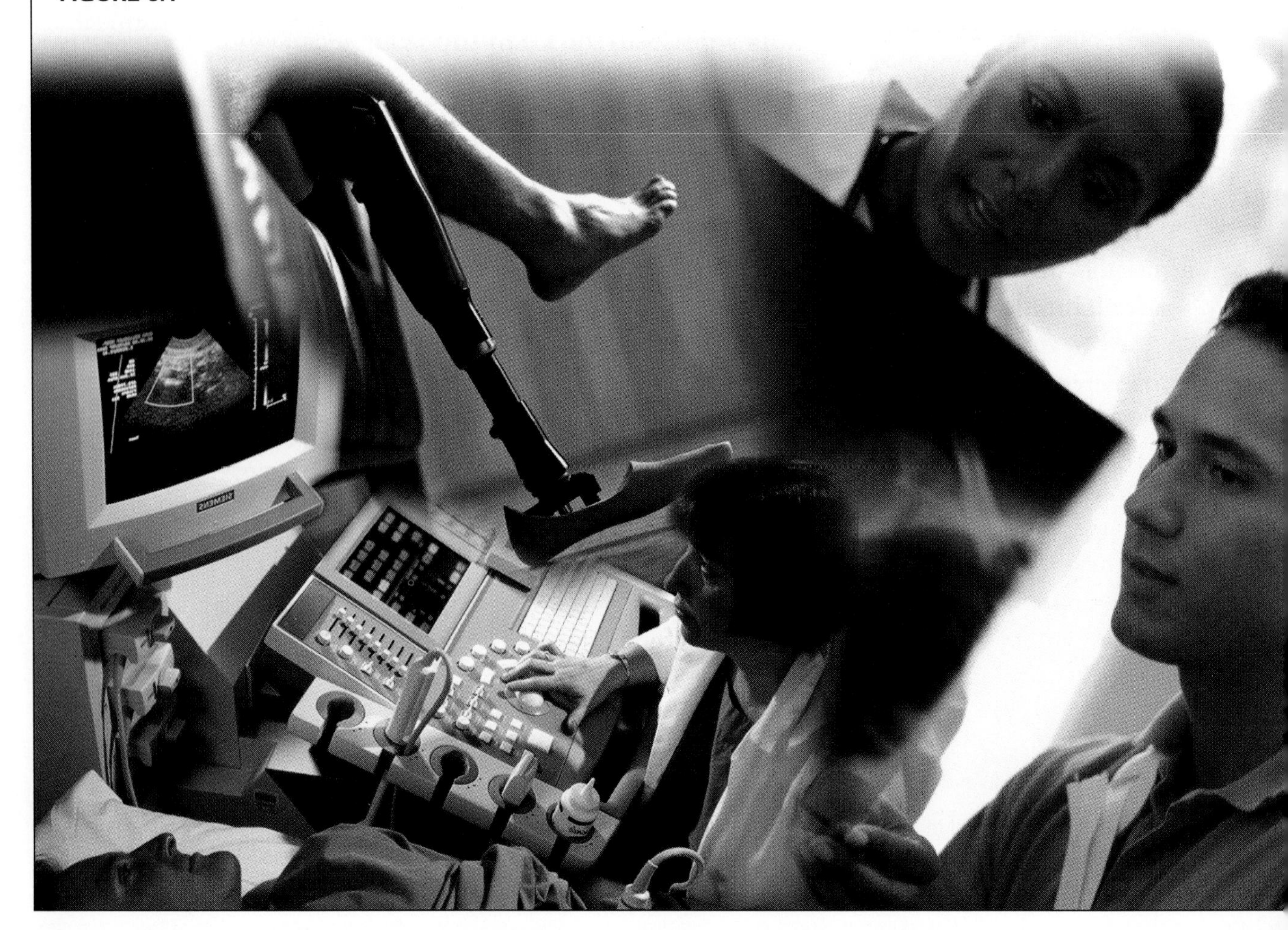

What You Will Learn

After completing this chapter, you will be able to:

- Understand some terms related to medical technology (6.1, 6.2, 6.3)
- Explain the use of technology for diagnosing medical conditions (6.1)
- Describe the use of technology for biomedical repair (6.2, 6.3)
- Provide examples of how science and technology have influenced the diagnosis and treatment of human illness (6.2)

What You Will Do

- Research and present information on the role of a medical technology in daily life (Activities 6A, 6C)
- Conduct a laboratory experiment simulating the process that occurs during kidney dialysis (Lab 6B)
- Select appropriate instruments and use them effectively in collecting and observing data (Lab 6B)
- Select and use appropriate methods to communicate scientific ideas and results (Activity 6A, Lab 6B, Activity 6C)
- Locate, select, analyze, and integrate information, using library and electronic resources (Activities 6A, 6C)
- Compile, organize, and interpret data in an appropriate format (Activity 6A, Lab 6B, Activity 6C)
- Communicate lab results and research for specific purposes (Activity 6A, Lab 6B, Activity 6C)
- Research careers related to medical technology (Activity 6C)

Words to Know

amniocentesis
chemotherapy
cochlear implant
computed axial tomography (CAT)
corrective lenses
defibrillator
diagnose
dialysis
echocardiogram
hearing aid
hemodialysis
joint replacement
magnetic resonance imaging (MRI)
medication
pacemaker
prosthesis
radiation therapy
sonogram
ultrasound
X-ray radiation

A puzzle piece indicates knowledge or a skill you will need for your project, The Future of Medical Technology, at the end of Unit 3.

6.1 Types of Diagnostic Technology

You will need to know about diagnostic technologies for your project, The Future of Medical Technology, at the end of Unit 3.

When a person seeks treatment for a medical condition, the doctor's first job is to **diagnose**, or determine the specific cause of, the condition. Depending on the condition, a diagnostic technology may be used. In this section, you will learn about some technologies used to diagnose medical conditions.

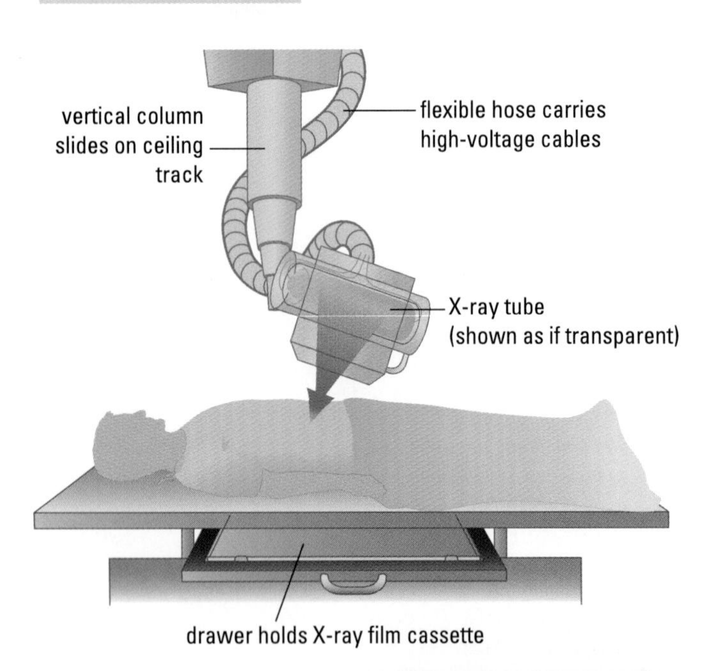

FIGURE 6.2 Important features of X-ray technology. Why does the X-ray technologist leave the room when the X-ray is given?

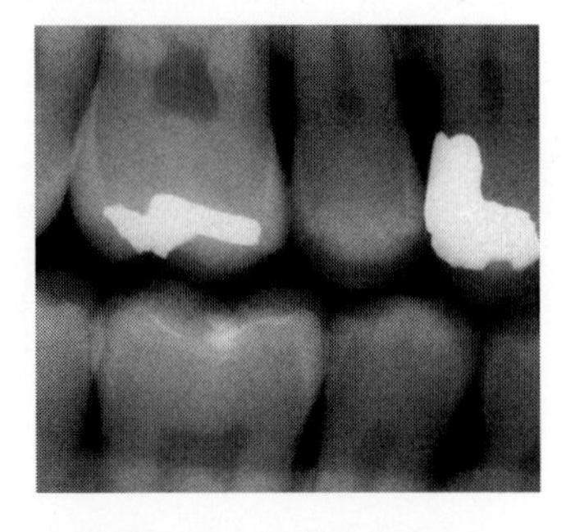

FIGURE 6.3 An X-ray may be taken as a part of a dental check-up. In this example, the lightest areas are fillings.

X-Ray Radiation

X-ray radiation is a diagnostic technology in which high-energy beams (X-rays) are directed on a part of the body. X-rays pass through soft tissues, such as muscle, but are stopped (absorbed) by the bones of the skeleton. You can find out more about X-rays in Section 4.1. Any areas in which X-rays passed through the body are detected by changes on a piece of film. An X-ray machine is therefore similar to a camera, but it uses X-rays instead of light rays (Figure 6.2).

Different parts of the body absorb different amounts of X-rays, and can be identified on the film. The lightest grey areas on the film are those where the most X-rays were absorbed. Some common reasons a patient might be sent for X-ray radiation are for dental examinations (Figure 6.3) or to determine if a bone is fractured.

X-rays can damage the DNA of cells, so parts of the body that are not to be diagnosed must be covered with a material that does not allow the X-rays to pass through, such as lead.

Amniocentesis

Amniocentesis is a procedure in which a doctor removes fluid from around a developing baby, by inserting a needle into the mother's uterus. This removes cells that can then be used to diagnose genetic abnormalities (Figure 6.4). The procedure is safer than removing cells from the fetus itself, which could damage the developing child. The cells of the amniotic fluid are identical to the cells of the fetus. Doctors can therefore detect chromosomal abnormalities in the fetus without removing any of the fetus's cells directly. For example, amniocentesis can detect the genetic disease Down syndrome, which you read about in Section 5.2.

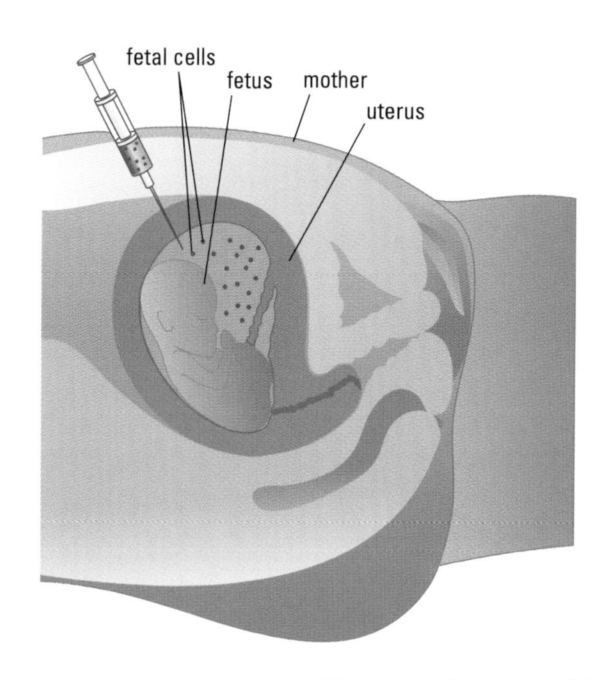

FIGURE 6.4 Amniocentesis is a medical technology used to diagnose genetic disorders. Could a karyotype be obtained from a fetus? Why or why not?

Cells for diagnosis of genetic disorders in an unborn child can also be obtained from the blood in the umbilical cord. A tiny needle is inserted directly into the vein of the umbilical cord (Figure 6.5). A karyotype can then be prepared from the cells in the blood. The blood sample can also be used to detect non-genetic diseases, such as anemia. This can allow treatment while the fetus is still in the uterus. For example, if anemia is diagnosed, medical personnel can give the fetus a blood transfusion while the needle is still in place.

Like all medical procedures, there are some risks associated with amniocentesis. In 1 out of 200 cases, the fetus is miscarried or the baby is delivered prematurely after amniocentesis is performed.

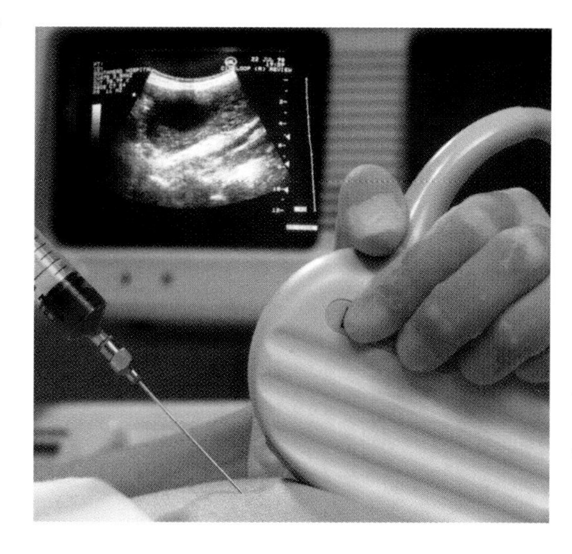

FIGURE 6.5 The needle is guided into the umbilical cord by an ultrsound scan, a technology you will read about on the next page.

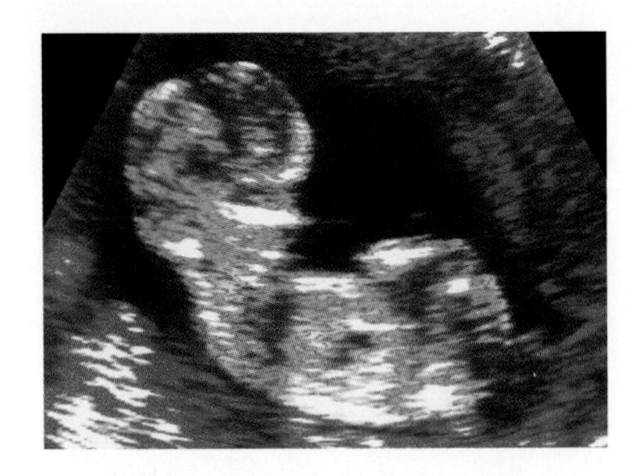

FIGURE 6.6 A sonogram can be used to determine the rate of growth of the developing fetus. What other information about an unborn child could a sonogram provide?

Ultrasound Scans

Sound is made up of waves of several different frequencies. You can find out more about sound waves and frequency in Section 3.2. **Ultrasound** waves are very high frequency sound waves that cannot be heard by the human ear. Ultrasound can be used to create images of the internal parts of the human body, including the soft tissues. During an ultrasound scan, a device that emits ultrasound waves is held against the body. The sound waves are reflected back at different speeds, depending on the type of tissue they encounter in the body. The reflected sound waves are then used to produce a **sonogram**, which is a picture of the pattern of sound waves. Figure 6.6 shows a sonogram of an unborn child. Ultrasound scans are commonly used to monitor the health of developing babies, and to diagnose medical conditions such as kidney stones.

ScienceWise Fact

Sonograms are also used for non-medical purposes. For example, sonograms can be used to find fish deep in lakes and oceans.

Echocardiogram

An **echocardiogram** is an ultrasound scan of the heart. An echocardiogram is often carried out when a patient is at rest and again during exercise (Figure 6.7). This diagnostic test provides the doctor with information about the overall condition of the heart, how hard the heart must work, or how quickly the heart can recover after exercise. Echocardiogram results can also diagnose irregular heart rhythms and other forms of heart disease.

A)

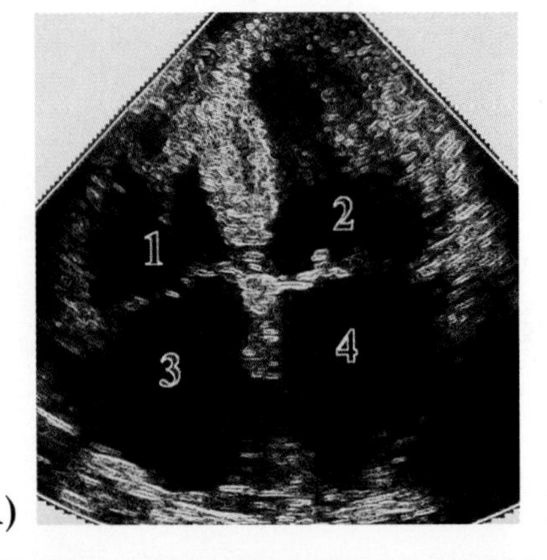

B)

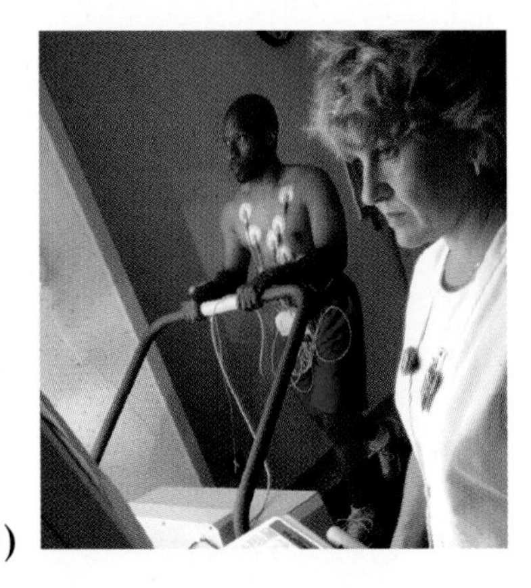

FIGURE 6.7 A) An echocardiogram, in which the four chambers of the heart are labelled. B) An echocardiogram is often done in conjunction with a "stress test," which is a measurement of the heart's functioning during exercise.

Magnetic Resonance Imaging (MRI)

Magnetic resonance imaging (MRI) is a diagnostic medical technology that uses magnetism, radio waves, and a computer to produce detailed images of different parts of the body (Figure 6.8). An MRI scan can detect disease anywhere in the body without the doctor having to open the body or remove tissues. For example, trauma to the brain, which is invisible from the outside, can be seen on an MRI scan. MRI scans are commonly used to determine if a patient has had a stroke, and to detect cancer in the brain or spine. An MRI scan cannot be performed on patients with metal in the body.

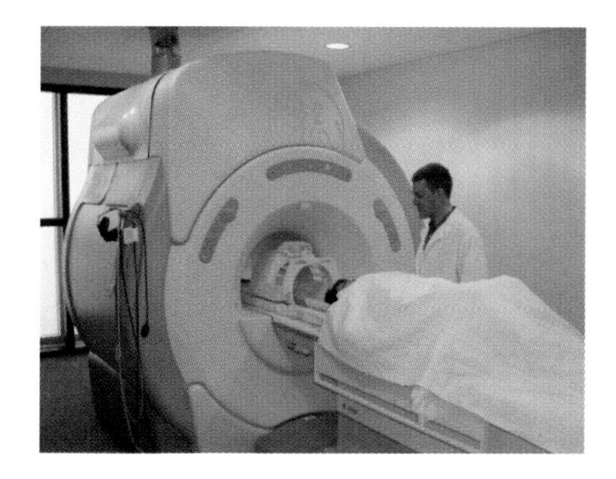

FIGURE 6.8 During an MRI scan, the patient is placed on a movable bed. The bed is then placed in the machine, and a large circular magnet moves around the patient's body.

Computed Axial Tomography (CAT)

Computed axial tomography (CAT) is a diagnostic technology that uses computers to combine many X-ray images together, and generate cross-sectional (cut-away) views of the body (Figure 6.9). Like MRI scans, CAT scans allow doctors to "see" inside the body without actually opening the body up. CAT scans can also generate three-dimensional (3-D) images of the internal organs and structures of the body. However, CAT scans expose the patient to X-ray radiation.

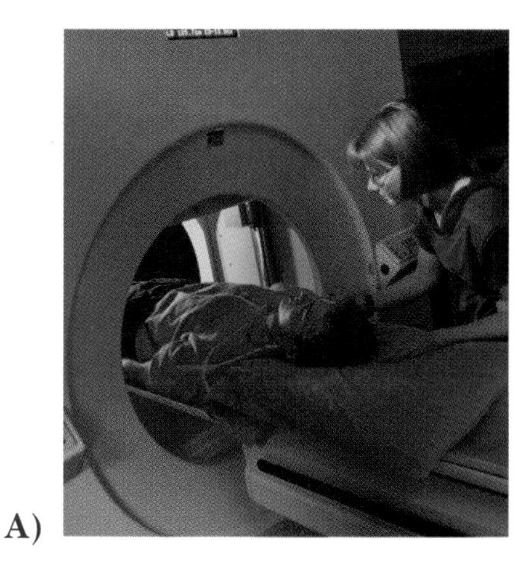

A)

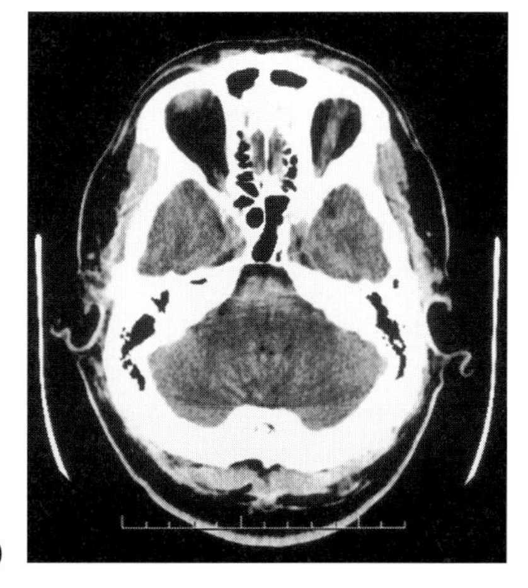

B)

FIGURE 6.9 A) In a CAT scan, a large doughnut-shaped X-ray machine takes X-ray images at many different angles around the body. B) An example of a CAT scan image. This image is of a patient's brain.

Surf the Web

Some tools that were developed for medical purposes are also used in other workplaces. For example, airport personnel scan passengers' luggage using devices based on technology similar to that of X-ray radiation and CAT scans. Find out how these machines work. Begin your search by visiting **www.science.nelson.com** and following the links for ScienceWise Grade 12, Chapter 6, Section 6.1.

ACTIVITY 6A

Optical Fibres in Medicine

Optical fibres are thin fibres of glass, usually about 120 mm in diameter, that carry pulses of light over distances. You can find out how optical fibres work in Section 4.2. Optical fibres have many different uses. For example, they can be used to transmit computer data or voice signals. Optical fibres are very flexible. In medicine, fibre optics are used in a number of diagnostic instruments, as well as some surgical tools.

FIGURE 6.10 The development of optical fibres gave doctors many new tools to help diagnose illness.

What You Will Need

- paper and pens or pencils
- access to library or Internet resources

What You Will Do

1. Choose one of the following instruments that employ optical fibres:
   - laparoscope
   - bronchoscope
   - cytoscope
   - proctosigmoidoscope
2. Using print and electronic resources, conduct research on the medical instrument you have chosen. Find out how this medical technology works, how it is used, and any risks associated with it.
3. Draw a sketch or create a model that clearly shows how optical fibres are used in the medical instrument you chose.

What Did You Find Out?

4. Write a summary of the information you gathered. Your summary should include a discussion of the advantages and disadvantages of using this instrument over other instruments used to diagnose the same condition.

Making Connections

5. The following occupations may also involve use of fibre optic technology: cable television service technician, telecommunications equipment technician, electronics assembler, and dental assistant. Survey your local newspaper or Internet job boards to find out if positions in these occupations are available, what qualifications are required, and the salary range. Summarize your research in point form.

Review and Apply

1. Briefly describe how X-ray radiation technology works.
2. What is the advantage of using amniocentesis to diagnose genetic disorders in a fetus? What are the risks of this procedure?
3. Prepare a chart that summarizes the similarities and differences between a CAT scan and an MRI scan.
4. An athlete has just been diagnosed with a torn knee ligament. Given that a ligament is a soft tissue, which of the following diagnostic technologies could NOT have been used to diagnose this condition: X-ray radiation, MRI scanning, or CAT scanning? Explain your choice.
5. Organize the concepts you have learned in this section in a graphic organizer.

Job Link

Medical Secretary

Medical secretaries are often the first person a patient sees, so they must be knowledgeable, friendly, and professional.

Responsibilities of a Medical Secretary

- schedule appointments and communicate messages between doctors, and between doctors and patients
- type and file confidential medical records, reports, and other correspondence
- prepare billing statements, order supplies, and maintain inventories

Skills for the Job

- computer skills, especially word processing
- friendly, outgoing personality
- familiarity with medical terminology

Education

- high school diploma, but may require additional training

FIGURE 6.11 Medical secretaries need to know the medical terms used by physicians.

6.2 Technology for Medical Treatment

Surf the Web

Diabetes affects the body's ability to process glucose, a sugar found in blood. Many diabetics have to monitor their blood sugar levels several times a day. To find out how diabetics can monitor their blood sugar levels, visit **www.science.nelson.com** and follow the links for ScienceWise Grade 12, Chapter 6, Section 6.2.

The goals of medical treatment are to return patients to health and to reduce suffering due to illness. Some illnesses can be cured by medical treatments, leaving a patient with no lasting change to his or her health. Other medical conditions cannot be cured, such as genetic disorders or lost limbs. For these conditions, medical technology can help patients live longer or more active lives.

Medication

The most common technology used to treat illness is a medication. A **medication** is any substance used for medical treatment. Medications can cure a disease, treat disease symptoms, or prevent a disease (Figure 6.12). For example, antibiotics are medications that cure disease by killing disease-causing micro-organisms. Injected insulin is a medication that treats the symptoms of the disease called diabetes. A diabetic person does not produce enough insulin, and injected insulin can keep his or her insulin levels within a healthy range. Vaccines are an example of a medication that can prevent disease. Vaccines contain weakened or dead cells of a disease-causing micro-organism. When a patient is injected with a vaccine, the injected cells stimulate the production of protective antibodies.

A)
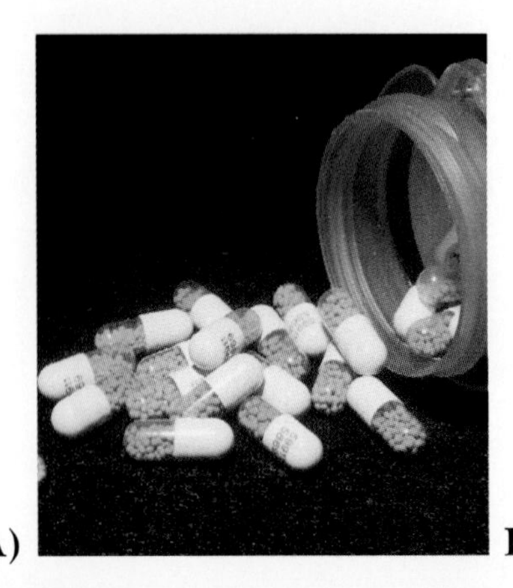

B)
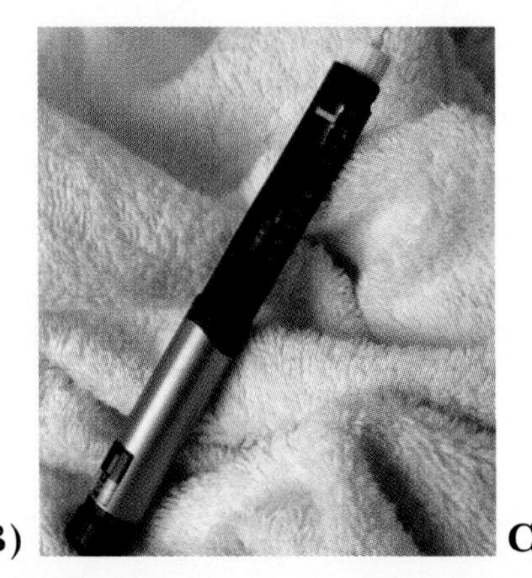

C)
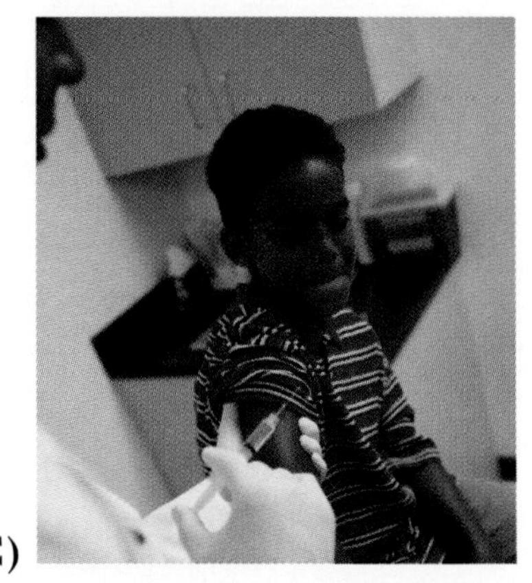

FIGURE 6.12 A) Antibiotics are medications that can cure a disease; B) Insulin is medication used to treat the symptoms of a disease; C) Vaccines are medications used to prevent disease.

Technology for Cancer Treatment

Cancer is a disease that will cause death if untreated. Cancer cells are cells of a person's own body that have started to divide uncontrollably. Cancer cells can spread throughout the body and interfere with its normal functioning.

Chemotherapy is the treatment of a disease, usually a cancer, with chemical medications. Many chemotherapy drugs stop cancer cells from dividing. Unfortunately, since cancer cells are formed from the body's normal cells, chemotherapy drugs also affect the normal cells in the body. Chemotherapy therefore often has significant side effects, such as hair loss, loss of red blood cells, and extreme nausea. Nevertheless, chemotherapy saves many lives.

Radiation therapy is a cancer treatment that uses high-energy radiation to kill cancer cells (Figure 6.13). The radiation used for cancer treatment has more energy than that used for X-ray radiation and CAT scans, so it can penetrate deeper into the tissues. Like chemotherapy, radiation therapy also affects non-cancer cells, and so has significant side effects.

You will need to know how medical technologies are used to treat human illness for your project, The Future of Medical Technology, at the end of Unit 3.

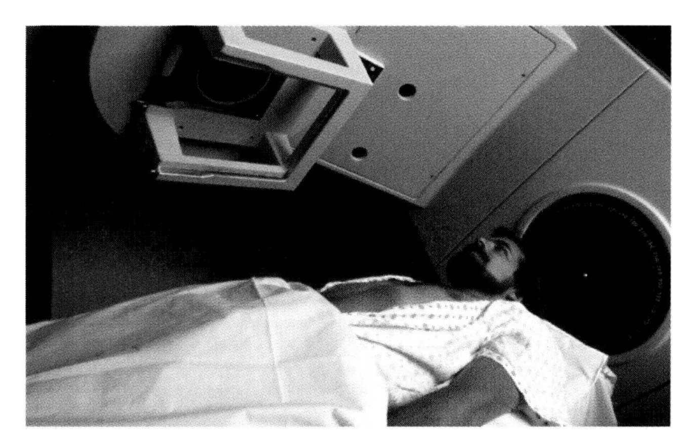

FIGURE 6.13 What does radiation therapy have in common with CAT scans and X-rays?

ScienceWise Fact

Canadian scientist Harold E. Johns and machinist John MacKay developed the first radiation therapy machine. This machine, called the "cobalt bomb," used the chemical cobalt 60 to deliver high-energy radiation to cancer cells. The first cobalt 60 unit was installed in London, Ontario, in 1951.

Hemodialysis

The kidneys are organs of the human body that filter excess fluids and wastes from the blood. These wastes are then eliminated in the urine. If, due to illness or accident, the kidneys stop or slow down, then the levels of toxic chemicals can build up and cause severe illness or even death.

Dialysis separates large particles in a solution, such as cells and proteins, from small particles, such as salts. Dialysis requires some sort of filter that only allows particles below a certain size to pass through (Figure 6.14A). **Hemodialysis** (dialysis of blood) is a medical technology in which a machine purifies the blood through a series of steps, which are shown in Figure 6.14B. The dialysis machine mimics the functioning of the kidneys. Blood from the patient is pumped through a filter unit composed of many thin glass tubes. The walls of the tubes contain tiny holes, or pores. These pores are too small to allow the cells of the blood to pass through, but large enough to allow through wastes such as urea and salts. The cleaned blood is then returned to the patient's body.

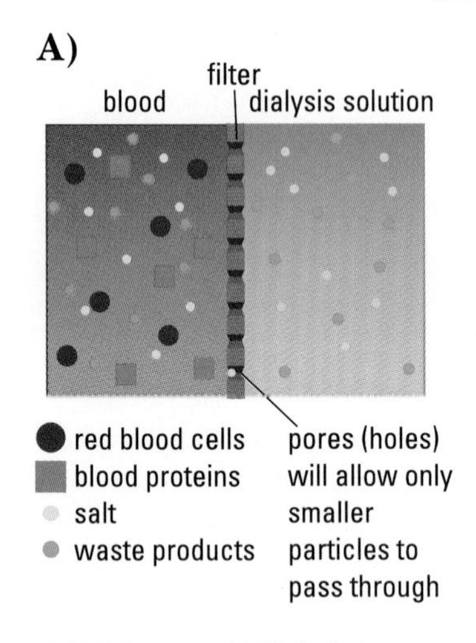

FIGURE 6.14 A) Dialysis is the filtration of particles in a solution, according to their size, across a filter membrane.

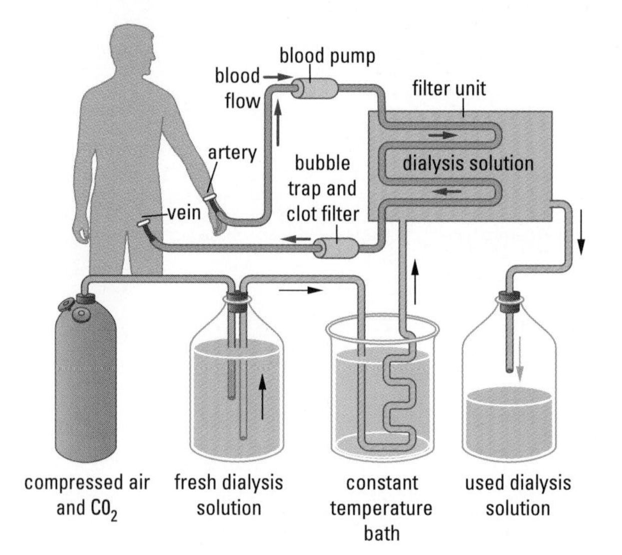

FIGURE 6.14 B) A hemodialysis machine carries out the functions represented here, inside a closed computer-controlled unit. The appearances of the parts and solutions are simplified in this illustration. The dialysis solution is first treated with compressed air and carbon dioxide gas to ensure the concentration of dissolved gases is equal to that in blood. Next, the dialysis solution is passed through a constant temperature bath, which brings the temperature to that of the human body (37°C), and then into the filter unit. The patient's blood is pumped from the artery into the filter unit. The filter removes excess salts and wastes by dialysis. The cleaned blood is then returned to the patient's body through a vein, and the used dialysis solution is removed from the system.

Technology to Assist the Heart

Heart disease is one of the leading causes of early death in Canada. Medical technologies to treat heart disease include those used during an emergency, such as when a patient has a heart attack, and those used to maintain a patient's health.

Defibrillators

A **defibrillator** is a device used to apply an electric shock to the heart (Figure 6.15). Defibrillators are used when a patient is experiencing irregular heartbeats. The electric shock stops the heart momentarily. When it begins to beat again, the heart will beat regularly. Without defibrillation, the heart of a person with an irregular heartbeat could be severely damaged or even stop. Defibrillators are standard equipment in ambulances, emergency rooms, and even on many airlines.

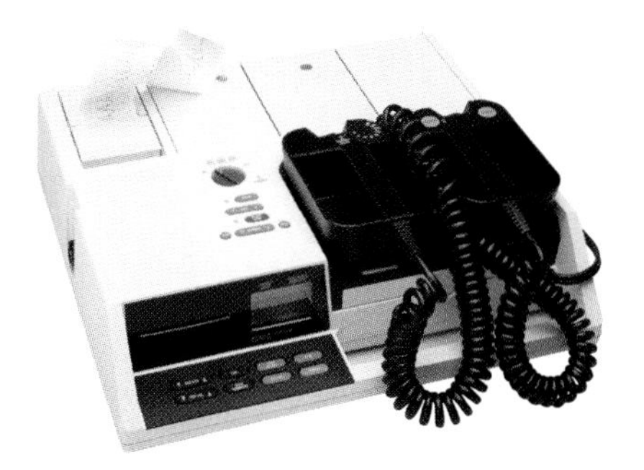

FIGURE 6.15 Flight attendants and life guards, who work with the public, often need to know how to operate a defibrillator.

Some patients who suffer frequent episodes of irregular heartbeats are now being fitted with implanted defibrillators. These patients are able to activate the defibrillator themselves, correcting the problem and avoiding damage to the heart.

Pacemakers

A **pacemaker** is an electronic device that sends timed electrical signals to the muscles of the heart, to keep the heart beating regularly. Pacemakers are usually implanted directly in the body (Figure 6.16). They have two components: a generator that produces the electrical signals, and an insulated electrical lead that connects the generator to a specific part of the heart. The generator of the pacemaker is also able to track the heartbeat and make adjustments, as needed, to the timing of the signals.

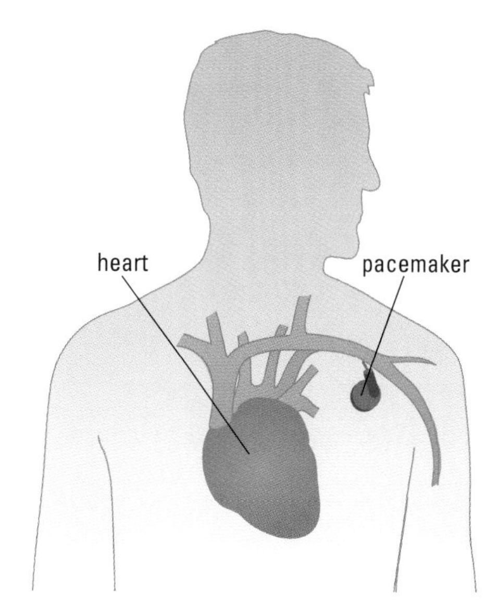

FIGURE 6.16 Pacemakers are implanted just under the collar bone. If a patient needs the device only for a short time, such as after heart surgery, the pacemaker's generator may be left outside the body.

Corrective Lenses

Corrective lenses, which are glasses or contact lenses, are a technology that many of us take for granted. Take a minute and look at your classmates. How many students are wearing glasses or contact lenses? How important is this medical technology to our society in general?

A person may need lenses to correct either nearsightedness or farsightedness. These conditions occur when the eyeball is either too long or too short, and so cannot bend light rays properly to focus the image sharply on the retina. This is illustrated in Figure 6.17. A nearsighted person can see only close-up objects sharply. A farsighted person can see things sharply only at a distance.

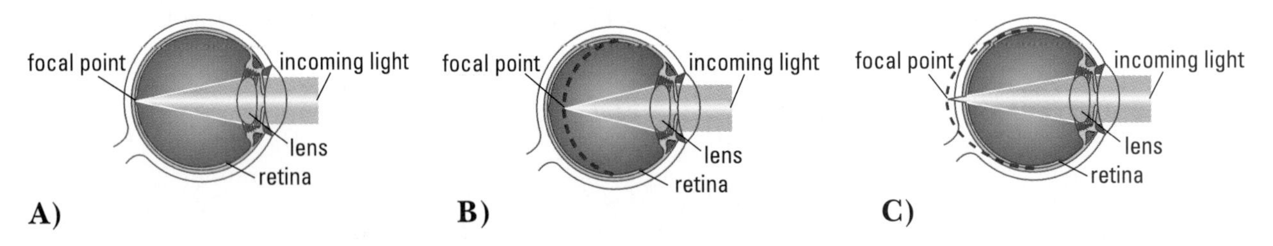

FIGURE 6.17 A) In a person with normal vision, the lens of the eye focuses light directly on the back of the retina. B) In a person with nearsighted vision, the light is focused just in front of the retina. C) In a person with farsighted vision, incoming light is focused on a point just behind the retina. Glasses or contact lenses correct vision by focusing light directly onto the retina.

A)

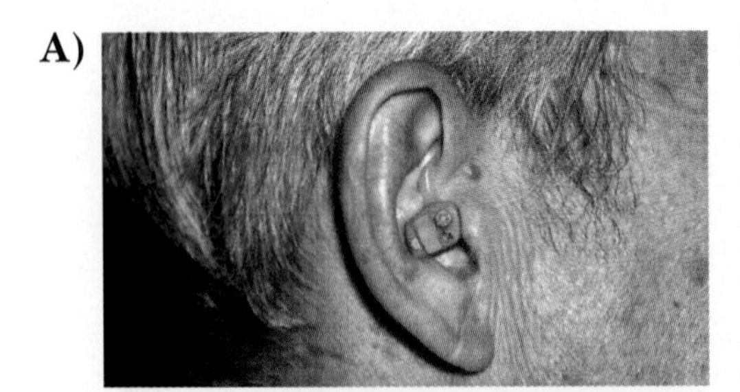

B)

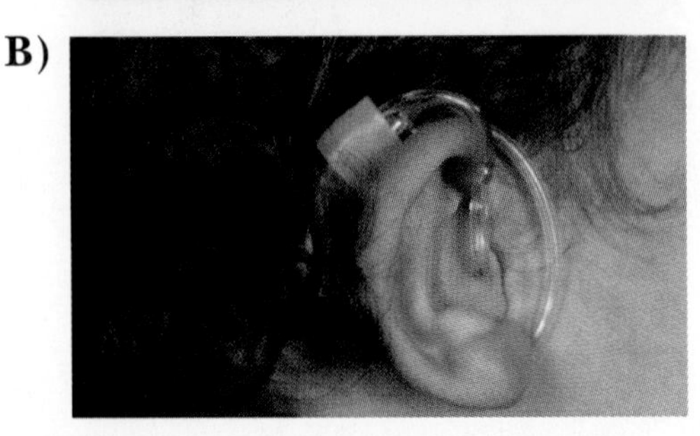

FIGURE 6.18 Two examples of hearing aids. A) Many hearing aids today are extremely small; B) Digital hearing aids deliver clearer sound.

Hearing Aids

A **hearing aid** is a medical device that is placed in the ear to amplify sound. People who have some hearing are able to detect the amplified sound. You can find out more about how sound is produced in Section 3.1.

The first hearing aids were trumpet-like devices that directed sound into the ear. The first wearable hearing aids were developed in the 1930s. These were large devices that ran off a battery. The patient wore a microphone around the neck, which was connected to an earpiece by wires. Today's devices are extremely small, and some are almost invisible (Figure 6.18A). The latest advance is digital hearing aids, which deliver very clear sound (Figure 6.18B).

Review and Apply

1. Copy each term in column 1 in your notebook, then match each term with the correct phrase in column 2.

Column 1	Column 2
Dialysis	A device that applies an electric shock to the heart
Medication	Refocus light on the retina
Pacemaker	A cancer treatment
Corrective lenses	Any substance used for medical treatment
Defibrillator	A method of separating particles in a solution by their size
Radiation therapy	A device that sends timed electrical signals to the heart

2. Explain why both chemotherapy and radiation therapy affect normal cells as well as cancer cells.

3. Why do many airlines carry defibrillators?

4. What is the difference between dialysis and hemodialysis?

5. Using flip charts, a storyboard, or graphic software, create an animated presentation showing what happens during the process of dialysis.

6. With a partner, brainstorm jobs related to the treatment of disease. Write your ideas on Bristol board. When you have thought of as many jobs as you can, prepare a list to share with the class. What is the salary range of these occupations?

7. Add the new concepts from this section to the graphic organizer you started in Section 6.1.

8. **Seeing Your Future?**

   Brainstorm some occupations related to treating vision problems. Choose one of these occupations to investigate further. Use print and electronic resources to find out what training is required for the position. Is this an occupation you might enjoy? Why or why not?

LAB 6B

Dialysis

Be Safe!

- Wear gloves, goggles, and a lab coat throughout this lab.
- Wash your hands before leaving the lab.
- Dispose of all chemicals according to your teacher's instructions.

Dialysis is a method of separating the particles in a solution according to their sizes. Hemodialysis uses this process to filter out small toxic substances from the blood.

Purpose

In this lab, you will simulate the process that occurs in a hemodialysis machine.

Materials

- 250-mL beaker
- dialysis tubing
- 0.1% albumin (w/v), 1% glucose (w/v) aqueous solution
- Biuret solution
- glucose test strips
- test tube
- scissors
- transfer pipette

Procedure

1. You will use dialysis on a solution that contains a protein (albumin) and a sugar (glucose) in water. Proteins are large molecules, whereas glucose is a small molecule. Predict and record the outcome of the dialysis.

2. Fill a beaker $\frac{3}{4}$ full with water. Soak the dialysis tubing in the water for about one minute, until it is soft and pliable. Tie a knot at one end of the dialysis tubing or fold the tubing over on itself and tie it tightly with thread or string (Figure 6.19).

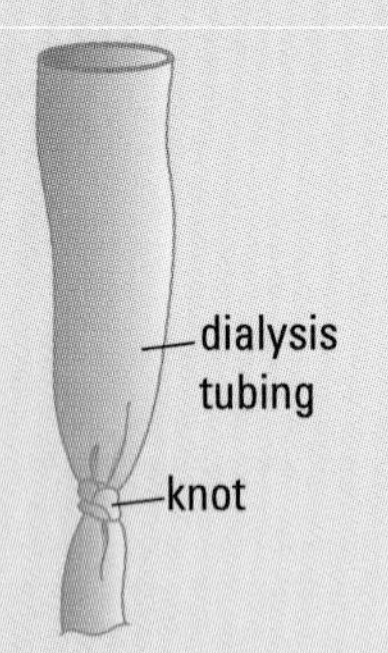

FIGURE 6.19 Tie off one end of your dialysis tubing.

3. Open the untied end of the dialysis tubing. You may need to rub the tubing between your fingers. Using the transfer pipette, fill the dialysis tubing with the albumin and glucose solution. Be careful not to spill any of the solution on the outside of the tubing (Figure 6.20).

4. Carefully tie another knot in the top of the tubing (or tightly tie off with string or thread), as shown in Figure 6.21. Rinse the dialysis tubing in distilled water to remove any spilled albumin and glucose solution.

5. Fill the second 250-mL beaker $\frac{3}{4}$ full with water. Place the dialysis tubing in the water, and leave it for 10 to 20 minutes.

6. While your solution is dialyzing, you will perform tests for the presence of protein and the amount of glucose. Design a data table to record this data for the albumin-glucose solution and the water, before and after dialysis.

7. Follow the manufacturer's directions on the Biuret solution to test for protein in the initial albumin-glucose solution and the water. Record your data.

FIGURE 6.20 Step 3

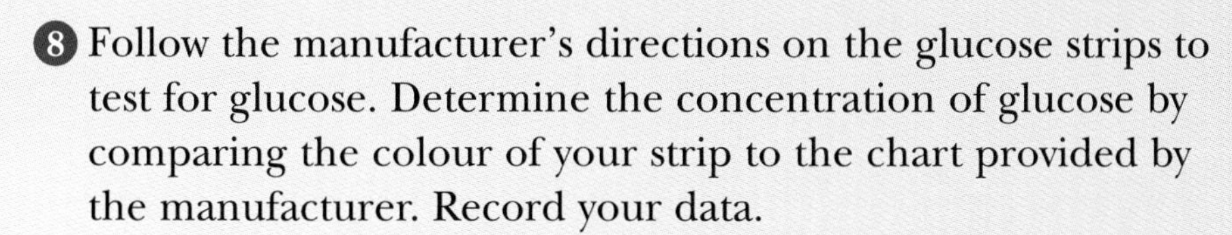

8. Follow the manufacturer's directions on the glucose strips to test for glucose. Determine the concentration of glucose by comparing the colour of your strip to the chart provided by the manufacturer. Record your data.

9. After dialysis is complete, carefully cut the knot off one end of the dialysis tubing. Empty the solution into the test tube. Repeat steps 7 and 8 for the dialysed solution, and for the water in the 250-mL beaker.

FIGURE 6.21 Step 4

Analysis and Conclusion

1. Was the outcome of the dialysis as you predicted in step 1? If not, rewrite your prediction so it is supported by your data.

2. Did the albumin pass through the dialysis tubing? Did the glucose pass through the dialysis tubing? Explain how you know.

Extension and Connection

3. Explain how the movement of the glucose and albumin molecules relates to the process of hemodialysis.

4. With a partner, use the library or the Internet to find out what causes kidney disease. Under what circumstances is it necessary for patients to undergo hemodialysis? What role do kidney transplants have in the treatment of patients with kidney disease?

CASE STUDY

Laser Eye Surgery

Instead of corrective lenses, patients with nearsightedness may undergo laser eye surgery. Laser eye surgery reshapes the lens of the eye (the cornea) with a cool, computer-controlled, precise beam of light (a laser). The procedure is completed in a few minutes and patients can resume their daily activities in one to three days. After laser eye surgery, most patients no longer need corrective lenses.

How Is Laser Eye Surgery Carried Out?

Before surgery begins, the shape and size of the patient's eye is very accurately measured. These measurements allow the surgeon to determine exactly what is wrong with the eye shape. The surgeon constructs a "map" of the eye's surface from the measurements, and then enters this information into the computer that controls the laser. Laser surgery removes some tissue from the cornea, reshaping it so that incoming light is focused directly on the back of the eye. Figure 6.22 outlines the steps of one type of laser eye surgery.

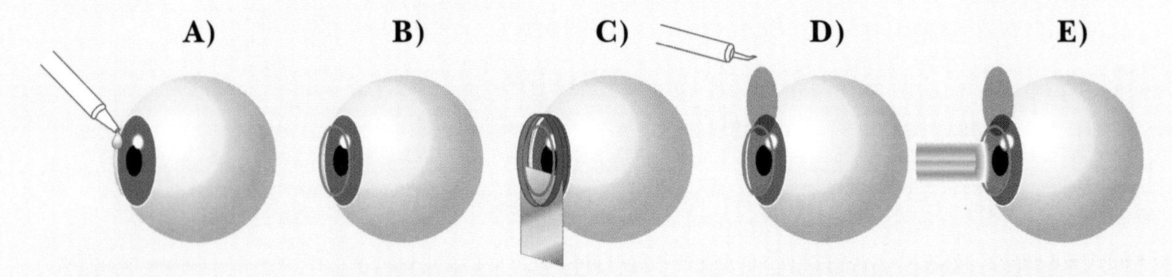

FIGURE 6.22 A) Anaesthetic drops are used on the eye to numb the area. The doctor then marks the area to be cut. B) A special device cuts a tiny hinged flap on the surface of the cornea. C) and D) The flap is flipped open. A portion of the cornea surface remains attached. E) The laser reshapes the cornea by removing some of the underlying tissue. The flap is then replaced on the eye.

a) Use illustrations to show the approximate position of the focal point of an eye before and after laser eye surgery. Remember that laser eye surgery can correct only nearsighted vision.

Who Is a Candidate for Laser Eye Surgery?

- Patients must be at least 18 years of age. In patients younger than 18, the eyes are still undergoing change.
- Patients' eyes must be healthy and free of disease.
- Patients must have mild to moderate nearsightedness. Farsightedness cannot be corrected by laser surgery.
- Patients must be prepared to cover the cost themselves.

What Are the Risks?

As with any medical procedure, there are some risks involved with laser eye surgery. Patients have reported the following problems after having laser eye treatment:

- corneas became infected after the surgery;
- vision was either over-corrected or under-corrected, so that corrective lenses were still required;
- vision became slightly worse; or
- the presence of a "halo" effect affected night vision.

There is also a risk that the flap created during the procedure can become distorted or come off completely, causing it to become damaged or lost. Another small risk is that the computer controlling the laser could malfunction during the procedure. Such events, however, are extremely rare.

b) People who have undergone laser eye surgery may not be permitted to work in occupations such as security guards or police officers. Suggest a reason for this restriction. Do you think the restriction is justified? Explain your answer.

Analysis and Communication

1. What are the advantages of laser eye surgery over corrective lenses? What are the disadvantages?

2. Do you consider laser eye surgery to be risky? State your reasons.

Making Connections

3. You are a welder, and you have just been diagnosed as nearsighted. You are told you could wear glasses or contact lenses, or have laser eye surgery. Create a chart listing the advantages and disadvantages of glasses, contact lenses, and laser eye surgery, referring specifically to the occupation of a welder.

6.3 Spare Parts

FIGURE 6.23 Terry Fox depended on a prosthesis to replace his amputated leg.

You will need to know about prostheses for your project, The Future of Medical Technology, at the end of Unit 3.

Have you heard of the Terry Fox Run, an annual charity run to support cancer research? Terry Fox was a young Canadian who, at age 18, had his right leg amputated above the knee to treat bone cancer. Terry responded to this challenge by running over 5000 km across Canada on an artificial leg, to raise money for cancer research (Figure 6.23). Unfortunately, Terry died before he could complete his planned run across Canada. The annual Terry Fox Run is part of the legacy of this remarkable person.

Terry could not have made his run without a prosthesis. A **prosthesis** is an artificial body part, such as Terry's artificial leg. The plural of prosthesis is prostheses. In this section, you will learn about some medical technologies that replace damaged or missing body parts.

Limb Prostheses

Missing limbs may be the result of birth defects, accidents, or surgical amputations due to cancer, infection, or circulatory diseases. In some cases, doctors can provide prostheses that perform some of the functions of the natural limb.

Limb prostheses have a long history, and many different devices have been invented over the years. Some of these are shown in Figure 6.24. Various materials have been used to create artificial limbs, including wood, copper, steel, and plastics. Modern prostheses take advantage of human-made materials, such as plastics, that are stronger and lighter than natural materials. Some prostheses are coated with a silicon material to make them look as real as possible. You can find out more about some of these materials in Section 1.3.

There are two general types of limb prostheses: static and dynamic prostheses. Static prostheses do not move, whereas dynamic prostheses can. For example, some arm and leg prostheses respond to nerves in the wearer's skin, which send signals to electronic components in the prosthetic device, causing it to move in a manner similar to the missing natural limb.

A) 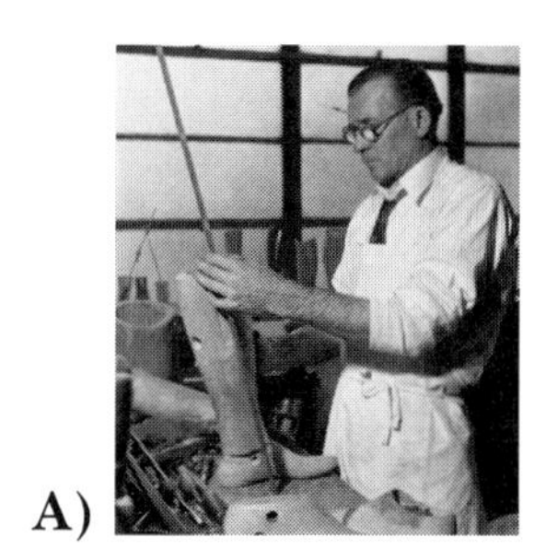

B) 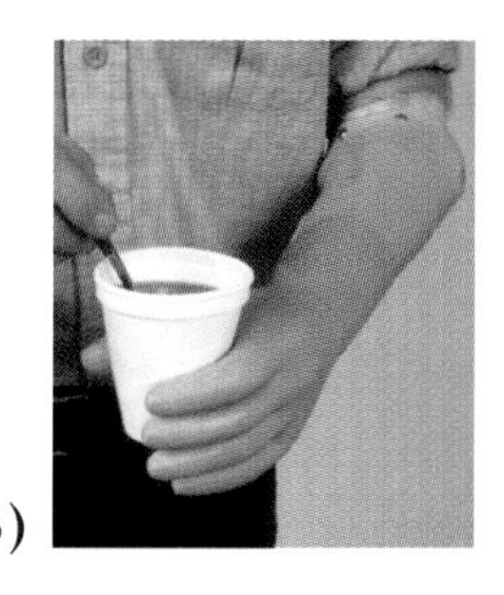

C) 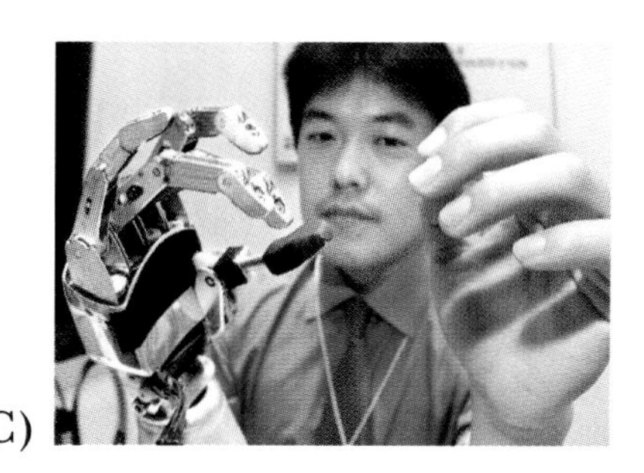

FIGURE 6.24 A) The earliest prostheses were made of heavy materials, like this wooden leg, and allowed little movement. B) The invention of plastics allowed the design of prostheses that allowed some movement, were lighter, and looked more natural. C) Advances in robotics technology permit prosthetics that can move in response to nerve impulses, much like a natural limb.

Joint Replacements

Joint replacements are prostheses that replace joints. Human joints (for example, knees, wrists, or elbows) can become damaged by an accident or by disease. When a joint is so damaged that it does not function or causes severe pain, the doctor may recommend a joint replacement.

Natural joints are composed of bones, which provide strength, and soft tissues such as cartilage, ligaments, and muscles. The ligaments and muscles allow the bones to move in particular ways, such as when a knee bends backward. They also prevent the bones from moving in ways that would cause harm, such as a knee bending too far forward. The cartilage protects the bones from wearing against each other. Figure 6.25A shows the arrangement of bones and ligaments in a natural knee joint.

Most joint prostheses are made in a ball-and-cup design (Figure 6.25B and C). The cup is moulded from plastic. The ball is made of metal or ceramic and is attached to a metal stem. Even large joints, such as those of the hip, can be replaced. In fact, hip replacement is now a common procedure among elderly patients.

A) 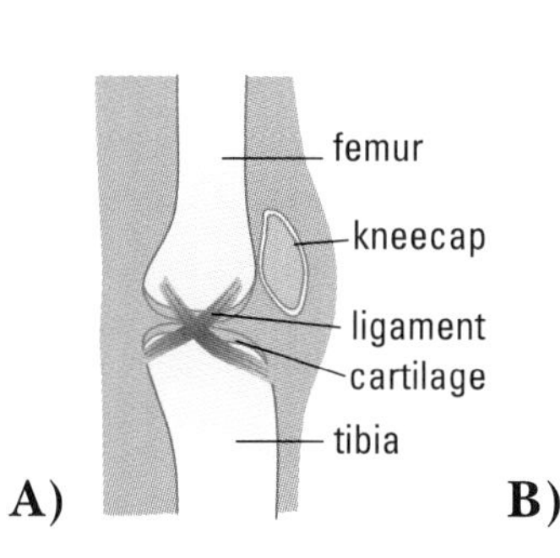

B) 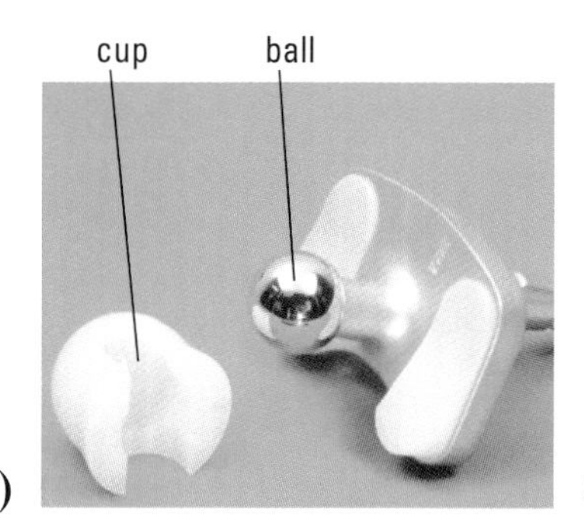

C) 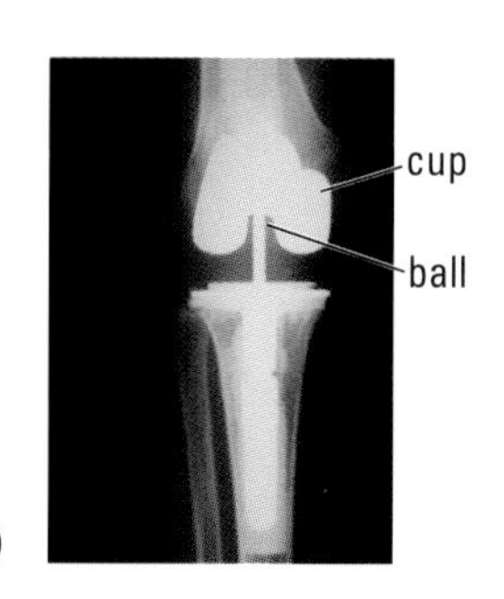

FIGURE 6.25 A) A natural knee joint, B) an artificial knee joint, C) an inserted artifical knee joint (X-ray).

FIGURE 6.26 Patients who have never had hearing can detect sound using a cochlear implant.

Cochlear Implants

A **cochlear implant** is a small electronic device consisting of a microphone within a headset, which picks up sound (Figure 6.26). Unlike hearing aids, cochlear implants are surgically inserted into the patient. The implants act by stimulating the nerves involved in hearing. This is very different from the mechanisms in hearing aids, which only amplify sounds.

When a cochlear implant detects sound, a processor in the implant translates the sound waves into digital code with the help of a computer. The digital code is then sent to a radio receiver implanted under the skin. The receiver then stimulates the auditory nerve, which sends impulses to the brain. These impulses are interpreted as sound (Figure 6.27).

Surf the Web

Some deaf activists say that cochlear implants threaten deaf culture. What do you think about this issue? Find out more about the debate over cochlear implants and deaf culture. Start your research by visiting **www.science.nelson.com** and following the links for Science Wise Grade 12, Chapter 6, Section 6.3.

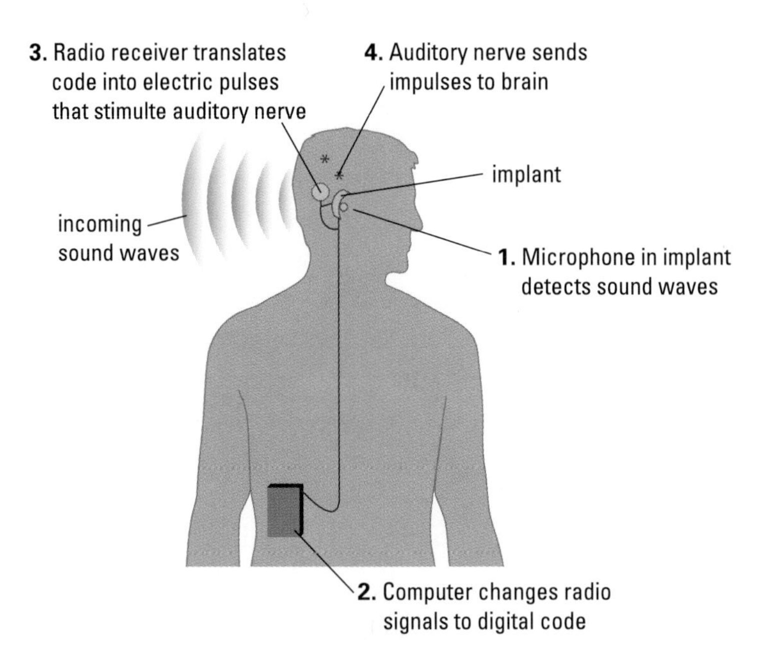

FIGURE 6.27 A natural ear translates sound waves into nerve impulses in the brain. A cochlear implant uses electronics to perform the same function. You can find out more about devices that receive and transmit sound waves in Section 4.2.

Artificial Organs

Although the technology to transplant organs is available, there is a worldwide shortage of donor organs. To help patients affected by this shortage, medical scientists have developed some artificial organs. Most artificial organs available today can be used only temporarily, to extend a patient's life while he or she waits for a donor organ to come available. However, research is underway to develop artificial organs that can permanently replace a damaged natural organ.

Artificial Hearts

The artificial heart is the most advanced artificial organ today. The first successful artificial heart was implanted in a human patient in 1982. Although the patient lived for only 112 days, this event was nevertheless an important milestone. The first artificial heart functioned similarly to a natural heart. It had two pumps that acted like the two sides of the heart, and a disk-shaped mechanism that acted like a heart valve. However, this early artificial heart required a large external energy source, which prevented the patient from moving freely.

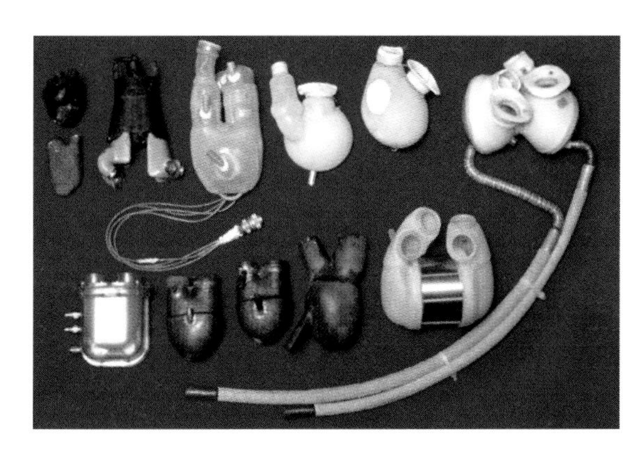

FIGURE 6.28 Some examples of artificial heart designs. Older designs are on the left side of the photograph.

Since that first success, researchers have developed better, more efficient, and smaller artificial hearts (Figure 6.28). Newer designs have an internal battery as well as a small external battery. This design allows the patient more freedom of movement. Patients with an artificial heart today also live much longer. Most medical researchers expect that eventually an artificial heart will be available that can permanently replace a natural human heart.

ACTIVITY 6C

People in Motion

Many people in our society use a kind of medical technology called a mobility aid to assist them in moving from place to place. For example, people who tire easily may use a cane while walking, to reduce the amount of effort required by their legs. Paraplegics (people who cannot use their legs) may use a wheelchair to get around.

At one time, medical technology could only offer devices with basic functions. Today, medical technology can offer a wider range of possibilities (Figure 6.29). Some technologies still provide only basic functions, but others are designed to simulate or complement the natural abilities of the human body. For example, the wheelchair design shown in Figure 6.30 allows the user to compete in high-level athletic competitions.

In this activity, you will conduct research on the different types of mobility aids available today. You will also prepare a list of companies in your area that are involved in producing or providing these mobility aids, the types of job opportunities that may be available in these companies, and the qualifications needed for these jobs.

What You Will Need

- paper and pen or pencils
- access to library and Internet resources

What You Will Do

1. Working in a group, brainstorm a list of mobility aids that are available.
2. When you have finished brainstorming, assign each person in your group a mobility aid or aids to research. More than one person may share responsibility for a mobility aid, if needed.
3. Use print and electronic resources to find out as much as possible about the mobility aids you were assigned. Keep a record of your research, and collect copies of illustrations and photographs you find helpful.

FIGURE 6.29 How do these devices assist people in their daily lives?

Your research should address the following questions:

- What are the features available on different models of this device?
- What are the advantages of each of these features?
- What are the disadvantages of each of these features?
- Who is most likely to use the features?

4. When you have completed your research, meet with your group members and share your findings. Decide on the one mobility aid your group finds of greatest interest. Using the Internet and the library, find at least two companies in your province that are involved in producing or providing this mobility aid. Record the name, address, and telephone number of these companies in your notebook.

5. Conduct additional research to find what job opportunities may be available at these companies. You may need to access the company's website, conduct library research, write a letter to the company, or conduct a brief phone interview with the human resources department. If you contact a person, ensure you have planned your questions in advance, and have asked for their time. Emphasize that you are researching job opportunities, not looking for employment.

6. For each job opportunity, conduct additional research to find out the duties of the job, the training required, and the salary range.

What Did You Find Out?

1. Working together, prepare a summary of the information your group collected on the types of mobility aids that are available. Use an appropriate format to present your summary. For example, you may wish to prepare a poster, a web page, or a presentation using software.

2. Write a newspaper-style job listing for each of the occupations you found in step 5, using all the information you researched.

Making Connections

3. Canada is expected to have a higher proportion of people over the age of 60 over the next 20 to 30 years. How will this affect the number of opportunities available in industries that produce medical devices such as mobility aids? Give reasons for your answer.

FIGURE 6.30 Some modern wheelchairs are designed for specific purposes, such as these racing models.

Review and Apply

1. What are the two general types of limb prostheses?
2. Prostheses are very expensive. With a partner, discuss the fairest way for everyone (rich and poor) to get needed prostheses. Share your ideas with another pair of students.
3. Every artificial limb or artificial joint must be designed for the specific patient. Explain why.
4. How is a cochlear implant different from a hearing aid?
5. Human heart transplants are not common, but are usually successful. Why is the development of an artificial heart still an important goal of medical research?
6. Add the new concepts from this section to the graphic organizer you started in Section 6.1.

Surf the Web

To find out more about what procedures of plastic surgery are available, how they are performed, and what risks are involved, visit **www.science.nelson.com** and follow the links for ScienceWise Grade 12, Chapter 6, Section 6.3.

7. **Plastic Surgery**

Plastic surgery is surgery to reconstruct parts of the human body. Some people undergo plastic surgery to reconstruct deformed or injured parts of the body. For example, the genetic disorder known as cleft palate can be corrected by plastic surgery. Plastic surgery can also be used to change the appearance of a person's body even when there is nothing medically wrong. For example, some people choose to have liposuction, a procedure that removes fat, to make their abdominal muscles more noticeable.

As a class, discuss the following issues:

- What are some of the benefits of plastic surgery? What are some of the risks?
- Do you think it is acceptable for teens to undergo plastic surgery in some circumstances? What are those circumstances?

6.4 Chapter Summary

Now you can...

- Demonstrate an understanding of terms related to medical technology (6.1, 6.2, 6.3)
- Explain the use of technology for diagnostic medical applications (6.1)
- Describe the use of technology for biomedical repair (6.2, 6.3)
- Provide examples of how science and technology have influenced the diagnosis and treatment of human illness and have made medical technology an integral part of our lives (6.1, 6.2, 6.3)

Concept Connections

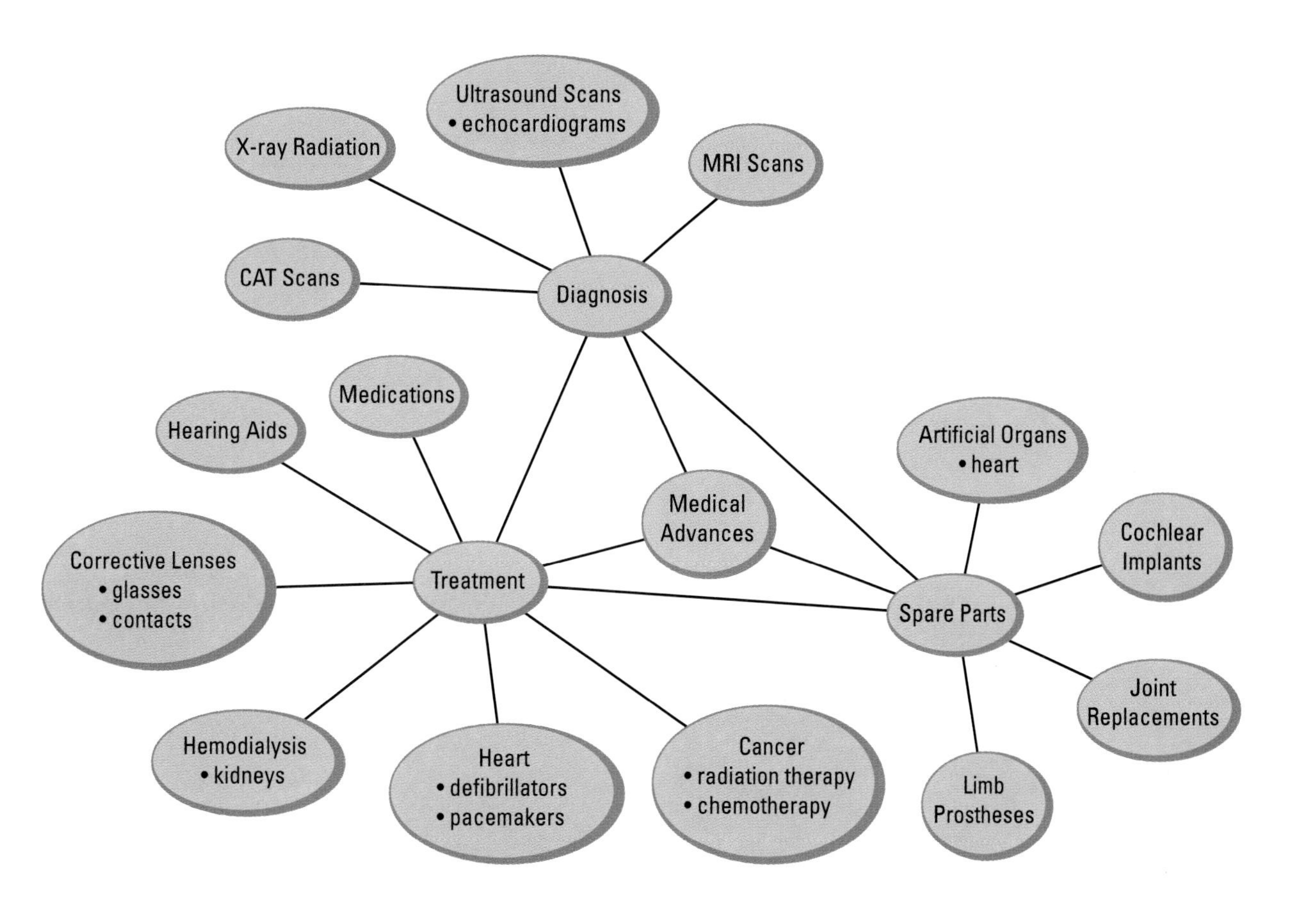

FIGURE 6.31 Compare your completed graphic organizer to the one on this page. How did you do? What new links can you add to your organizer?

CHAPTER 6 *review*

Knowledge and Understanding

1. In your notebook, match the term in Column A with the information in Column B.

Column A	Column B
CAT scan	Sound waves used to show a picture of the heart
Amniocentesis	Shows images of the body in various shades of grey
X-ray radiation	Removal of amniotic fluid for genetic testing
MRI	Uses magnetism, radio waves, and a computer to produce images of body structures
Ultrasound	An image produced by sound
Echocardiogram	Very high frequency sound inaudible to the human ear
Sonogram	A process used to clean waste from the blood
Chemotherapy	Uses computers to combine many X-ray images to form cross-sectional views
Hemodialysis	A treatment that uses anti-cancer medications

2. Compare MRI and CAT scans.
3. Briefly describe two diagnostic tests that a pregnant woman may need to have.
4. Give three examples of medical technology that has been developed to treat heart disease.
5. Describe three devices used to replace or repair non-functioning body parts.
6. Write each of the following in a corner of a piece of paper: MRI scan, CAT scan, ultrasound scan, X-ray radiation. With a partner, brainstorm reasons why a doctor may suggest each of these tests. Summarize your work in point form.

Inquiry

7. A student fills a small plastic bag with a solution of household salt in distilled water. The bag is then placed in a large container of pure water. The next day, a small sample of the water from the large container is placed on a watch glass and allowed to evaporate. After the water has evaporated, there is a small amount of salt on the watch glass. Explain what happened.

Making Connections

8. A doctor diagnoses a patient with kidney failure. What medical technology will this patient need to use?

9. Most multi-storey buildings are required to have elevators, in part to allow people using medical technology such as wheelchairs and walkers to use the building safely. Give two other examples of common modifications to human-built structures that accommodate people using these medical technologies.

10. Twenty to 30 years from now, Canada's population is expected to have a higher proportion of people aged 50 years or more. Predict how this change will affect the number of jobs related to medical technology. Give reasons for your answer.

11. Some people think that medical technology may someday allow humans to live forever. If this occurred, what problems might result?

Communication

12. Create a sketch of a human body. Using arrows and labels, identify as many structures as possible that can be enhanced, repaired, or replaced by medical technology.

13. Society must make accommodations for the use of mobility aids such as wheelchairs. One such accommodation is wheelchair access to buildings. With one or two classmates, survey the accessibility of your school to people in wheelchairs. Make sure you include all the facilities in the school, including washrooms, libraries, and gymnasiums. When you are finished, write a letter to your school board that discusses the wheelchair accessibility of your school and how it might be improved.

PUTTING IT ALL TOGETHER

The Future of Medical Technology

In this unit, you have learned about the diagnosis and treatment of human medical conditions, including genetic disorders, diseases such as cancer and heart disease, and injuries. You have developed an understanding of some diagnostic tools, including karyotypes, pedigrees, CAT scans, and ultrasound scans, among others. You will now use this knowledge to create an information presentation for a developing medical technology, to be presented to an audience of medical workers.

FIGURE C1 Medical technology has changed human life in many ways.

The Plan

You and several classmates are part of a marketing team for a company that designs and sells medical technology to clinics and hospitals. This company is a leader in new designs. Your job is to create an information presentation that will help medical workers understand the advantages of this new design, and let them know what your company is likely to produce in the next few years.

What You May Need

- paper, pencils, and pens
- reference sources such as magazines, books, or Internet access
- other materials to make an effective presentation

What You Will Do

1. In groups of two to four, choose one medical condition that you would like to learn more about. Read through the rest of the project to find out what you will need to research. Decide how you will divide the work among group members.

2. Using print and electronic resources, begin your research. Find out how the condition arises, whether genetics has a role, and what percent of the population suffers from the condition.

3. One way of demonstrating the value of an improved technology is to compare it to past technology. Find out about technologies that were used to diagnose or treat the condition in the past. Sketch or copy any diagrams or photographs of devices. Make notes on the advantages of each new technology over the previous technology.

4. Identify the most commonly used technology for diagnosing or treating the medical condition today. Make note of its advantages and disadvantages. Consider its effectiveness in diagnosing or treating the condition, how easily patients can gain access to the technology, how much it costs to use, and any side effects.

5. Identify the advances in medical technology that are currently being developed. Choose one advance on which to base your information presentation. Remember, you will need to clearly present the advantages of this technology, so be sure you know as much as possible about its advantages and disadvantages compared to present technology. Include as an advantage any jobs that will be created or made easier by the use of this technology.

6. Meet as a group to discuss the results of your research. Review the information and determine whether you need any additional facts and figures. Carry out any further research that is needed.

CONTINUED

7. Create your information presentation. You will have only 15 minutes to discuss your technology, so your presentation has to make an impact. Some ideas that might enhance your presentation are as follows:
   - a pamphlet to hand out
   - a multimedia presentation
   - a graphic timeline on a banner
   - a model of the latest technology
   - a poster presentation

FIGURE C2 How will you make sure your presentation creates interest?

What Did You Find Out?

1. Show time! Make your presentation to your classmates.
2. Have many advances been made in the field you researched? What problems did these advances solve?
3. Does medical technology benefit only the patients that use it directly? Why or why not?

Assessment

1. Comment on what you felt other groups did well. Which presentations did you find the most interesting? Why?
2. Not including your own group, which group do you think had the most effective presentation? Justify your choice.
3. What did you like best about your own group's project? Based on your classmates' presentations, what improvements could you make to your presentation if you had more time?

UNIT 4

Gardening, Horticulture, Landscaping, and Forestry

CHAPTER 7: Plants
CHAPTER 8: Working with Nature
PUTTING IT ALL TOGETHER: Planting a Garden

CHAPTER 7

Plants

Plants are among the most important organisms in our everyday lives. Through photosynthesis, they provide us with food and oxygen. Plants are also an important source of building materials, clothing, fuel, and medicines. They can provide us with a pleasant environment or a hobby. Since we use plants in so many ways, many occupations involve growing plants. The photographs in Figure 7.1 show people working in industries that require knowledge of plants. Describe some of the knowledge that all these workers must have to do their jobs. What specialized knowledge would a forestry worker require? How is this different from the knowledge required to work as a florist, or as a fruit seller?

FIGURE 7.1

What You Will Learn

After you have completed this chapter, you will be able to:

- Describe, with examples, the differences among common plants according to their life cycle or method of culture (7.1)
- Identify the general conditions necessary for healthy plant growth (7.2)
- Describe the basic steps in growing plants from seeds (7.1, 7.2)
- Identify and collect information on careers related to growing plants (7.2)
- Identify evidence of plant problems (7.3)

What You Will Do

- Test the effects of environmental conditions on seed germination (Activity 7A)
- Design and conduct an experiment to determine the effects of changes in light intensity, moisture level, and temperature on plant growth (Lab 7B)
- Demonstrate an understanding of safety practices (Lab 7B)
- Select appropriate instruments and use them to collect data (Lab 7B)
- Demonstrate the skills required to plan and carry out an investigation (Lab 7B)
- Carry out soil tests to determine optimal conditions for plant growth (Lab 7C)
- Select and use appropriate modes of representation to communicate results (Activity 7A, Labs 7B, 7C)
- Compile, organize, and interpret data using data tables and graphs (Labs 7B, 7C)
- Communicate the results of lab activities (Labs 7B, 7C)

Words to Know

annual
biennial
bulb
cloning
cultivated plants
cutting
dormant
germination
horticulturalist
native plants
optimal growth conditions
perennial
photosynthesis
plant propagation
respiration
seedling
variable
vegetative propagation
wilt

A puzzle piece indicates knowledge or a skill that you will need for your project, Planting a Garden, at the end of Unit 4.

7.1 Types of Plants

You will need to know about optimal growth conditions for your project, Planting a Garden, at the end of Unit 4.

Whether you live in a big city or a small rural community, you are surrounded by many different types of plants. There are trees, grass, flowers, ferns, and vines. Some plants will be **native plants**, or plants that grow naturally. Some will be **cultivated plants**, or plants produced by humans for their specific qualities, such as large flowers. These different types of plants can be grouped according to certain shared characteristics. Understanding the characteristics of each group can tell us how to care for its members. This section presents some of the ways in which people who work with plants classify them.

Seeds or Spores?

Seeds and spores are reproductive structures of plants. Since a particular type of plant will always produce only one or the other of these, plants can be classified by whether they produce seeds or spores. Figure 7.2 shows several examples of plants in each of these two groups. The inset shows one example of seeds and one example of spore cases. Although plants in these two groups may look very different from one another, all seed plants reproduce only by seeds, and all spore-bearing plants reproduce only by spores.

FIGURE 7.2 Examples of plants that bear seeds include A) petunias, B) wheat, and C) white pine. Examples of plants that produce spores are D) ostrich fern, E) marsh horsetail, and F) maidenhair moss.

Flowers or Cones?

More than 80 percent of all plants produce seeds. Seed plants can be divided further into two groups: those that produce seeds on flowers (flowering plants) and those that produce seeds on cones (cone-bearing plants). Figure 7.3 shows examples of these two classes of seed plants.

FIGURE 7.3 Examples of flowering plants include A) snapdragon, B) apple, and C) squash. Examples of cone-bearing plants include D) blue spruce, E) cedar, and F) Sago palm.

Classifying by Life Cycle: Annual, Biennial, or Perennial?

Another way of classifying a plant is according to the length of its life cycle. The life cycle of a plant includes three general stages. **Germination** is the first stage of growth, from a seed or spore to a **seedling** (a young plant with only its first few leaves). The second stage is the growth stage, in which a seedling increases in size. The final stage is the reproductive stage, in which the plant produces seeds or spores. Classifying plants according to their life cycles is especially important to those who grow plants for their seeds, flowers, or fruits, since these are produced only during the reproductive stage.

An **annual** is a plant that lives for one year only. During this time, the plant germinates, grows, reproduces, and eventually dies (Figure 7.4).

A **biennial** is a plant that lives for two years only, during which time it will germinate, grow, reproduce, and eventually die. Biennials reproduce only once, in the second year of their growth (Figure 7.5).

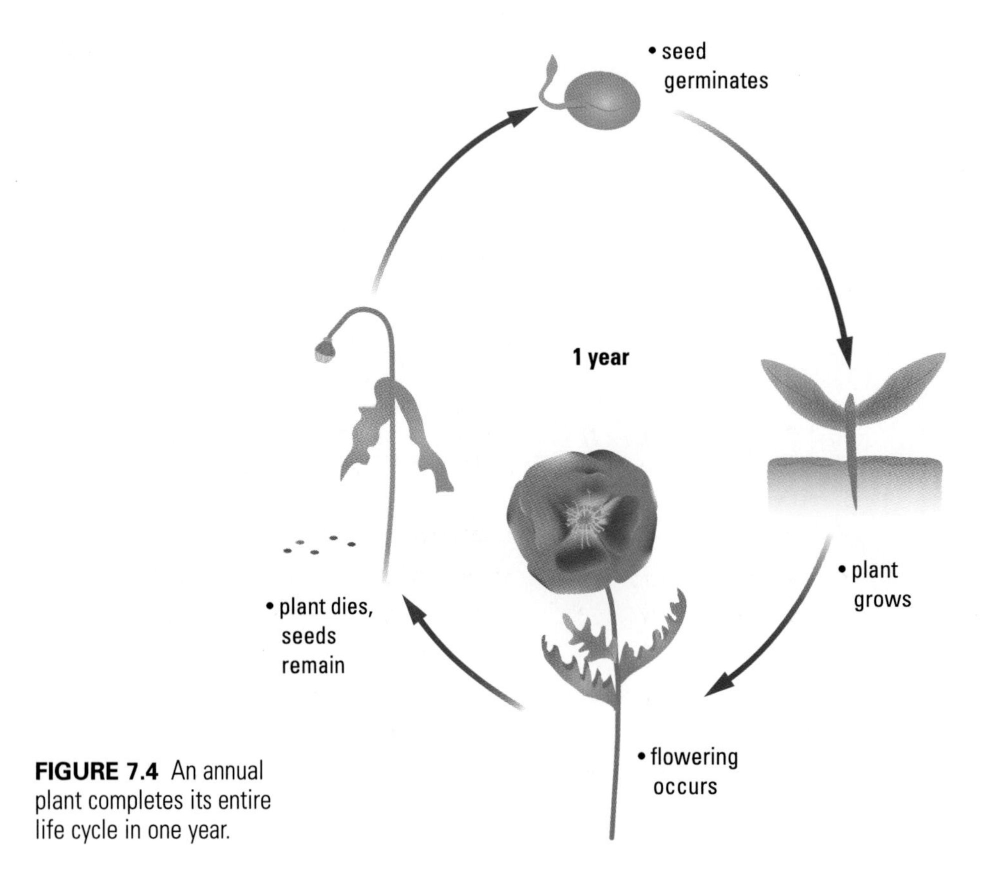

FIGURE 7.4 An annual plant completes its entire life cycle in one year.

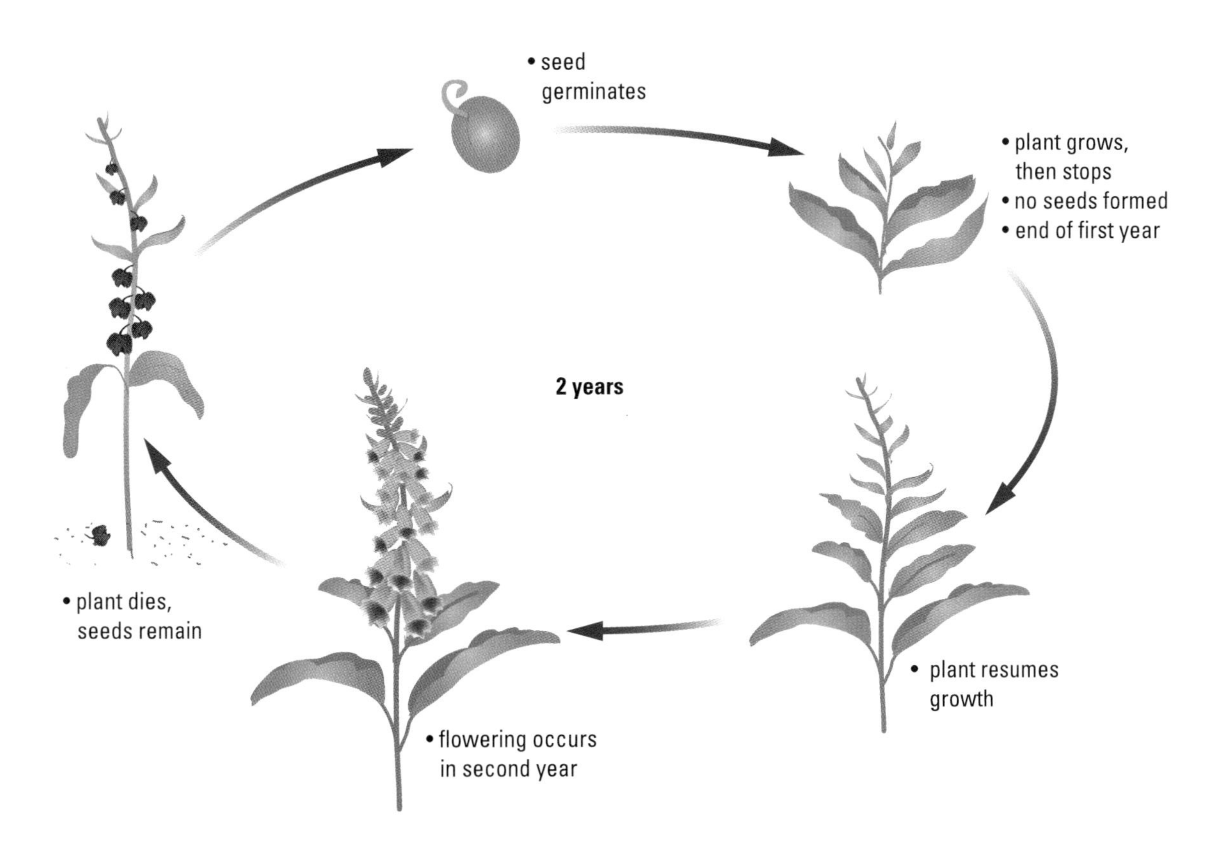

FIGURE 7.5 A biennial plant requires two years to complete its life cycle.

A **perennial** is a plant that lives for more than two years. Figure 7.6 shows some examples of perennials that are commonly grown in Canadian gardens. Some perennials, such as trees, can survive hundreds of years. Perennials can reproduce more than once.

FIGURE 7.6 A) Aquilegia, B) irises, and C) lilacs are examples of common perennials. How long do perennials live? Do they die after they reproduce? Explain.

Classifying by Propagation Method

You will need to know about the different ways to propagate plants for your project, Planting a Garden, at the end of Unit 4.

A common task for people who work with plants is to make more individuals of a certain type of plant. For example, someone who sells ready-to-plant rose bushes may need to produce thousands of a particular type of rose bush. **Plant propagation** refers to the production of new individuals of a particular plant. Another useful way to classify plants is by propagation method. Most plants can be propagated by more than one method, but a grower must know the best way to propagate the particular plants she or he needs.

From Seed

The least expensive way of propagating a seed plant is to plant and grow its seeds. One individual plant can produce thousands of seeds, so this method of propagation can increase the number of plants very quickly. The basic steps in growing plants from seeds include the following:

1. Collect the seeds.
2. Sow the seeds.
3. Provide conditions that allow the seeds to germinate.
4. Provide conditions that allow the seedlings to grow.
5. Allow the plants to flower. This may also require provision of particular conditions.
6. Allow the seeds to develop and ripen.
7. Collect the seeds.

You will learn more details about these steps throughout this unit. Many nursery plants, particularly annuals, are started from seed in greenhouses, early in the year (Figure 7.7).

FIGURE 7.7 Annuals are often started from seed in greenhouses well before the winter is over. Why are they not started outdoors in spring instead?

Plants that are produced from seeds are not always identical to their parent plants. This can be a disadvantage to the grower. Unless seed production is carried out very carefully, new plants may inherit different characteristics from the parent plants. You can find more information about inheritance of characteristics in Section 5.1 of this book. Many plants are also difficult to grow from seed. Some species will germinate only under specific conditions. For example, orchid seeds need, among other conditions, to be infected by a certain type of fungus (Figure 7.8).

FIGURE 7.8 Orchid seeds are very difficult to germinate.

Vegetative Propagation

Vegetative propagation is propagation of plants using any part of the plant (root, stem, or leaf) other than the reproductive structures (seeds or spores). Plants that are propagated by vegetative propagation are identical to the parent plant. Vegetative propagation is therefore a way of **cloning** plants. You can read more about cloning in Section 5.3. Perennials and spore-producing plants are often propagated by these methods. Perennials may take years to begin producing seeds, and growing plants from spores requires particular conditions that can be difficult to meet. You will read about two methods of vegetative propagation: bulbs and cuttings.

Some perennial plants can be propagated by bulbs. A **bulb** is an underground structure that stores energy and nutrients for the next year's growth. Many common spring flowers, such as tulips, are propagated by bulbs (Figure 7.9).

ScienceWise Fact

Not all the items available in the bulb section of a nursery are in fact bulbs. Some plants produce other underground structures that can be used for propagation. For example, the "bulbs" of gladiolas are more correctly called corms, and the "bulbs" of irises are more correctly called rhizomes.

A) 

B) 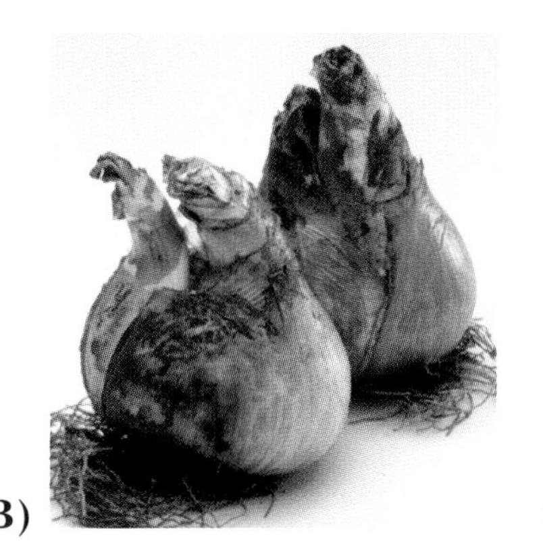

C) 

FIGURE 7.9 Bulbs can have many different shapes. These bulbs are A) tulip, B) daffodil, and C) onion.

FIGURE 7.10 Bulbs can be divided to produce new plants.

As a plant matures, it will produce either one large bulb or many smaller bulbs, depending on the species. To propagate plants that produce one large bulb, the grower digs up the bulb and cuts it into smaller sections. Each section, when replanted, will then produce a new plant. If a plant produces many smaller bulbs, the grower can separate the new bulbs and replant them to form new plants (Figure 7.10).

Cutting off a small piece of the plant and encouraging it to grow roots can also propagate plants. A **cutting** is a section of a plant, usually about 10 to 15 cm long, comprising a section of stem and at least one leaf. Any leaves at the bottom end of the stem are removed, and then the stem is placed in water, soil, or another suitable substance (Figure 7.11). The stem may also be dipped into a commercially available rooting compound before planting, which speeds up the rooting process. The cut stem eventually forms new roots. Once roots have appeared, the new plant may be transplanted to soil. The method used to produce cuttings varies with the type of plant. Most commercially available houseplants, such as coleus plants and rubber trees, are propagated from cuttings.

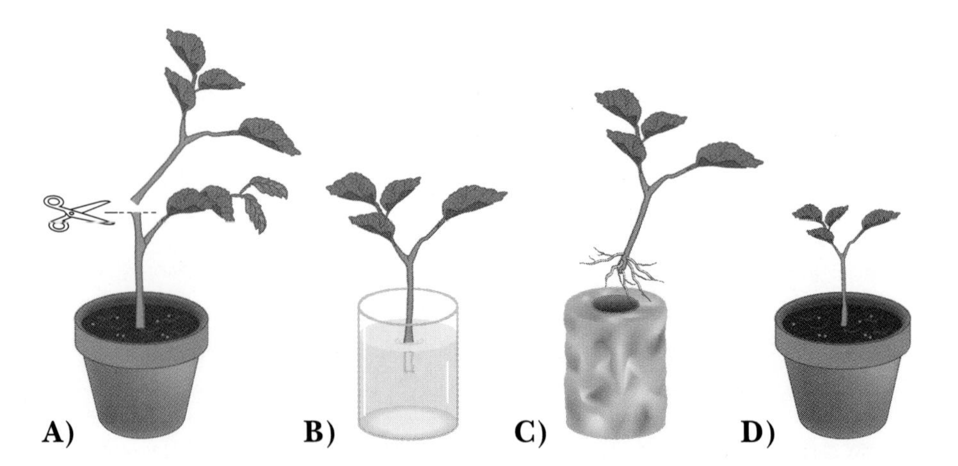

FIGURE 7.11 One method of rooting a cutting. A) Make a clean cut in an area of new growth. Include at least one leaf. B) Place the stem into a small container of water. There should be no leaves in the water. If necessary, cover the container with foil or paper to prevent light from entering. C) Lift the cutting and check for new roots every few days. D) When new roots appear, transplant the new plant to soil.

Review and Apply

1. In your notebook, match the terms in Column A with the definitions in Column B.

Column A	Column B
Biennials	Plants that live for more than two years
Germination	A 10- to 15-cm length of the plant, comprising a section of stem and at least one leaf
Annuals	Plants that grow for two years only
Cutting	The first stage of growth of a seed or spore to a seedling
Perennials	Underground structures that store energy and nutrients for the next year's growth
Bulbs	Plants that grow for one year only

2. Give two examples of plants that do not produce seeds.

3. Plants that produce seeds can be separated into two groups. Identify these groups, and give two examples of members of each group.

4. Agree or disagree with the following statement: All plants produce flowers. Explain your answer, and include examples of specific plants in your response.

5. You are working at a farm that produces seeds of flowering plants for growers. You have been asked to keep track of when the biennial species were planted. Explain why this information would be important to the seed producer.

6. You have started your own landscaping service. One of your clients wants a large area to have flowers that bloom in the first year. Would you recommend that your client plant annuals, biennials, or perennials? Justify your answer.

7. Describe two ways that plants can be propagated. Include an example of a plant that is commonly propagated by each method you describe.

8. Organize the concepts you have learned in this section in a graphic organizer.

ACTIVITY 7A

Propagating Plants

Anyone in the business of producing plants will likely need to propagate them. In this activity, you will conduct research to find out more about how plants may be propagated. Based on your research, you will select a plant and choose an appropriate method by which to propagate it. You will then propagate your plant and follow its growth.

What You Will Need

- paper and pens or pencils
- access to library or Internet resources
- plants and other materials (according to your research)

What You Will Do

1. In a small group, conduct research on plant propagation. There are many methods not presented in this textbook. Be sure to find as many different methods as you can.
2. Write a summary of your research. For each propagation method, describe how the propagation is carried out, list the materials that are needed, and name some examples of plants that may be propagated by that method.
3. Choose a plant to propagate, using one of the methods that you researched. Write a step-by-step procedure of how you will propagate your plant. Your procedure should include a list of the materials you will need.
4. Gather your materials and carry out your procedure. Observe your plant every two or three days, and record any changes you see. Design a chart that will allow you to record your observations.

What Did You Find Out?

1. Write a brief report outlining the advantages and disadvantages of the method of propagation you carried out. Your report should discuss the results you achieved and any problems you encountered.
2. Based on your experience, would you use the same propagation method to produce 100 plants? How about 1000 plants? Explain your answer.

Making Connections

3. Visit a garden centre and find out the cost of all the materials you would need to produce 100 plants by the propagation method you carried out. Based on the cost you calculate, how much would you need to charge in order to make a profit of $1.00 per plant? Would a profit of $1.00 per plant give you a reasonable wage? Explain your answer.

7.2 Plant Growth

To produce healthy growth, growers must provide plants with specific conditions. These conditions vary for different types of plants. They can also change with the age of the plant or the type of growth that is wanted. For example, when growing a Christmas cactus, the conditions required to produce flowers are not the same as those required to produce the biggest possible plant.

Plant Growth Basics

All plants require light, water, nutrients, and a source of carbon dioxide gas just to stay alive. Figure 7.12 summarizes how a plant uses these substances:

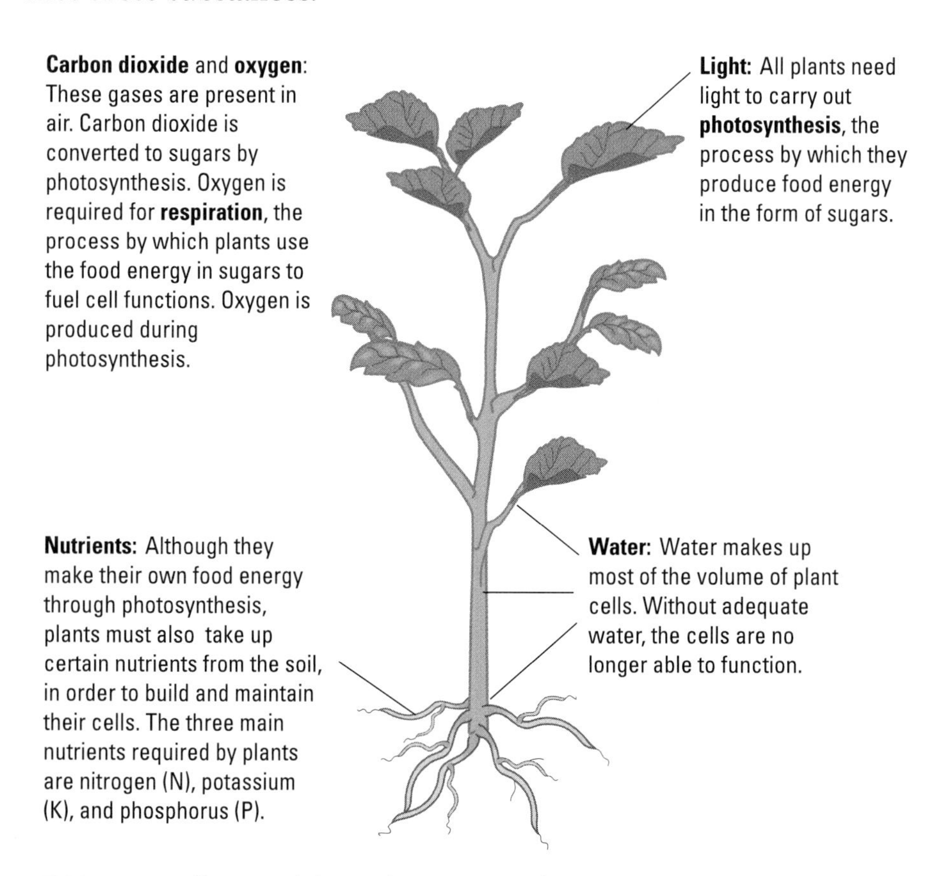

FIGURE 7.12 Plants need these substances to survive.

Optimal Growth Conditions

You will need to know about optimal growth conditions for your project, Planting a Garden, at the end of Unit 4.

Optimal growth conditions are the conditions under which a plant will grow the strongest and healthiest possible, and produce the desired type of growth. By providing optimal growth conditions, growers can produce plants that have the characteristics needed to make them suitable for sale or use. For example, a berry grower could produce more berries of higher quality by providing optimal growth conditions. Optimal growth conditions are related to the basic requirements of plants, but they also include some other factors.

Light

Since they use light to produce food energy, plants would die in complete darkness. However, the amount and quality of light is also very important in determining the plant's appearance. Some plants, such as sunflowers and coleus, need high amounts of light. Other plants prefer to grow in partially shaded areas. When a plant receives inadequate light, its leaves tend to be pale and small, and its stem (or stems) thin and very long. Plants that receive too much light may have leaves that appear bleached or burned. Figure 7.13 shows the appearance of an Earth Star plant grown under optimal and less-than-optimal light conditions.

A) 

B) 

FIGURE 7.13 A) Earth Star grown under optimal light conditions, and B) under low light conditions. Under low light, the plant no longer makes green pigments, and so it appears red.

Light also affects plant development. Many plants require specific light conditions in order to germinate or to flower. People who produce flowers (**horticulturalists**) must know the light requirements of the plants with which they work, to be able to produce flowers when they are needed. Home gardeners can also observe the effect of light on flowering. For example, the lettuce plant is grown for its edible leaves. When lettuce flowers, it produces a tall stalk and stops producing new leaves (Figure 7.14). The gardener is then unable to get the desired crop. Lettuce will only flower when it receives 12 or more hours of sunlight per day.

A) 

B) 

FIGURE 7.14 Lettuce plants grown under A) short days and B) long days.

Temperature

Temperature can affect seed germination, the rate of plant growth, and the timing of flowering. Each type of plant has a particular optimal temperature range for each of these stages. For example, lettuce seeds will not germinate if the temperature is too high, but red pepper seeds will not germinate if the temperature is too low. Most plants will grow more quickly in higher temperatures, providing they have sufficient water. Many plants stop growing, or become **dormant**, when the temperature falls below a certain level. This is very important here in Canada, because dormancy helps plants survive our long winters.

FIGURE 7.15 Many bulbs can be "forced," or made to bloom out of season, by artificially exposing them to cold temperatures.

Temperature also affects the flowering of some plants. Most plants will not flower until they have achieved a certain size. This ensures that the plant is able to provide the developing seeds with enough nutrients. Since plants often grow more quickly at higher temperatures, many plants flower earlier at warmer tempatures. Other plants require exposure to cooler temperatures before they will produce flowers. For example, spring-blooming bulbs such as daffodils will not bloom unless they are first exposed to a period of cold temperatures (Figure 7.15). Most biennials also require a period of cold before blooming.

Water and Oxygen

The optimal growth conditions of a plant include a certain amount of water. Some plants, such as the Boston fern, need to be in moist conditions all the time, whereas others, such as the aloe plant (Figure 7.16), grow best in relatively dry conditions. When plants have too little water, they tend to droop, or **wilt**. If a plant has too much water, however, its roots can become water-logged and no longer be able to take up the oxygen it needs for respiration. The plant may die as a result. Some plants, such as water lilies, are adapted to grow with their roots in water. Aquatic plants can grow only if they are fully submerged in water.

A)

B)

FIGURE 7.16 A) Boston fern and B) aloe require dramatically different levels of water for optimal growth.

Nutrients

FIGURE 7.17 The biggest and best rose blooms are achieved only when all nutrients are kept at optimal levels.

Optimal growth conditions for a plant include specific amounts of the main plant nutrients: nitrogen (N), phosphorus (P), and potassium (K). Different plants require different levels of these nutrients. For example, roses require higher levels of phosphorus than grass, which tends to require more nitrogen. If roses are not supplied with the optimal level of each of these nutrients, the plant's growth rate may slow or the plant may become sick. Plant problems caused by nutrient deficiency will be discussed in Section 7.3. The relative amounts of these nutrients can also affect plant development. For example, when roses are given too much nitrogen, they tend to produce more leaves and stems and fewer flowers (Figure 7.17).

In nature, all the nutrients plants need are usually provided by natural decomposition of dead plants and animals and of animal waste, and by the minerals in soil. The decomposed matter is an important part of the soil. Soil that is deficient in nutrients can be improved by adding either compost, manure, or chemical fertilizer. Both compost and manure improve the soil structure, but it is difficult to add specific amounts of nutrients with these fertilizers. Chemical fertilizers add specific amounts of nutrients, but they do not improve the general soil conditions.

Nutrients and Soil pH

Most plants take up nutrients through their roots, along with water. For this to occur, the nutrients must be dissolved in the water. Unfortunately, these nutrients are not always present in a form that will dissolve. If the medium in which a plant is growing (for example, soil) is too acidic or too basic, these nutrients are chemically changed to forms that cannot dissolve. Optimal growing conditions therefore also include a particular range of acidity or pH.

LAB 7B

Optimal Growing Conditions

Whether plants are grown for their beauty, for food, or for other materials, obtaining the desired characteristics depends on providing optimal growing conditions. Plant growers determine these conditions by carefully testing the effects of different variables on plant growth.

Be Safe!

Always wash your hands after handling soil.

Purpose

In this lab, you will design and carry out an experiment to test the effect of different variables on the growth of seedlings of one type of plant. From your data, you will determine some of the conditions that are needed for optimal growth of that plant type.

Materials

- seedlings
- six foam cups or similar containers
- potting soil
- pencil
- water
- ruler, scale, or other measuring devices

Procedure

Part One: Transplanting Seedlings

FIGURE 7.18 Make sure you do not damage the root when you transfer your seedlings.

1. Fill the foam cups about $\frac{3}{4}$ full with moist potting soil. With the pencil, make a small hole in the soil surface. The hole should be no more than 1 cm deep.
2. Obtain seedlings of one species only.
3. Gently transfer one seedling from the container to a foam cup. Place the seedling root down into the hole, making sure all of the root is in the soil, as shown in Figure 7.18. Gently fill in the hole with soil so that the root is buried. Do not press the soil down.
4. Add a small amount of water to the transplanted seedling.
5. Continue this process until you have transplanted six seedlings.

Part Two: Design an Experiment

1. You will use these transplanted seedlings to test the effect on growth of the following variables: light intensity, moisture level, and temperature. A **variable** is an experimental condition that is purposely changed during an experiment. You will design your own experiment. To conduct a fair test, you must make sure only one variable is changed at a time. Write a procedure describing how you will test the effect of these variables. Your procedure must include the following:
   - **a)** A step-by-step description of how you will test each variable, including how you will make sure that all other factors in the experiment remain the same.
   - **b)** A description of how you will measure changes in the growth of your seedlings. Will you measure changes in height? Will you determine the change in mass, or count the number of leaves?
   - **c)** A list of all the materials you will need.
   - **d)** A data table that will allow you to record your observations.
   - **e)** A description of any safety procedures you will need to follow.
2. Label your foam cups and carry out your procedure.

Analysis and Conclusion

1. Did all plants survive? If not, describe the conditions that did not support the life of the seedlings.
2. Using the data you collected, construct a graph to show the effect of each variable on the growth of your plants.
3. Based on your graph, describe the optimal growth conditions for the type of plant you tested.

Extension and Connection

4. Prepare a handout naming your seedling species and describing the optimal growth condition for the type of plant you tested. Exchange handouts with other groups.
5. Compare your results with those of other groups. Could you plant all these different species in the same area of a garden and provide them all with optimal growth conditions? Explain your answer.

Review and Apply

1. State the four basic requirements for plant life.
2. Explain what is meant by optimal growth conditions. How are optimal growth conditions different from the basic requirements for plant life?
3. Are the optimal conditions for seed germination the same as the optimal conditions for seedling growth? Explain and give examples.
4. Why is it important for a grower to know the optimal growth conditions for the plants that he or she produces?
5. Give two examples of how light can affect the growth of plants.
6. Discuss why you might not be able to grow all the same types of plants in North Bay that you might in Hamilton.
7. Describe what happens if a plant receives too little water. Is it possible to give a plant too much water? Explain.
8. Why is the pH (acidity) of a soil important?
9. Add the new concepts in this section to the graphic organizer you started in Section 7.1.
10. **Favourite Flowers**

    In a small group, choose at least two flowering plants that you would enjoy growing in a garden. Using print and electronic resources, research the optimal growing conditions for the plants you chose. Prepare a chart that summarizes the similarities and differences in the growing conditions of the plants. Could you grow these plants under identical conditions? Explain your answer.

Surf the Web

In addition to light, water, temperature, and nutrients, seed producers also have to provide conditions to ensure that flowers are pollinated (pollen is transferred from one flower to another). If pollination does not occur, seeds do not form. Research some of the ways that seed producers encourage flower pollination. Begin your search by visiting **www.science.nelson.com** and following the links for ScienceWise Grade 12, Chapter 7, Section 7.2.

7.3 Plant Problems

Like any other living things, plants can develop health problems. Problems can occur if plants become infected by micro-organisms or are attacked by insects. Plants can also develop problems if they are not provided with optimal growing conditions. In this section, you will learn about some common plant problems and how to identify them.

Evidence of Plant Disease

Plant diseases are caused by micro-organisms such as viruses, fungi, or bacteria. These organisms may use nutrients that the plant needs, or use the plant as a source of food. Figure 7.19 shows some common symptoms of plant disease.

Symptoms of Plant Disease

Symptom	Example
Dusty white or grey coating on leaves	Willow infected with powdery mildew
Spot of dead or discoloured tissue on leaves or fruit	Apple infected with apple scab
Swellings or deformed growth	Corn cob infected with smut
Rotting of roots and stems	Tobacco stem with black shank disease

FIGURE 7.19 The symptoms of disease vary with the type of plant and the particular micro-organism.

Evidence of Insect Pests

Plants serve as a source of food and shelter for many different organisms, including insects. Unfortunately, insects can cause significant damage, sometimes enough to kill a plant. Insects can also change the appearance of a plant, making it less attractive. This can cause economic harm to those whose business involves selling or using plants. To minimize insect damage, a grower must be able to recognize the evidence that an insect is attacking a plant, and take appropriate action. Figure 7.20 shows some of the signs of insect attack.

Signs of Insect Attack

Evidence	Example
Holes in leaves, fruit, or bark	Weevil damage on alfalfa leaves
Plants cut off at stems at soil surface	Cut worm damage to corn seedlings
Changes in leaves, such as appearance of lines or curling of edges	Leaf miner tunnels on velvet leaf
Presence of webs or casings	Tent caterpillar webs on apple tree
Presence of insects or their larvae	Aphids on rose buds

FIGURE 7.20 Insects can cause damage to any part of a plant.

Evidence of Nutrient Deficiency

Soil rarely provides the optimal levels of all nutrients for all types of plants. Plants can also become nutrient deficient when too many plants are growing in an area. When plants are deficient in a nutrient or nutrients, they will develop certain symptoms. Some examples of symptoms are shown in Figure 7.21. The symptoms vary with the type of plant, so if a plant develops any of these symptoms, the nutrient level in the soil or other growth medium must be analyzed in order to identify the problem.

Symptoms of Nutrient Deficiency

Symptom	Example
Leaves begin to yellow, or there is yellowing between leaf veins	Nitrogen deficient corn plant
Plant growth slows down and plant becomes stunted	Normal and phosphorus deficient wheat plants
Stems and/or leaves turn red, purple, or brown	Magnesium deficient rose leaf
Leaves start turning brown on the edges	Potassium deficient cotton leaf
Seeds, flowers, or growing points die or become deformed	Calcium deficient beans

FIGURE 7.21 Plants can often survive a long time even when nutrients are deficient. Why would it be important for a landscaper to correct a nutrient deficiency?

Review and Apply

1. In your notebook, state whether each of the following is evidence of nutrient deficiency, disease, or insect pests:
   a) Pea plants form only deformed or empty pea pods.
   b) The leaves of a rhododendron plant are yellowish instead of deep green.
   c) The fruit on a pear tree has dark spots on the skin.

2. Describe or sketch three examples of nutrient deficiency in plants.

3. During a summer job in an apple orchard, you notice that leaves on the apple trees have small dead spots. Are these spots likely to be caused by nutrient deficiency, disease, or insect pests?

4. You run a small company that maintains the plants in office buildings. One very expensive potted plant has a tiny web on one of its shoots. Should you be concerned? Explain your answer.

5. Add the new concepts in this section to the graphic organizer you started in Section 7.1.

6. **From Seed to Seedling**

   Working with a partner, choose one type of seed plant. Brainstorm the basic steps you would need to follow to produce 100 seedlings from seed. The seedlings must be ready for transplanting into the garden. Use print and electronic resources to find the optimal growing conditions and any other information you need. Present your work as a pamphlet, video, or multi-media presentation.

Job Link

Fruit Grower

Fruit growers produce fruit for sale in grocery stores or to food processors.

Responsibilities of a Fruit Grower

- caring for fruit-bearing plants (planting, pruning, watering, fertilizing, and controlling weeds, harmful insects, and birds)
- thinning, harvesting, and packaging fruit for shipment
- operating and maintaining related machinery

Where do they work?

- orchards and vineyards

Skills for the Job

- ability to work outdoors and perform heavy lifting
- knowledge of plants, especially conditions needed for flowering and fruiting

Education

- high school diploma required
- may require specialized training

FIGURE 7.22 Fruit growers try to provide the plants they care for with the optimal conditions for fruit production.

LAB 7C

Nutrients and Plant Growth

In outdoor landscapes or gardens, most plant nutrients must be supplied by the soil. Landscapers and gardeners often test soils for nutrient levels before they plant. If any nutrients are lacking, they can be added by using fertilizers.

Purpose

In this lab, you will test the effects of different amounts of nitrogen, phosphorus, and potassium on the growth of plants.

Materials

- seedlings
- garden soil (not potting soil)
- full-strength liquid fertilizer (according to manufacturer's instructions)
- half-strength liquid fertilizer (according to manufacturer's instructions)
- soil-testing kit
- three small pots or other containers
- water
- masking tape
- waterproof marker
- small trowel or spoon
- ruler

Procedure

1. You will transplant seedlings into three pots. The first pot (pot 1) will contain garden soil only, the second pot (pot 2) will receive half-strength liquid fertilizer, and the third pot (pot 3) will receive full-strength liquid fertilizer. You will then monitor the growth of the plants over several weeks. You will keep track of the following to compare growth: height, number of leaves, and leaf colour. Design a data table in your notebook or in a spreadsheet file to record this data.

2. Choose one species of plant. Note the conditions of light intensity, temperature, and moisture level that give optimal growth for this species, either using the results from Lab 7B or the information provided by the supplier.

3. Using the masking tape and waterproof marker, label the pots (pot 1, pot 2, pot 3). Include your name and the date on the label.

4. Add soil to within 1 or 2 cm from the top of each container.

5. Using a trowel or spoon, make a hole in the soil in each container. The hole should be about the size of the containers in which the seedlings are growing. Add a small amount of water to the hole.

6. Place a seedling on its side, and gently tap the container to loosen the soil and roots. Grasp the stem of the plant close to the soil surface, and gently pull the plant out (Figure 7.23).

FIGURE 7.23 When removing the seedlings, pull on the lowest part of the stem to avoid breaking the roots.

7. Place the roots of the plant in the hole, and then bury the roots. Add more soil if needed. Gently tamp down the soil around the plant.

8. Repeat steps 5–7 until you have transplanted seedlings into each of the three pots.

9. When all seedlings are transplanted, moisten the soil in pot 1 with additional water. Moisten the soil in pot 2 with half-strength liquid fertilizer, and the soil in pot 3 with full-strength liquid fertilizer.

10. Place your seedlings in a place where they will have optimal conditions of light and temperature.

11. Check your plants daily. Moisten the soil as needed with water (pot 1), half-strength liquid fertilizer (pot 2), or full-strength liquid fertilizer (pot 3).

12. Every two to three days, observe and record the height, number of leaves, and leaf colour of each of your plants. Record your data in your data table.

CONTINUED

13. Remove a small amount of the soil from each pot, being careful not to disturb the seedling. Following the instructions on the soil test kit, determine the levels of nitrogen, phosphorus, and potassium in each soil sample. Record your data.

Analysis and Conclusion

1. On paper or using spreadsheet software, create graphs to show the changes in height and number of leaves of each of your plants over time.
2. From the information on the graph and any changes in the leaf colour you observed over time, which seedlings had the healthiest plant growth? Explain your choice.
3. What were the nutrient levels in the pot that supported the healthiest plant growth? What were the nutrient levels in the pot that supported the least healthy growth?

Extension and Connection

4. Prepare a handout naming your plant species and describing the nutrient levels that provided optimal growth conditions for the type of plant you tested. Exchange handouts with other groups.
5. Compare your results to those of classmates who grew a different species. Write a short summary of the similarities and differences between their results and yours.
6. A chemical fertilizer company prepares two different products for lawn care. One product is for spring application, and has high levels of nitrogen. The other product is for fall application, and has lower levels of nitrogen. Suggest reasons why the levels of nutrients are different for these two products.

7.4 Chapter Summary

Now You Can...

- Describe and give examples of spore- and seed-bearing plants; of flowering and cone-bearing plants; of annuals, biennials, and perennials; and of plants propagated by seeds, bulbs, or cuttings (7.1)
- Describe how specific conditions of light, temperature, water, and nutrients are required for healthy plant growth (7.2)
- Describe the basic steps in growing plants from seed (7.1, 7.2)
- Identify evidence of plant problems caused by diseases, insect pests, and nutrient deficiency (7.3)

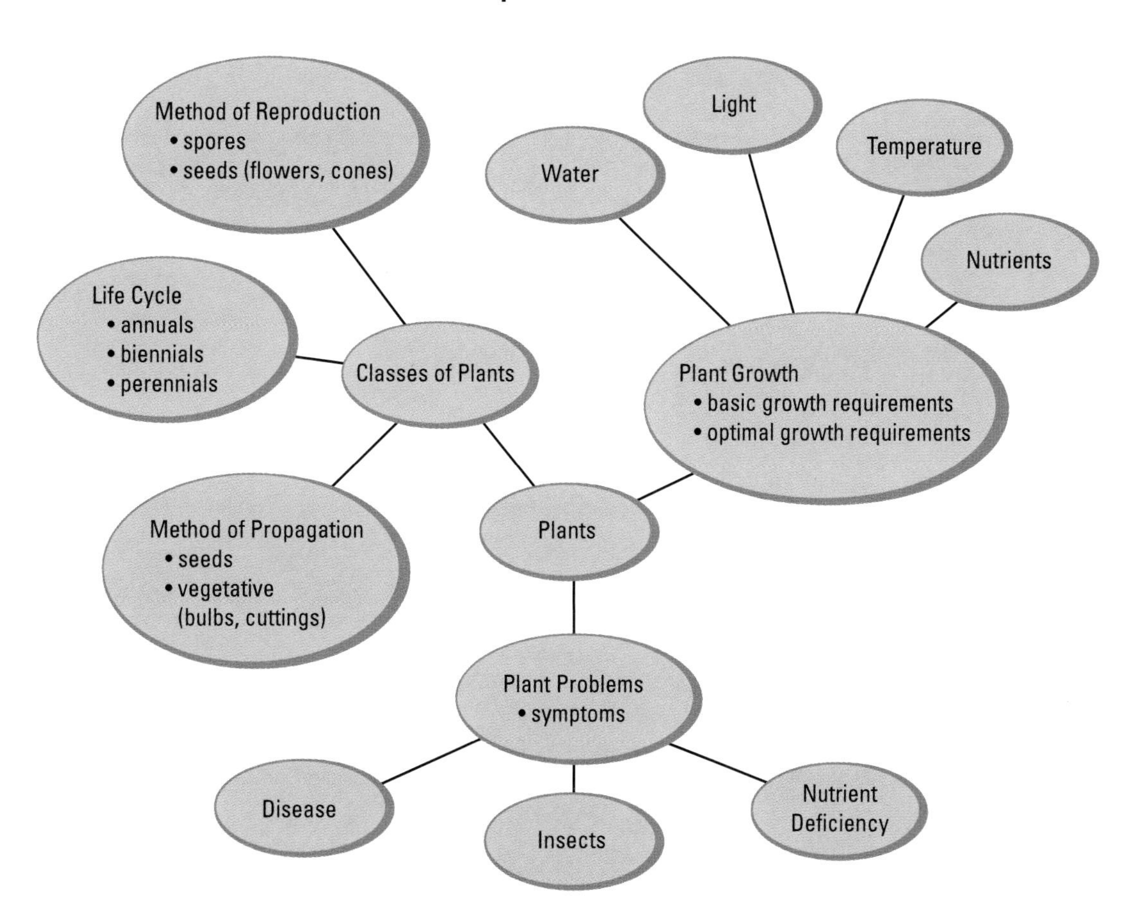

FIGURE 7.24 Compare your completed graphic organizer to the one on this page. How did you do? What new links can you add to your organizer?

CHAPTER 7 *review*

Knowledge and Understanding

1. In your notebook, identify whether each of the following statements is true or false. Rewrite any false statements to make them true.
   - **a)** The reproductive structures of plants are either flowers or cones.
   - **b)** Plants produce spores or seeds every year.
   - **c)** Perennials reproduce by bulbs only.
   - **d)** Annuals reproduce every year.
   - **e)** All seed plants can be propagated easily by germinating their seeds.
   - **f)** Plants produce all their own nutrients.
2. Give two examples of vegetative propagation.
3. Explain the difference between an annual and a biennial.
4. Give two examples of how a high level of light might affect a plant.
5. Identify two stages of plant growth that can be affected by temperature.
6. Name the three main nutrients required by plants.
7. Look at the photographs in Figure 7.25 below. Identify whether the plant problems shown are most likely due to nutrient deficiency, disease, or insect pests.

A)

B)

C)

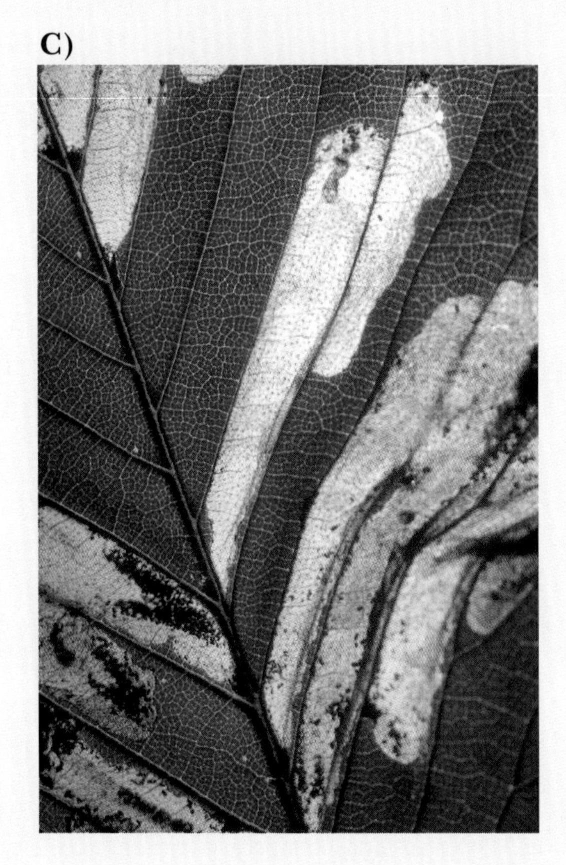

FIGURE 7.25

Inquiry

8. You have discovered an unopened package of seeds with no date on it. Seeds stored for a long time may no longer germinate. Describe how you could test the seeds to determine whether or not they are worth planting.

9. You work as an assistant to a greenhouse grower who produces carnations for florists. One of your jobs is to determine the optimal conditions for each type of carnation. The graph in Figure 7.26 shows the result of a test of number of hours of light per day on flowering. What is the optimal number of hours for each of the carnation varieties tested?

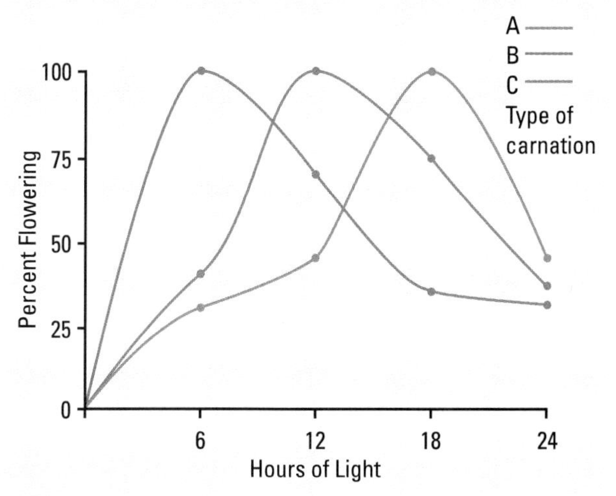

FIGURE 7.26

Making Connections

10. A local charity wants to sell houseplants for a fundraiser. Unfortunately, they have received only 10 plants from donors. Explain how they could increase the number of plants without buying more.

11. You work as an assistant groundskeeper with a municipal parks department. The plants in one of your flower beds have not flowered. The same plants in another flower bed close by are all in bloom. Given that the entire area is in the shade most of the day, what factors in the environment could be responsible for this difference?

12. You own a hardware store that sells gardening tools, pots, and other supplies for home gardening. You do not sell any plants. Why would it be an advantage for you to know about the needs of plants?

Communication

13. The Ontario government offers several apprenticeships in areas related to growing plants. Use print or electronic resources to identify one apprenticeship program related to growing plants that interests you. Research what the program involves, and the kinds of jobs for which you would be qualified after completing the apprenticeship. Use the information you find to write a newspaper-style advertisement for this apprenticeship program.

14. Select one of the following: roses, tomatoes, or a plant of your own choosing. Use print and electronic resources to find out the optimal growing conditions for the plant, and any diseases or pests that commonly cause problems for it. Present a summary of the information in the style of a garden catalogue entry. Include photographs or drawings.

CHAPTER

8

Working with Nature

Plants benefit humans in many ways. For example, they offer us food to eat, and materials with which to make clothing, housing, medicine, and other goods. However, humans are not the only organisms that depend on plants. All life on Earth ultimately depends on plants for food energy, nutrients, and oxygen. Plants also help to create environments in which other organisms can live, such as nesting areas for birds. The methods we choose to grow and harvest plants can have negative effects on the environment. Figure 8.1 shows some methods of growing plants. What occupations might be involved with growing plants by the method shown? For each situation shown, identify all the advantages to humans. What are the environmental consequences? How could we reduce negative consequences?

FIGURE 8.1

What You Will Learn

After completing this chapter, you will be able to:

- Describe the diversity of environments that must be maintained to provide habitats for a wide variety of plants (8.1)
- Describe different methods of gardening and how each controls conditions of growth (8.1)
- Describe the design elements and materials used in landscaping (8.2)
- Describe some common forest-management practices (8.3)
- Demonstrate an understanding of the role of forests as essential habitats for other plants and animals, including threatened and endangered species (8.3)
- Demonstrate an understanding of various ways in which humans depend on healthy plant populations (8.1, 8.2, 8.3)
- Analyze the social, economic, and environmental factors involved in gardening, horticulture, landscaping, and forestry (8.1, 8.2, 8.3)

What You Will Do

- Investigate various methods used to control the conditions of growth for plants (Activity 8A)
- Identify the features of a good landscape architecture site, and prepare a plan to scale for an outdoor garden (Activity 8B)
- Analyze the social, economic, and environmental factors that determine the different approaches and methods required in gardening, horticulture, landscaping, and forestry (Activity 8C)

Words to Know

balance
clay
clear-cutting
climate
colour
commercial thinning
contrast
forestry
greenhouse
habitat
hard materials
hardiness
harmony
humidity
humus
hydroponics
landscape gardening
landscaping
repetition
sand
selective cutting
silt
soft materials
sustainable development
texture
topography

A puzzle piece indicates knowledge or a skill that you will need for your project, Planting a Garden, at the end of Unit 4.

8.1 Gardens Provide Habitats

A **habitat** is an area with environmental conditions that allow a particular organism to survive. Gardening can be thought of as creating a habitat for plants, by providing suitable growing conditions.

Plant Habitats

FIGURE 8.2 Does a habitat always provide optimal growing conditions? Explain.

A plant's habitat usually provides a range of environmental conditions that the plant can tolerate (Figure 8.2), not just optimal growing conditions. The main factors that determine if an area provides habitat for particular plants are that area's climate and soil characteristics, and the presence of other organisms.

You will need to know about plant hardiness zones for your project, Planting a Garden, at the end of Unit 4.

Climate

The conditions of light, moisture, and temperature of an environment affect how well a plant will grow. **Climate** is the average conditions of temperature, precipitation (rain and snow), and sunlight of an area. **Hardiness** is the relative ability of a plant to survive extreme climate conditions. Most plants sold commercially are assigned a number that describes their hardiness. Plants that are the hardiest have the lowest numbers. A map of plant hardiness zones for Canada is available online or on paper. Each zone is an area in which plants of a certain hardiness rating, or above, can survive.

Surf the Web

Agriculture and Agri-Food Canada provides an electronic map of Canada's plant hardiness zones. To access this map, go to **www.science.nelson.com** and follow the links for ScienceWise Grade 12, Chapter 8, Section 8.1. Which hardiness zone do you live in?

Soil Characteristics

Whether an area can provide habitat for a plant also depends on the characteristics of the soil. Soil is made up of particles of sand, clay, and silt, along with various amounts of humus. **Sand** is very small particles of rock. **Clay** is small particles of minerals called silicates. **Silt** is a mixture of sand and clay particles that have been carried by water and then deposited. **Humus** is partially decomposed material from living organisms. Humus is the main source of nutrients in soil, and helps soil to hold water. Soils are classified by the relative amounts of these materials (Figure 8.3).

Soil Types

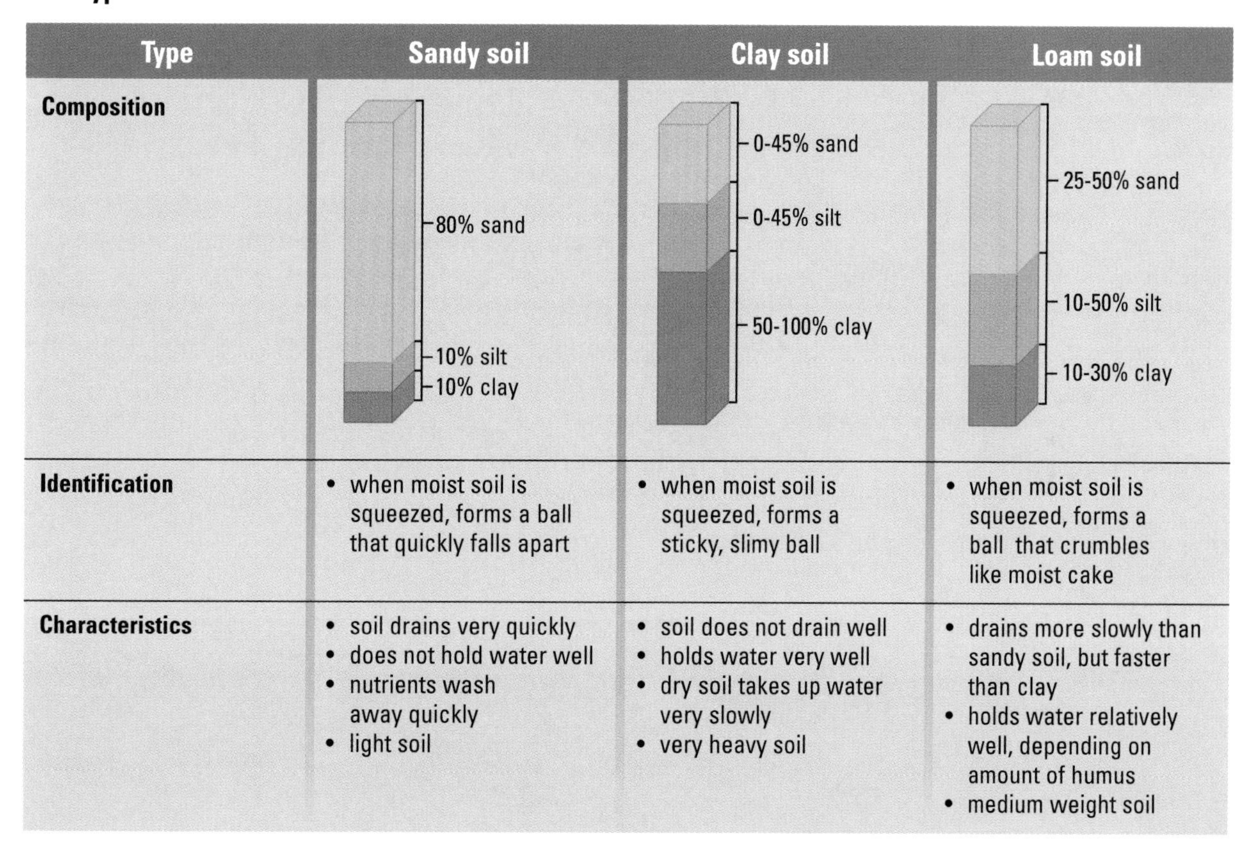

Type	Sandy soil	Clay soil	Loam soil
Composition	80% sand 10% silt 10% clay	0-45% sand 0-45% silt 50-100% clay	25-50% sand 10-50% silt 10-30% clay
Identification	• when moist soil is squeezed, forms a ball that quickly falls apart	• when moist soil is squeezed, forms a sticky, slimy ball	• when moist soil is squeezed, forms a ball that crumbles like moist cake
Characteristics	• soil drains very quickly • does not hold water well • nutrients wash away quickly • light soil	• soil does not drain well • holds water very well • dry soil takes up water very slowly • very heavy soil	• drains more slowly than sandy soil, but faster than clay • holds water relatively well, depending on amount of humus • medium weight soil

FIGURE 8.3 The amount of humus, the soil pH, and the level of nutrients can vary in each soil type. How might this affect the ability of a soil to provide habitat for a particular plant?

FIGURE 8.4 How are plants affected by the presence of these organisms? Are all types of plants affected in the same way?

Other Organisms

The other organisms present in an area may also determine if that area can provide suitable habitat for a plant. For example, many large trees will block out sunlight, creating conditions that would not allow sun-loving plants to survive. Similarly, if an area contains many organisms that eat a particular plant, very few individuals of that plant will be able to survive (Figure 8.4).

Controlling Growing Conditions Outdoors

To successfully grow plants outdoors, a gardener must know the climate and soil type of the area to be used. Using this information, a gardener can either choose plants that suit the existing habitat, or modify the environmental conditions to provide a more suitable habitat for desired plants.

You will need to know how growing conditions can be controlled for your project, Planting a Garden, at the end of Unit 4.

Climate

It is not possible to change the climate conditions of an outdoor garden dramatically. The easiest approach to outdoor gardening, therefore, is to identify the hardiness zone and choose only plants that suit that zone. However, it is possible to control some climate conditions in an outdoor garden. Figure 8.5 shows some common methods. What are the conditions of climate that are being controlled?

Adding water | Preserving soil moisture | Adding shade | Adding insulation

FIGURE 8.5 Examples of methods that can change the climatic conditions in small areas of an outdoor garden. What other methods are available? Give two occupations in which you would need to know how to control climate conditions in an outdoor garden.

Soil

Soil characteristics can be changed significantly in an outdoor area. Coarse sand can be added to clay soil, to lighten it and increase its drainage. Humus can be added in the form of compost (partially decomposed material from living organisms) or manure (animal waste). Soil pH can also be changed by adding materials such as lime, which increases pH, or bone meal, which decreases pH. Sometimes, gardeners make more dramatic changes to their soil (Figure 8.6). For example, people may buy commercial topsoil (loam soil) and layer it on existing soil, or use it to replace soil altogether.

FIGURE 8.6 Many people choose to alter the soil characteristics of their gardens. How does this affect habitat? What organisms are affected?

Other Organisms

Controlling other organisms is another important part of gardening outdoors. Weeds (undesirable plants) can easily invade cultivated areas. Insects can cause problems by eating the plants, as you saw in Section 7.3. Animals can damage gardens, by eating plants, trampling, or digging in a cultivated area. Figure 8.7 shows some examples of methods used to control other organisms in outdoor gardens.

FIGURE 8.7 How are unwanted organisms being controlled in these situations? Name a workplace that would be involved in producing, selling, or using the products shown in each situation pictured.

Surf the Web

Organic gardens are gardens in which no human-made chemical products are used. Conduct research to find the costs and benefits of organic gardening. Start your research by visiting www.science.nelson.com and following the links for ScienceWise Grade 12, Chapter 8, Section 8.1.

Costs and Benefits of Outdoor Gardening

The principal benefit of outdoor gardening is that it takes advantage of the natural climate. Outdoor gardening uses the Sun's energy and the natural precipitation in an area. It also uses the natural soil of an area, although gardeners may change the soil's characteristics to suit particular plants. However, the gardener is limited to specific types of plants in an outdoor garden (Figure 8.8).

FIGURE 8.8 How do these two gardens demonstrate of the limitations of outdoor gardening?

Outdoor gardening can also have economic, social, and environmental benefits. Many businesses have a stake in gardening, including horticulture and landscaping businesses, and businesses that produce fertilizers, gardening tools, or pesticides. Gardening provides a source of pleasure, can be good exercise, and also be helpful to people recovering from injury or disease (Figure 8.9). Gardening can also result in environmental benefits. For example, gardens can provide habitat for insects such as moths and butterflies, or birds.

FIGURE 8.9 Gardening can offer exercise and enjoyment. What other social benefits can you suggest?

However, some practices used to modify growing conditions in an outdoor area have environmental and economic costs. Figure 8.10 shows examples. When people water their gardens during hot, dry weather, instead of soaking into the soil, the water evaporates and brings dissolved salts up toward the surface of the soil. These salts can become so concentrated that they poison plants (Figure 8.10A). Excess nutrients from chemical fertilizer or manure can cause water pollution (Figure 8.10B). Using pesticides can reduce the number of insects in an area, which will also reduce the number of organisms that depend on insects for food (Figure 8.10C). By changing the environment of an area to suit plants we choose to grow, we also reduce the amount of habitat for native plants (Figure 8.10D).

A)

B)

C)

D)

FIGURE 8.10 Each of these photos shows an example of the potential costs of outdoor gardening. In your notebook, explain the connection between gardening and the photo.

Controlling Growing Conditions Indoors

Plants may be grown indoors if they require growing conditions significantly different from those of the natural environment, if a gardener wishes to provide optimal growing conditions, or just for pleasure (Figure 8.11). Plants may be grown indoors in homes (houseplants), or in **greenhouses**, which are structures designed specifically to grow plants.

FIGURE 8.11 Houseplants can make our indoor spaces more attractive.

Surf the Web

Horticulture (growing plants for flowers) is a growing field in more than one sense. Horticulturalists are expected to be in demand for some time to come. Using the Internet or library, find out what kinds of work are available in your area. What qualifications are needed? Begin your search at **www.science.nelson.com**. Follow the links for ScienceWise Grade 12, Chapter 8, Section 8.1. Would you like to work in horticulture? Why or why not?

Greenhouses

Greenhouses are constructed to provide a controlled environment for plants. They provide an alternative environment to the natural environment of Earth. You can find more information about alternative environments in Unit 5. Greenhouses can control some or all growing conditions, depending on their design.

Most greenhouses are large enough to allow people to enter and work. They usually have flooring, which helps to insulate the structure so that temperature can be more easily controlled. Most greenhouses also have a venting system to allow excess heat to escape. Many large commercial greenhouses have specially designed shades to help cool certain areas of the structure. Some greenhouses also have heating and cooling systems.

FIGURE 8.12 Not all designs of greenhouses control the same growing conditions. What factors would affect the choice of greenhouse design?

Greenhouses are built entirely or partially of transparent materials (Figure 8.12) to take advantage of natural sunlight. They may also have artificial lighting, to provide plants with optimal light conditions. The moisture levels of plants in greenhouses may be controlled by simple watering with a hose or watering can, or by more sophisticated computer-controlled automatic watering systems. Greenhouses can also have a different **humidity** (moisture content of the air) than the outside environment. Plants in greenhouses are usually grown in containers, so greenhouse growers can easily adjust the characteristics of the soil they use for different plants. Nutrients can also be added as needed.

Hydroponics

Hydroponics is the cultivation of plants in nutrient-enriched water. Plants are usually held up in a non-nutritive medium such as vermiculite, peat moss, or coarse sand (Figure 8.13). Their roots are then bathed in a liquid containing nutrients at concentrations that will encourage optimal growth. Plants grown by hydroponics always have optimal nutrient and water levels, so they tend to grow larger and healthier than non-hydroponically grown plants. For example, hydroponically grown tomatoes tend to be juicier and more tender than tomatoes grown in soil. Hydroponic plant production usually occurs in greenhouses, where conditions of light, temperature, and humidity can be controlled, but plants also may be grown hydroponically on a small scale at home.

FIGURE 8.13 Hydroponically grown vegetables are often more expensive. Why?

Costs and Benefits of Indoor Gardening

Indoor gardening has the same economic and social benefits as outdoor gardening. However, indoor gardens do not interact with the natural environment to the same extent as outdoor gardens, and so do not have many environmental benefits. The principle benefits of indoor gardening, compared to outdoor gardening, are as follows:

- Plants produced indoors can be healthier than those grown outdoors, if they receive close to optimal growing conditions.
- Plants can be grown in an area, or at a time of year, in which they would not normally survive.

In general, indoor gardening costs more money than outdoor gardening. The structures and environmental control systems of a greenhouse or hydroponics operation are expensive to build and run, and use a lot of energy. Indoor gardens cannot use natural precipitation or soil, so the grower must also supply all the water and nutrients. Indoor gardens do offer some protection from other organisms, particularly other plants. However, insects and disease-causing micro-organisms can cause many problems in greenhouses. The greenhouse environment also protects these organisms, and may supply them with a ready-made banquet. Pest and disease control is therefore extremely important, and can have significant environmental and economic costs.

ScienceWise Fact

In 2000, Canadian growers produced more than $494 million dollars worth of vegetables in greenhouses.

ACTIVITY 8A

How Does Your Garden Grow?

You and a few friends have decided to earn some money this summer by growing and selling fresh tomatoes. Of course, the most important step in any project is good planning. In this activity, you will assess the costs and benefits of the following methods of producing tomatoes:

a) growing the plants in an outdoor garden, using chemical products to provide nutrients (fertilizer) and to control unwanted organisms (for example, herbicides and pesticides);

b) growing the plants in an outdoor garden, using only natural organic products (for example, manure and compost) to add nutrients and control unwanted organisms, and;

c) growing the plants indoors by hydroponics.

After you have compared the costs and benefits of each method, you will use this information to decide which method you will use for your summer business.

What You May Need

- pen or pencil
- paper
- advertising flyers from grocery, hardware, and garden supply stores
- reference books on vegetable gardening
- library and Internet access

What You Will Do

1. Conduct research to find information you think would help you to grow healthy tomatoes. For example, you will need to know the optimal growing conditions for tomatoes, and what pests and diseases commonly affect tomatoes.

2. With your group, make a list of all the materials you would need to provide suitable growing conditions for tomatoes, using each of the three gardening methods listed in the introduction to this activity.

3. Use advertising flyers to calculate how much it would cost to get sufficient materials to plant and grow 12 tomato plants. Assume that you do not need to pay for any land you might use in an outdoor garden, or space for an indoor garden.
4. Find out the selling price per kg of tomatoes grown by each of the three methods described in the introduction to this activity. These will most likely be described as "vine-ripened," "organically grown," and "hydroponically grown" tomatoes (Figure 8.14). You may need to visit a grocery store to get accurate prices.
5. When research is complete, make a chart showing the costs and benefits of each method of growing tomatoes. Consider the environmental, social, and economic costs and benefits. Include an estimate of how much it will cost you to produce 1 kg of tomatoes and the selling price per 1 kg for each method.

What Did You Find Out?

1. Create a report outlining which method of producing tomatoes you think would be best for your business. Use the information from your research to justify your choice.

FIGURE 8.14 How will the cost per kilogram affect your business?

Making Connections

2. Of all the luck! During the summer that you are running your business, there is a drought in July. Then, in late August, there is an early frost. How do these two events affect your business? Which method of growing tomatoes would be least affected by these events?

Review and Apply

1. Why is it important for a gardener to know about the climate of the area in which plants are to be grown?
2. How could you determine if a soil sample was sandy soil, clay soil, or loam soil?
3. Give two examples of how climate conditions can be controlled in an outdoor garden.
4. Create a storyboard or skit that shows how gardeners can help to provide habitat for particular plants by controlling the environmental conditions of an area.
5. Describe how greenhouses control the conditions of light and temperature under which plants are grown.
6. Look through the business pages of your local phone book for businesses related to gardening. Make a list of the companies you find and what they do. What jobs are available at these businesses?
7. Organize the concepts you have learned in this section in a graphic organizer.

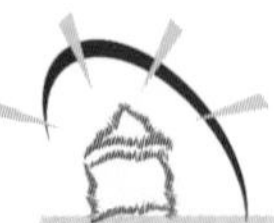

Try This at Home

An Interior Landscape

How would you like to create a landscape of potted plants for your home? Start by conducting a brief survey of the environmental conditions of the rooms in your home. Make notes on the amount of sunlight, the average temperature, and the humidity in each area. When you have finished your survey, choose one area that would be suitable for an indoor garden. Measure the area that you want to use.

Set yourself a reasonable budget for your garden. Visit a store that sells houseplants, and make a list of plants that would grow in the conditions of your indoor garden and that are within your budget. Include on your list plants that would have a pleasing mixture of colour, texture, and size.

8.2 Landscaping

Landscaping or **landscape gardening** is the activity of laying out or designing gardens. Landscaping creates pleasant environments for us, and creates habitat for plants and other organisms. In this section, you will learn about some of the elements a landscaper considers when designing a landscape, and some of the materials that are used.

You will need to know about landscape design for your project, Planting a Garden, at the end of Unit 4.

Landscape Design

Before any work begins, a landscaper usually draws a scale plan, similar to a blueprint, of the proposed design. Landscape design must consider the existing features of the site and any features that will be added. Some of these are outlined in Figure 8.15.

Features to Consider in Landscape Design

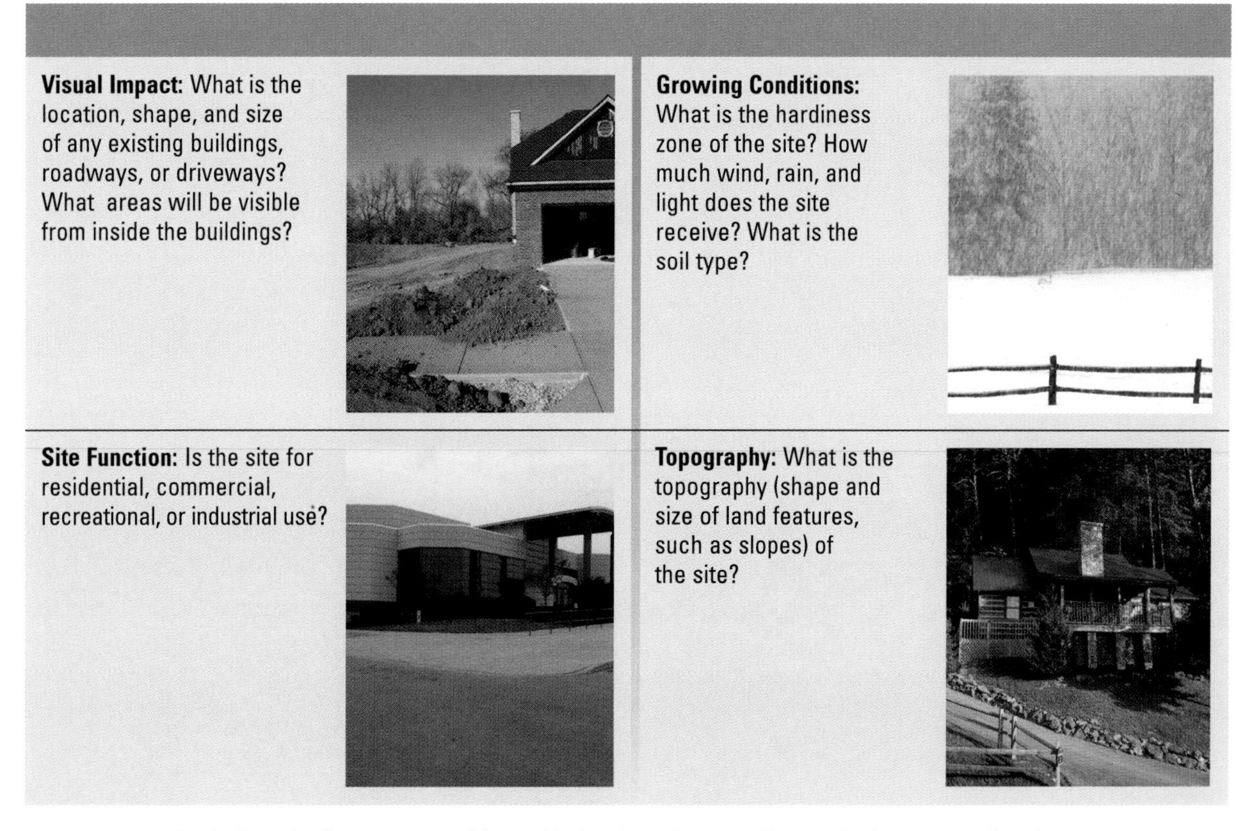

FIGURE 8.15 Designing a landscape starts with considering these features. Explain the importance of each.

Elements of Design

Every landscape plan must also consider elements of design. These elements, outlined in Figure 8.16, are considered in all creative designs, from graphic art used in advertising to fine art displayed in museums. Landscapers often put these design elements together in a way that creates a central theme or style. In many ways, landscape design is a form of art.

Elements of Design

Colour: Features (plants, rocks, buildings) may be organized according to colour.

Texture: Features may be organized according to texture (a characteristic of the surface of an object, such as softness, roughness, smoothness).

Balance: The visual impact of each feature in a landscape is considered in relationship to others on the site.

Repetition: Features are incorporated that repeat shapes, colours, or patterns.

Harmony: Features are selected that match in colour, size, shape, or texture.

Contrast: Features are chosen that have contrasting shape, size, colour, or texture.

FIGURE 8.16 A landscape design may incorporate any of these elements.

Landscaping Materials

Many different materials are used in landscaping. Plants such as trees, shrubs, bushes, and bedding plants are referred to as **soft materials**. **Hard materials** are non-living objects used in the design, such as walls, paving stones, rocks, decks, and gazebos. Figure 8.17 shows some of the many choices of materials available for landscaping.

Examples of Landscape Materials

Soft Materials	Hard Materials
Trees and Shrubs: Plant nurseries often supply young trees and shrubs with wrapped roots (a root ball) ready to place into soil.	**Concrete Building Materials:** Concrete products are available that are specifically designed to build pathways, patios, walls, and other landscape features.
Flats: Annual flowers, such as these petunias, are available in flats containing many individual plants.	**Decorative Containers:** Plant containers are available in many sizes, shapes, and colours.

FIGURE 8.17 Employment opportunities related to landscaping can be found in companies that make or sell any of these products. What are some of the companies in your area?

Landscape Maintenance

After landscapers have designed and planted a new site, they move on to new locations. Landscape maintenance workers look after the plants and other features of a landscaped property. Many gardeners enjoy doing this work themselves, but paid maintenance workers usually maintain large areas, such as municipal parks and schoolyards. Many landscape maintenance workers run their own businesses.

ACTIVITY 8B

Designing a Garden

Landscapers usually prepare a scale drawing, similar to a blueprint, which shows the features of the proposed design. This drawing can be used as a guide by landscape construction workers.

Landscape plans can be extremely simple, such as the one shown in Figure 8.18. In this plan for a vegetable garden, the area to be planted was drawn to scale on graph paper. Sections of the garden were then blocked out on the plan, and the plants to be placed in each block were noted.

Even this simple garden plan required careful research and thought. Here are some of the questions that the designer had to answer to plan this garden:

- What area on the site has climate and soil conditions suitable for this garden?
- What are the measurements of the garden area?
- What plants would I like to grow?
- Of these, which can grow well in the climate and soil conditions of this garden?
- How large will each plant grow?

In this activity, you will prepare a plan to scale for an outdoor garden. Your plan must consider the features of the site and the growing conditions needed by the plants you choose.

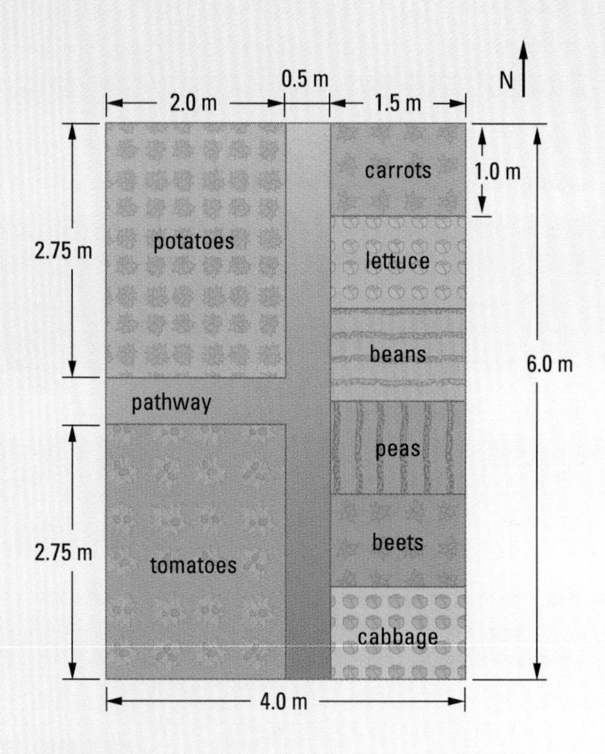

FIGURE 8.18 A simple landscape plan.

What You Will Need

- measuring tape
- pencils or markers
- ruler
- graph paper
- Computed Aided Design (CAD) software (optional)
- catalogues, seed packages, or other information sources on growing conditions of plants

What You Will Do

1. Choose a site on your school property that would be suitable for a garden.
2. Make notes on the features of the area. Examine the soil and identify the soil type, using the criteria in Figure 8.3.
3. Identify the hardiness zone of the outdoor site by referring to the map at www.science.nelson.com under the link for ScienceWise 12, Chapter 8, Section 8.2. If your site is indoors, record the air temperature of the site.
4. Choose a particular area that you think would be suitable for a garden. Note your reasons for choosing that area.
5. Using the measuring tape, measure the dimensions of the area you have chosen. Make a rough sketch of the area, noting the dimensions.
6. On graph paper or with software, draw the outline of your garden using an appropriate scale. For example, you might use a scale of 1 cm = 1 m. Include any existing features, such as buildings.
7. Make a list of the plants you would like to grow. Find out the optimal growing conditions for each plant (Figure 8.19).
8. Think about the elements of design and the plants you have chosen, and create a garden plan. Use labels and colours to show where you would place each element. Remember to draw all elements of the garden to scale.

FIGURE 8.19 Find out about optimal growing conditions before you choose the plants for your garden plan.

What Did You Find Out

1. What were the advantages in the site you chose for your garden? What were the disadvantages?
2. Exchange plans with a partner. Would you be able to construct a garden based on your partner's plan? Make a list of suggestions as to how your partner's plan could be improved. Make a second list of what you liked about the plan.

Making Connections

3. Which is the most important in planning a garden: choosing the site of the garden or choosing the plants that are to be grown? Justify your answer, using your experiences in completing this activity.
4. Would you prefer to work as a landscape designer or a landscape construction worker? Why?

You will need to know how to draw a garden plan to scale for your project, Planting a Garden, at the end of Unit 4.

Review and Apply

FIGURE 8.20

1. Explain why the local climate must be considered when planning an outdoor garden.
2. You work in a garden centre, and a customer brings in the sketch shown in Figure 8.20. Consider the features of the plan and recommend the best site for a garden planted with sun-loving plants. Explain your choice.
3. Figure 8.21 shows a formal garden design. In point form, describe how the design elements are used in this landscape.
4. Give two examples of soft materials and two examples of hard materials used in landscaping.
5. With a partner, survey the landscape of your schoolyard. What type of landscape maintenance does your schoolyard require? Who performs this maintenance?
6. Add the new concepts from this section to the graphic organizer you started in Section 8.1.
7. **Tools of the Trade**

   You and a partner are starting your own landscape maintenance company. From a current hardware catalogue, select a set of tools that would be essential for your business. For each tool, describe why you would need it. Using the catalogue prices, calculate how much it would cost you to start your business. Would you change your list based on this cost? Why?

FIGURE 8.21

Job Link

Landscape Construction Worker

ScienceWise: How long have you worked in landscape construction?

Ian Davis: Five years.

ScienceWise: How did you get started?

Ian Davis: When I finished high school, I applied for a job at a local nursery.

ScienceWise: What sort of work do you do?

Ian Davis: I work mainly in landscape construction. I build retaining walls, garden walls, and water gardens. I also install interlocking brick and rocks, and plant and remove trees.

ScienceWise: What do you like most about your work?

Ian Davis: I love working with my hands. When I take a drawing from the designer and make it come to life, it is very rewarding. I also like working outdoors.

ScienceWise: What do you not like about your work?

Ian Davis: Working outdoors when it is cold and raining—I wish I were somewhere else! Sometimes there are difficulties with the customers, like when they ask for changes halfway through a job.

ScienceWise: Are there any safety concerns on the job?

Ian Davis: I use power tools like a stone-cutting saw. I wear safety boots, and eye and ear protection. I also wear a mask when I use the saw, because it kicks up a lot of dust.

FIGURE 8.22 Ian Davis constructs many different features included in landscape designs. Here, he is shown cutting stone for a client.

8.3 Forestry—Managing the Ultimate Garden

Canada is a country rich in forests. **Forestry** is the science and practice of planting, caring for, and managing forests. The forestry industry is a major contributor to our economy. Many products use materials harvested from forests, such as lumber and pulp and paper. Figure 8.23 shows some examples.

Examples of Forest Products

Building Products: Forests supply us with many of the materials used in the construction of buildings.

Furniture: Wood is used in the manufacture of furniture.

Pulp and Paper: Paper and paper products are manufactured from wood pulp.

Other Products: Many products other than wood, such as maple syrup, are harvested from our forests.

FIGURE 8.23 What forest products do you use at school? Name at least one occupation that involves making, selling, or using each of these products.

Forests Provide Habitat

As well as providing people with many useful products, forests also provide habitat for many other organisms. One of the greatest challenges to using our forest resources wisely is finding ways of harvesting the trees we need without destroying habitat. Habitat can be destroyed by removing trees, and by use of chemicals to control insects and disease. If organisms lose too much of their habitat, they face the danger of extinction. Figure 8.24 shows some of the plants and animals that are endangered due to loss of forest habitat.

FIGURE 8.24 A) A wolverine, B) a Kirtland's warbler, C) a blue racer snake, D) a wood poppy. These species are all endangered due to loss of their forest habitat.

Forest-Management Practices

There are three main ways in which trees are harvested in Canada:

1. **Clear-cutting** is harvesting forests by removing all of the trees, regardless of their age or type. Clear-cutting may be done in blocks, strips, or patches (Figure 8.25A). Clear-cutting is the easiest and least costly method of harvesting the forest. The exposed land is then replanted with seedlings of a particular type and age. This results in a forest composed mainly of one type of tree, each individual being of similar size and age.
2. **Selective cutting** involves removing some trees in all size classes, either singly or in groups (Figure 8.25B). Although trees are usually selected for their commercial value, this method of harvesting also maintains the age and size distribution of a natural forest. However, selective cutting is more expensive to carry out than clear-cutting.
3. **Commercial thinning** is the selection and removal of smaller trees from a forest so that the remaining trees will be better able to reach the desired size and shape. The cut trees are removed and used for commercial purposes. Commercial thinning is more likely to occur in planted forests than in natural forests.

Surf the Web

In 1999, 352 000 people across Canada were directly employed by the forestry industry. What kinds of work are available in this industry? What companies in your province offer employment in the forestry industry? Begin your research by visiting **www.science.nelson.com** and following the links for ScienceWise Grade 12, Chapter 8, Section 8.3.

A)

B)

FIGURE 8.25 A) A forest after clear-cutting; B) a forest after selective cutting. Which method of forest management will have the least negative consequences to organisms that depend on a forest for their habitat? Explain your choice.

Sustainable Development of Forestry Resources

The Canadian government is working with the forestry industry to develop and promote methods that encourage sustainable development. **Sustainable development** is development that can be continued over the long term. Sustainable development ensures that we use the resources we have today in a manner that will not damage the environment or economy for tomorrow. This includes making sure that habitat is protected, so we can all continue to enjoy and benefit from the many organisms with which we share planet Earth.

Although clear-cutting removes all the trees in an area, new trees are always planted to replace the cut trees. Do these replanted forests replace habitats? The answer to this question is both yes and no. Some species of plants and animals can survive quite well in the replanted forests. However, replanted forests tend to be made up of only one type of tree, such as white pines, and each individual tree is about the same age. A natural forest offers a mix of small and large trees. This mix of trees provides habitats that are not found in a planted forest and on which some species depend (Figure 8.26).

FIGURE 8.26 The eastern cougar is an endangered species. Eastern cougars are found mainly in forests that have a natural mix of trees of different ages and species.

Selective logging maintains the natural forest structure. However, since selective logging is more expensive to carry out than clear-cutting, most of Canada's forests are harvested by clear-cutting (Figure 8.27).

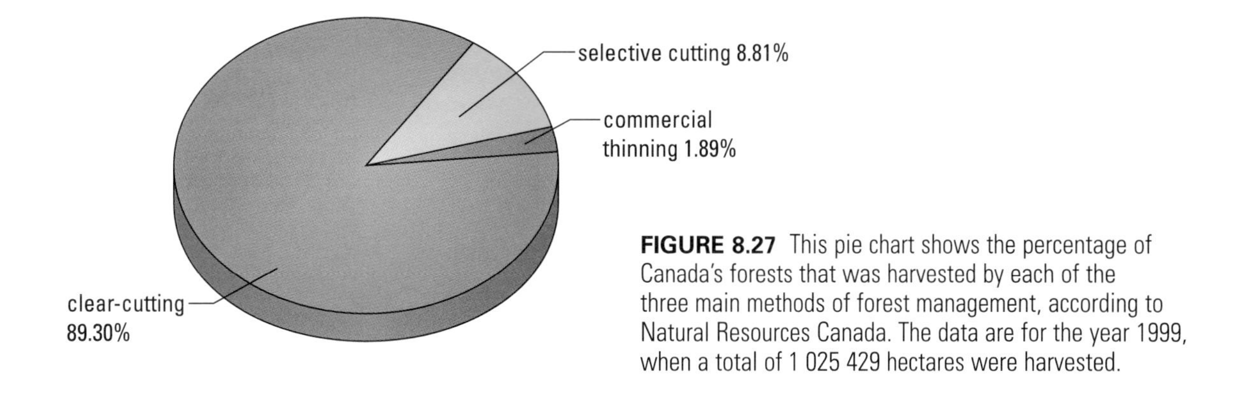

FIGURE 8.27 This pie chart shows the percentage of Canada's forests that was harvested by each of the three main methods of forest management, according to Natural Resources Canada. The data are for the year 1999, when a total of 1 025 429 hectares were harvested.

ACTIVITY 8C

Costs and Benefits of Forest Management

Currently, most of Canada's forests are harvested by clear-cutting. Why is clear-cutting so widely practised? What are the costs and benefits of clear-cutting? How do they compare to the costs and benefits of selective cutting and commercial thinning?

In this activity, you will conduct research into one method of forest management, and gather information on the costs and benefits of this method to Canada's environment, economy, and society.

What You Will Need

- pens or pencils
- notebooks
- reference books
- Internet access (if available)

What You Will Do

1. Working in groups of three or four people, decide on one method of forest management you would like to investigate: clear-cutting, commercial thinning, or selective cutting.
2. Using reference books and the Internet, research the costs and benefits of the method you chose to Canada's economy, environment, and society. Use the following questions to guide your research:
   - When and where in Canada is this method used?
   - What are the environmental costs of this method?
   - What is done to benefit the environment after or during cutting?
   - What are the economic benefits of this method in relation to other methods?
   - What are the economic costs of this method in relation to other methods?
   - Who in society benefits from this method of forest management?
   - Who in society is harmed by this method?
3. When your research is complete, meet with your group members and share the information you found. As a group, decide whether the benefits outweigh the costs.

What Did You Find Out?

1. Write a report that describes your views on this method of forest management. Your views must be supported by facts from your research.

2. In your notebook, prepare two tables, each with three columns. Title one table "Costs of Forest Management Methods" and the other "Benefits of Forest Management Methods." Label each of the columns in the tables as follows: clear-cutting, selective cutting, and commercial thinning. Enter all the costs and benefits of the method you researched in the appropriate table. In the columns for the other two methods, enter a check mark if that method shares each cost or benefit you listed. When you are finished, enter any additional costs and benefits for those two methods. You may use the information provided on the previous pages. Based on this comparison of the three methods of forest management, identify which method you think is best for the environment, the economy, and for society. Justify your choices.

3. If there is time, debate your views with class members. Remember to allow each person to finish speaking before you offer your views. Be prepared to defend your thoughts with facts.

Making Connections

3. Sustainable development is development that meets current needs without damaging our ability to meet future needs. Which method (or methods) of forest management do you think is an example of sustainable development?

4. Choose a forestry-related occupation that interests you. Conduct research to find out what skills and education are required to enter this occupation, and what companies in your province hire such workers. Summarize your answer in paragraph form.

FIGURE 8.28 What are the economic costs and benefits of the three different forest management methods?

Review and Apply

1. Name five forest products that you have used.
2. What is the relationship between sustainable development and habitat?
3. What is the most common method of forest management in Canada? Describe this method.
4. Do replanted forests provide the same habitats as natural forests? Explain.
5. Explain how sustainable development of Canada's forests helps our economy.
6. Add the new concepts from this section to the graphic organizer you started in Section 8.1.

Job Link

Logging Machine Operator

Logging machine operators work in many different phases of logging.

Responsibilities of a Logging Machine Operator

- operates machines such as harvesters, forwarders, and loaders

Where Do They Work?

- logging companies and contractors

Skills for the Job

- knowledge of maintenance and safe operating procedures of machinery
- ability to work as a team member

Education

- high school diploma may be required
- on-the-job training is usually provided

FIGURE 8.29 A logging machine operator works with very powerful, specialized machines.

8.4 Chapter Summary

Now You Can...

- Describe different gardening methods and how they control conditions of growth (8.1)
- Describe how different methods of gardening can affect habitat (8.1)
- Identify the features of a good landscaping site and create a scale plan for a garden, using the elements of design (8.2)
- Describe some common forest management practices (8.3)
- Discuss the necessity of maintaining sustainable forests (8.3)
- Discuss the various ways humans depend on healthy plants, and analyze the social, environmental, and economic factors related to how we grow and harvest plants (8.1, 8.2, 8.3)

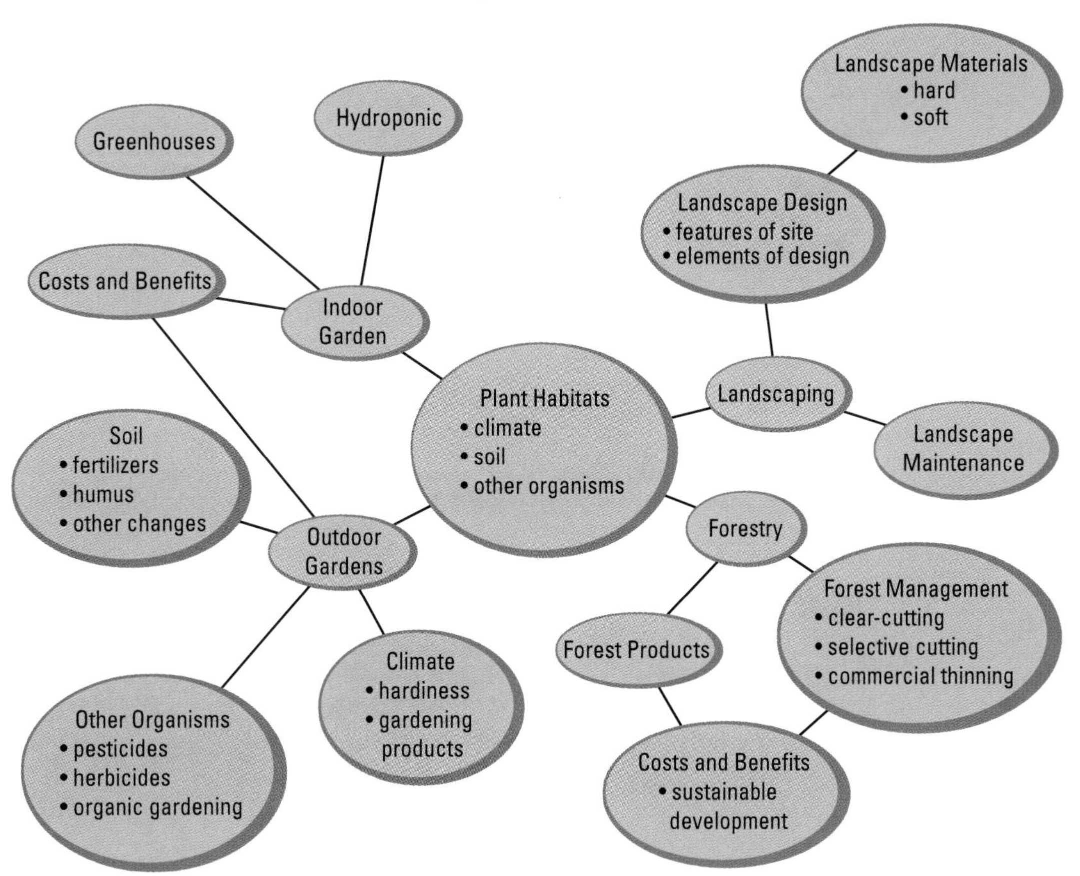

FIGURE 8.30 Compare your completed graphic organizer to the one on this page. How did you do? What new links can you add to your organizer?

CHAPTER 8 *review*

Knowledge and Understanding

1. What is plant hardiness? How is hardiness used in gardening?
2. Describe two ways in which soil can be modified in an outdoor garden.
3. A large tree in a local playground is removed. How does this affect the habitat of the area where the tree was growing?
4. What are the advantages of growing plants by hydroponics over growing plants in soil? What are the disadvantages?
5. Using an example, explain why a landscaper must consider the existing features of a site when designing a landscape.
6. Draw a sketch of a garden that uses harmonizing colours in its design and a second sketch of a garden that uses contrasting colours.
7. What is sustainable development?
8. In point form, describe the three main methods of forest management in Canada.
9. Give an example of how forestry affects the habitat of organisms within the forest.

Inquiry

10. You have your own lawn-care business. What soil tests would be important to your business? Explain your choices.
11. You want to create a landscape that provides habitat for native birds. Make a list of questions you would need to research before you started to create your landscape.

Making Connections

12. Explain how firefighting is important to the forestry industry. How does firefighting contribute to sustainable development of forestry resources?
13. You and a friend are planning to start a business making wooden tables and chairs. Should you be concerned with how forestry companies harvest the wood you use? Explain.

14. Many communities have banned or are considering banning the use of herbicides and pesticides in public parks and schoolyards. Others are switching to "environmentally friendly" products. Are these better choices? Justify your viewpoint, considering environmental, social, and economic factors in your answer.

15. The pie chart in Figure 8.31 shows the percentage of people in Canada employed directly and indirectly by the forestry industry. From this graph, how important do you think the forestry industry is to the Canadian economy? Identify two occupations that would be included in the pie chart under direct employment, and two that would be included under indirect employment.

Communication

16. Imagine that you have your own landscaping business. Create an advertising flyer to sell your services. Your flyer should list your services and prices, and point out anything that makes your business different from others.

17. Landscaping and horticulture in Ontario is a billion-dollar industry. Use print and electronic sources to find job announcements, and create a portfolio of advertised job opportunities in these fields.

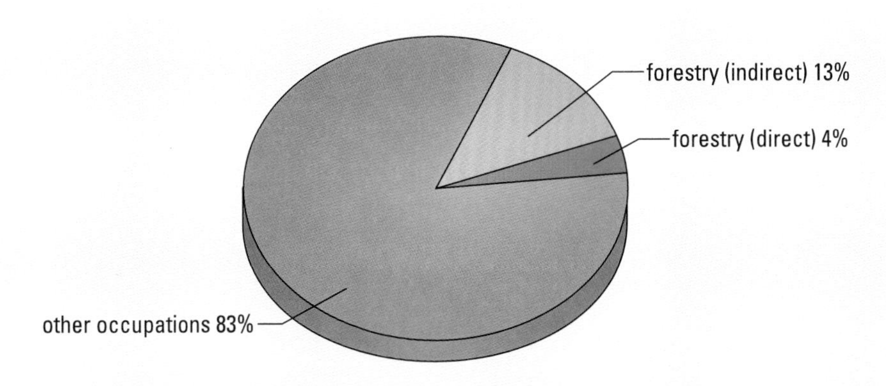

FIGURE 8.31 Employment in the forestry industry, as a percentage of employment in all industries in Canada, for 1998, according to Natural Resources Canada.

PUTTING IT ALL TOGETHER

Planting a Garden

Throughout this unit, you have been practising the skills needed to grow and propagate plants. You have also learned some of the skills required to plan a garden, using the design principles and materials used by professional landscapers. You will now use all this information to create a garden for the enjoyment of your community.

FIGURE D1 What effect does the landscape of a school have on the students, staff, and visitors? How might you improve the landscape of your school?

The Plan

You and your classmates have carried out lab exercises to determine some of the following information about various plant species:

- how the plants may be propagated
- the optimal conditions of light intensity, moisture level, and temperature for seedling growth
- the optimal levels of plant nutrients for seedling growth

You have also learned how to determine the type and pH level of soil, and the factors that must be considered in choosing a site suitable for a garden. You have created a garden plan.

You will use this knowledge to plan and plant an indoor or outdoor garden, working from a prepared garden plan. You will also prepare a maintenance schedule for your garden, to make sure that the plants stay healthy.

What You May Need

- seedlings (your own or commercial)
- a garden design
- seeds
- gardening catalogues and reference books
- Internet access (where available)
- gardening tools
- fertilizer (chemical, compost, or manure)
- water
- a suitable site
- camera (optional)

What You Will Do

1. Identify an area in your school that would make a good site for your garden. This may be the site you chose in Activity 8B, or a new site.

2. If you are using the same site you chose for Activity 8B, you and your classmates may vote to choose one garden plan that you would like to use. If you chose a different site, work with your classmates and follow the steps in Activity 8B to create a garden plan for this area. Make sure you have chosen plants that are suitable for the site of your garden.

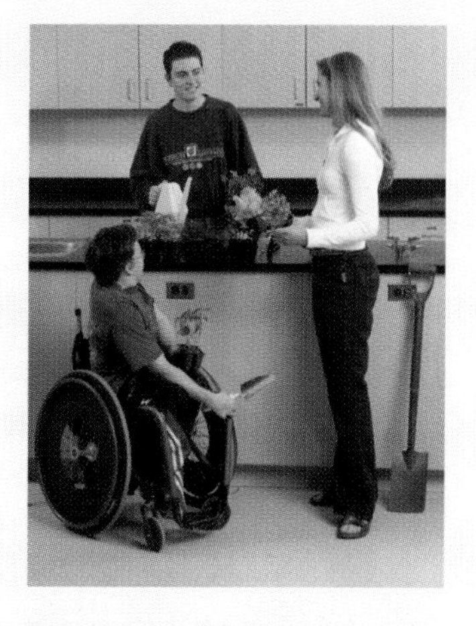

FIGURE D2 Make one last check that you have everything you need.

3. Obtain your plants. Look over the plants and check that they are all healthy. Use the information in Section 7.2 to identify any plant problems.

4. Using gardening books, the Internet, or information supplied by the grower, review how to transplant each of the plants to the garden, or how to plant the seeds. Make notes on the requirements for each type of plant you will grow, before you go to your garden site.

Be Safe!

Always wash your hands after handling soil.

5. Make a list of the tools you will need, and then gather them together. Assemble your plants, and carefully take all your materials to the garden site.

6. Prepare the soil in the area you are planting. If you are working outdoors, turn over the soil to loosen it. If you are planting an indoor garden, fill your containers with potting soil.

7. If you have access to a soil-testing kit, test the soil in the site you have chosen. Test for pH, potassium, nitrogen, and phosphorus levels. If necessary, add nutrients or adjust the pH of the soil to suit the needs of your plants.

CONTINUED

8. Transplant your plants to the appropriate area, according to the garden design. Water the plants. If you have access to a camera, take a photograph of your garden.

9. As a class, put together a schedule of activities that you will need to carry out to maintain your garden. Your schedule should include all of the following:
   - watering
   - weeding
   - checking for signs of problems (pests or disease)

   Your schedule should include the names of the students in your class, which task or tasks they will be responsible for, and the dates on which they are to carry out their tasks.

10. In a notebook, create a table in which the students in your class can record the growth of the plants in the garden. The records may be in the form of written descriptions, sketches, or photographs, recorded at one-week intervals.

What Did You Find Out?

1. After at least three weeks have passed, look over the plants once again, and read the records of their growth. Which plants are the healthiest? Identify any plants that are showing signs of problems.

2. Based on your answer to question 1 above, did your class choose a good site for the garden? What improvements could you make so that the growing conditions at the site are more suited to the plants?

Assessment

3. Which is more important to successful gardening: planning the garden or maintaining it? Justify your answer by referring to your experience in carrying out this project.

4. Most professional landscapers work with other people to complete a project, such as construction workers or nursery staff. Comment on the importance of teamwork in successful landscaping. Suggest one or two ways that your classmates could have improved their teamwork during this project. Give at least one example of your classmates working well together.

Alternative Environments

UNIT 5

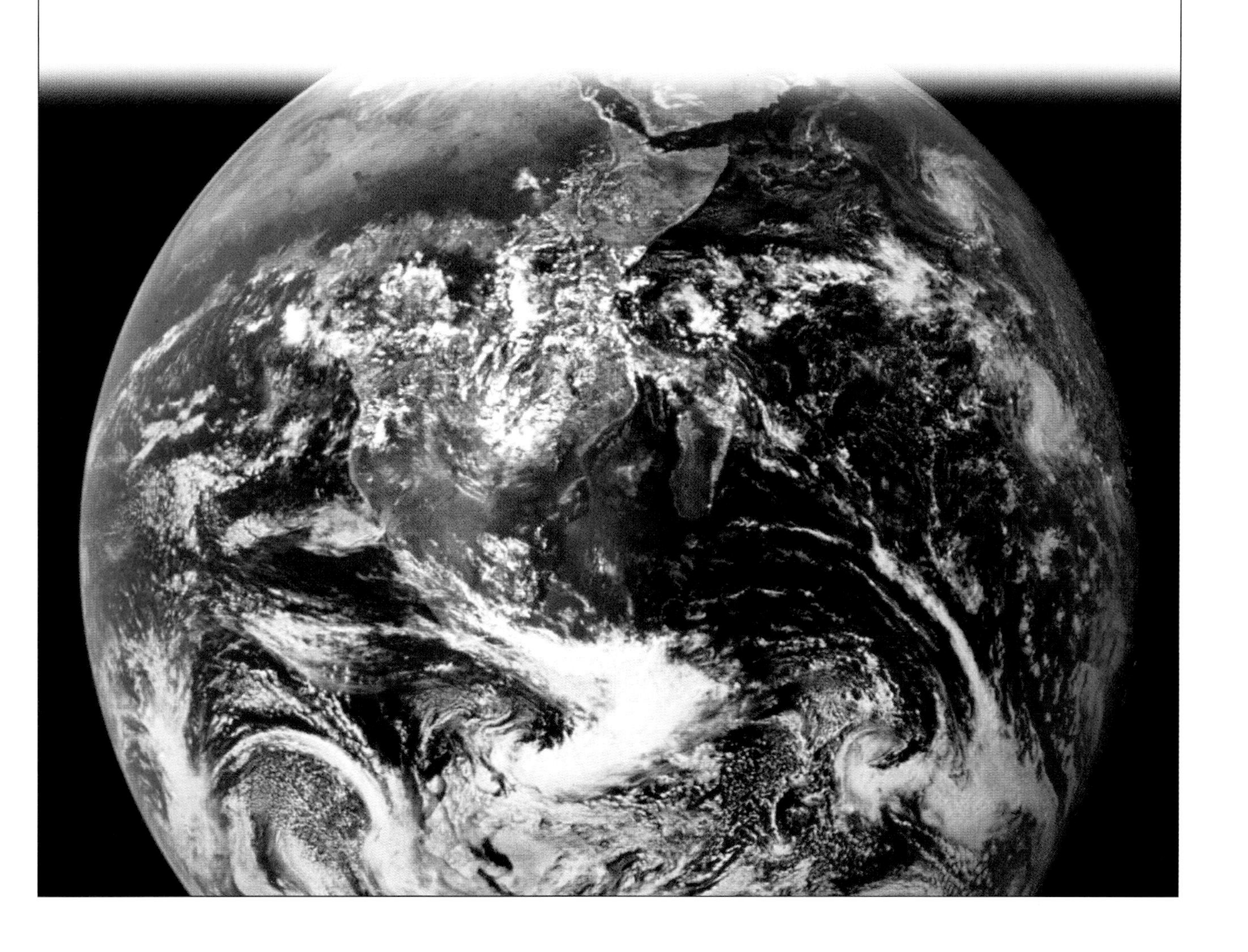

CHAPTER 9

Life on Earth

The **environment** is the specific conditions of an area in which an organism lives. In order to survive, humans need an environment that provides air to breathe, a certain temperature range, a source of energy, and water. Many places on Earth have an environment that can sustain human life (keep humans alive). An **alternative environment** is any environment other than the natural environment of Earth. The Space Shuttle is an example of an alternative environment. Your classroom is an alternative environment, especially on a cold winter day. Look at the photographs in Figure 9.1. How do the environments shown provide the conditions that sustain human life? What other examples of alternative environments can you suggest?

FIGURE 9.1

What You Will Learn

After completing this chapter, you will be able to:

- Identify the systems required to sustain human life in an environment (9.1)
- Describe the inputs of food, air, energy, and water needed to maintain a life-sustaining environment (9.1)
- Identify the components of a life-sustaining environment, and describe how they must interact to be successful (9.2)
- Relate life-sustaining processes in alternative environments to processes that sustain life in the natural environment (9.2)

What You Will Do

- Determine, through experimentation, what factors affect a controlled micro-environment (Lab 9A)
- Formulate scientific questions about the nature of alternative life-sustaining environments (Case Study)
- Demonstrate an understanding of safety practices consistent with WHMIS legislation (Lab 9A)
- Demonstrate the skills required to plan and carry out investigations safely and accurately (Lab 9A)
- Compile, organize, and interpret data using tables and graphs (Lab 9A)

Words to Know

abiotic factor
air
alternative environment
biotic factor
carbohydrates
carbon cycle
decomposers
ecosystem
energy
environment
matter
nitrates
nitrogen cycle
phosphorus cycle
photosynthesis
proteins
respiration
self-supporting environment
system
transpiration
water cycle

A puzzle piece indicates knowledge or a skill that you will need for your project, Another Way of Life, at the end of Unit 5.

9.1 An Environment to Support Life

FIGURE 9.2 This building provides an environment in which humans can survive more easily than in the natural environment.

FIGURE 9.3 What do you need to include in this environment to keep the lizard alive?

Why do we build alternative environments? Alternative environments are built to support human life in areas where people might not otherwise survive. For example, a heated building in a northern climate provides the temperature conditions that humans need to survive (Figure 9.2). Depending on its design, it might also provide running water and a means to prepare and store food.

Building a life-sustaining alternative environment is similar to preparing a suitable home for a pet. For example, suppose you had a pet lizard (Figure 9.3). What conditions would you need to provide in order to keep it healthy? You would, of course, have to provide food and water. Lizards breathe air, so you would need to make sure your lizard could get fresh air. Lizards can survive only at temperatures from 28°C to 40°C, so you would also need to provide a heater and a source of energy.

We learn how to build alternative homes for animals such as lizards by studying their natural environment. Learning how Earth sustains life is therefore a good starting point in understanding how to build alternative environments. How does our natural environment, planet Earth, provide the requirements for life: energy, food, air, and water?

Energy

All life requires energy. Energy is needed to fuel all processes carried out by living organisms and to maintain the temperature of the environment in a range that can support life. What is the energy supply for Earth?

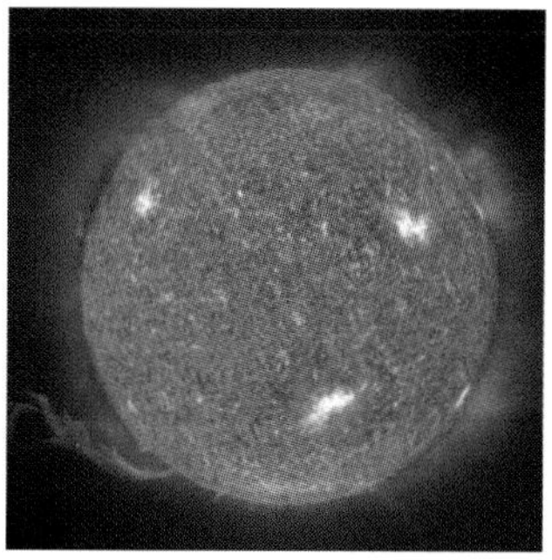

Virtually all the energy on Earth comes from the Sun (Figure 9.4). The Sun warms the air, water, and land of Earth, keeping the average temperature in a range that supports life. A small amount of this energy is captured by plants and converted to food energy.

FIGURE 9.4 Energy from the Sun warms Earth's atmosphere, land, and water. How is this energy converted to food energy?

Food

Food supplies the energy that the human body needs to function and move, and the nutrients it needs to build and maintain its cells. Plants ultimately supply all the food energy for all organisms on Earth, through the process of photosynthesis. **Photosynthesis** is the process by which green plants convert light energy and carbon dioxide into sugars. These sugars provide food energy for the organisms that eat plants. The process of photosynthesis is summarized in the equation shown in Figure 9.5.

Inputs and Outputs of Photosynthesis

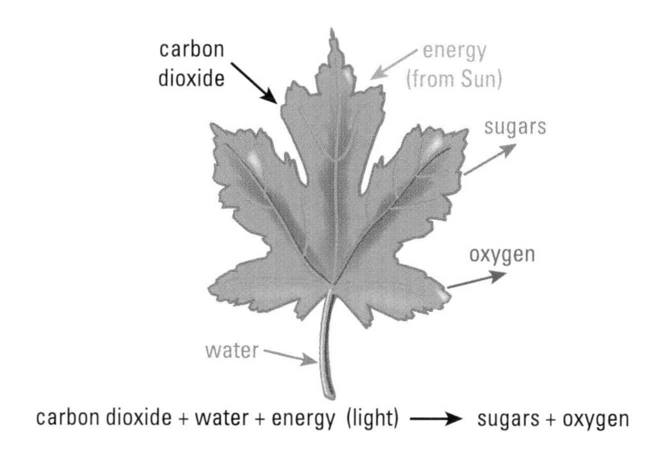

FIGURE 9.5 Photosynthesis produces food (sugars) and oxygen.

Plants also require nutrients, which they use for growth and maintenance. When we eat plants, our bodies use the energy and nutrients contained in the plant cells. Some plants, such as peas or carrots, we might eat directly. Others, such as grass, supply food to other organisms we might eat, such as cattle (Figure 9.6). When we eat animals, our bodies use the energy and nutrients that those animals obtained from eating plants. However, humans can survive by eating only plants, provided that a sufficient variety of plants is consumed.

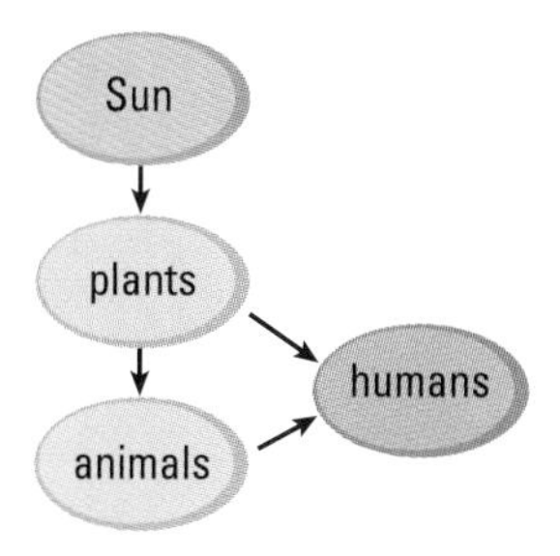

FIGURE 9.6 All the energy in our food starts out as energy from the Sun.

Air

Humans also need **air**, which is the mixture of gases that makes up Earth's atmosphere. Air is made mostly of nitrogen gas, but it also contains carbon dioxide, oxygen, and water vapour. Through photosynthesis, plants support human life by releasing oxygen to the atmosphere.

Humans need the oxygen in air to carry out respiration. **Respiration** is the process by which glucose, a type of sugar, is broken down to release energy that is used in our cells. Respiration is summarized by the equation in Figure 9.7.

You will need to know the factors that sustain human life in an environment for your project, Another Way of Life, at the end of Unit 5.

Inputs and Outputs of Respiration

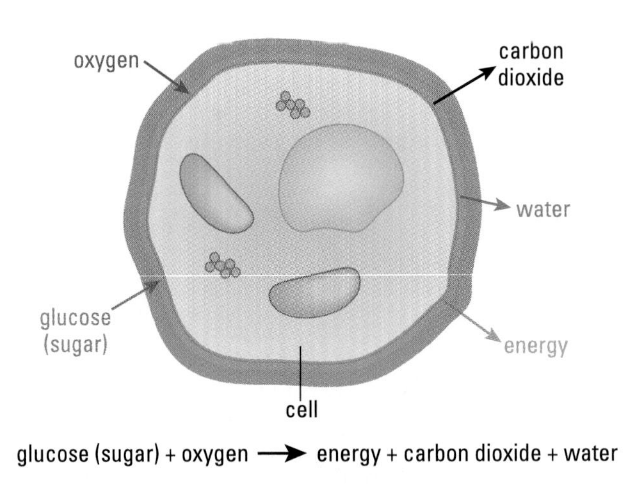

FIGURE 9.7 Respiration requires food (glucose) and oxygen.

The air also helps to maintain Earth's temperature in a range that supports life. The carbon dioxide in the atmosphere absorbs energy from the Sun and holds it, which helps to keep Earth warm. Carbon dioxide also helps to protect us from harmful rays in sunlight.

Water

Humans need to consume water daily to survive, so we require a fresh, clean, and continuous supply. About 70% of Earth's surface is covered with water.

Outputs of An Environment

FIGURE 9.8
An alternative environment must provide a means of handling wastes. What would be the result if waste management was not properly planned?

Think about the pet lizard discussed earlier in this section. If you provide the lizard with the food, energy, air, and water it needs to live, would that be enough to keep the lizard healthy? The answer, of course, is no. You would also need to take care of any outputs from that lizard. At a minimum, you would need to clean up any excrement and make sure that the water and air remain fresh and clean.

An alternative environment for human life would also require a means of handling the wastes, or outputs, that we produce. Human wastes include sewage and the carbon dioxide we produce during respiration. One way to handle wastes would be to remove them from an environment for some sort of processing, similar to how the wastes in chemical toilets are handled (Figure 9.8). This approach would take a lot of work and energy, and would be difficult for outputs such as carbon dioxide.

Sustainability and Alternative Environments

You will need to know what makes an environment sustainable for your project, Another Way of Life, at the end of Unit 5.

To support human life, an alternative environment must provide all the requirements for life: energy, food, air, and water. It must also provide a way of handling the outputs of the environment, or wastes. One approach to designing alternative environments is to build a structure, transport in all the requirements for life, and transport out all wastes. For example, we might build an alternative environment on Mars by constructing an airtight building, then regularly bring in energy, food, and water, and bring out wastes. However, this approach would require constant regular transport of the inputs and outputs of the environment between Earth and Mars. Such an alternative environment would not be sustainable (able to be continued) for any length of time.

A better, but more challenging, approach is to design alternative environments that are self-supporting. A **self-supporting environment** is one that provides all the requirements for life without any need for people to bring any material in or out of the environment. Earth is an example of a self-supporting environment.

Review and Apply

1. State the four basic requirements for maintaining life in an environment.
2. What is the source of energy on Earth? How does this energy source support human life?
3. Describe the relationship between photosynthesis and the food supply of humans.
4. For what cellular process do humans need air? What component of air is required?
5. Explain how plants help to maintain the oxygen in Earth's atmosphere.
6. Figure 9.9 shows two examples of alternative environments that can support the life of a caterpillar. The jars are both sealed, so air cannot enter or leave the jars. Describe how each environment supplies air, water, food, and energy. Is either environment sustainable? Explain your answer.

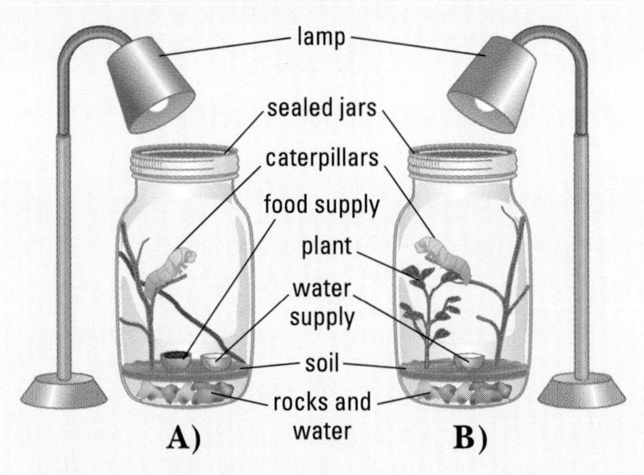

FIGURE 9.9 Is either of these environments self-supporting?

7. The distance from the Sun to Venus is 108 200 000 km, and the distance to Mars from the Sun is 227 900 000 km. Based only on these facts, which planet do you think would make the best site on which to build an alternative environment? Explain your choice.
8. Organize the concepts you have learned in this section in a graphic organizer.

Surf the Web

To find information on careers related to animal care, go to **www.science.nelson.com** and follow the links for ScienceWise Grade 12, Chapter 9, Section 9.1.

Job Link

Zookeeper, Gorilla Exhibit

FIGURE 9.10 Vanessa Phelan is a zookeeper at the Toronto Zoo, where she cares for the gorillas.

ScienceWise: What qualifications are needed to be a zookeeper?

Vanessa Phelan: The more hands-on experience you have with animals and animal care, the better. The Toronto Zoo mainly hires people with at least a diploma in veterinary technology.

ScienceWise: Can you explain how this exhibit is an alternative environment that meets the needs of the gorillas?

Vanessa Phelan: Gorillas require an environment very similar to that in which they would normally live. The environment simulates an African rainforest. Gorillas also need things to do, so we provide them with objects such as old tires to play with. They also have sprayers that turn on at times, and a small pond.

ScienceWise: I notice the climate in the exhibit is a lot different than outside. Can you tell us about that?

Vanessa Phelan: Gorillas do not have fur to keep them warm, so they could not survive in our climate here in Canada. These are lowland gorillas, so they need air temperatures of about 24°C to 27°C year round. Huge boilers downstairs heat the building, and the whole gorilla area has its own furnace and a humidifier to simulate the humidity of the rainforest. Without the humidifiers, the gorillas would suffer from dry skin and other ailments.

ScienceWise: What kind of food do you provide?

Vanessa Phelan: Gorillas are vegetarians, so they are fed with imported fruits, green vegetables, and "monkey chow." We also give them vitamin and mineral supplements, fruit and herbal teas, alfalfa hay (which is high in calcium and protein), and flax seed oil to prevent dry skin and keep their hair shiny.

LAB 9A

Factors That Affect a Controlled Micro-Environment

Through photosynthesis, plants convert light energy from the Sun into sugars, which then provide food energy for animals that eat plants. Photosynthesis also releases oxygen gas, which we need to breathe, to the atmosphere. Plants could therefore be very useful in an alternative life-sustaining environment. What factors in the environment affect the ability of a plant to provide oxygen?

Be Safe!

- Make sure the lamp is kept away from the water. Always turn off the lamp every time you need to handle your experimental set-up.
- Avoid touching the lamp bulb.
- Wash your hands before leaving the lab.

Purpose

In this lab, you will construct a micro-environment capable of supporting the life of an aquatic plant, Elodea. You will then determine how the amount of oxygen produced by the plant is affected by the following factors: light intensity, temperature, and level of carbon dioxide.

Materials

- thirteen 500-mL beakers
- seven test tubes
- seven retort stands
- seven ring clamps
- balance
- scissors
- waterproof marker
- Elodea
- water (distilled or de-chlorinated)
- sodium bicarbonate
- two lamps
- ruler
- thermometer

Procedure

1. You will be measuring the amount of gas produced by Elodea under the following conditions: high and low light, high and low temperature, and high and low carbon dioxide levels. You will be making observations every day. In your notebook, create a data table to record your observations.
2. Attach a clamp to each of the retort stands.
3. Fill seven of the beakers with water, to about 1 cm from the top.

4. Using the balance, weigh out six 5-g samples of Elodea (Figure 9.11). Cut the sprigs toward the root end to get the amount you need, being careful not to damage the plant.

5. Label the test tubes as follows: high light, low light, high temperature, low temperature, high carbon dioxide, low carbon dioxide, and control. Fill each tube with water at room temperature. Add a pinch of sodium bicarbonate to the test tube labelled high carbon dioxide. This chemical increases the amount of carbon dioxide in the water.

6. Working with one test tube at a time, place 5 g of Elodea into each test tube so that the root end is toward the bottom of the tube. Do not put a plant in the test tube labelled control. Quickly invert each test tube and place the open end under the surface of the water in the beaker. Make sure that the plant remains in the test tube. Avoid getting any air in the test tube.

7. Raise the test tube so that the lower 1–2 cm is still under water. Secure the test tube with the ring clamp on one of the retort stands (Figure 9.12).

8. Position a lamp 15 cm away from the test tubes labelled high light, high temperature, low temperature, high carbon dioxide, low carbon dioxide, and control. Position another lamp 30 cm away from the test tube labelled low light, making sure this tube will not receive light from the other experimental setup.

9. Fill five beakers with water. Place a water-filled beaker between the lamp and each of the beakers that contains a test tube, except for the one labelled high temperature. The water will absorb heat from the lamp, keeping the temperature in the test tubes at room temperature.

10. Measure and record the temperature in each beaker that contains a test tube. Turn on the lamps for 20–30 minutes.

11. Turn off the lamps. Using a ruler, measure and record the height of the gas in all the test tubes as shown in Figure 9.13. Measure and record the temperature in each beaker. Turn the lamps back on.

12. Repeat step 11 every day for up to 3 days.

FIGURE 9.11 Determine the mass of the plants.

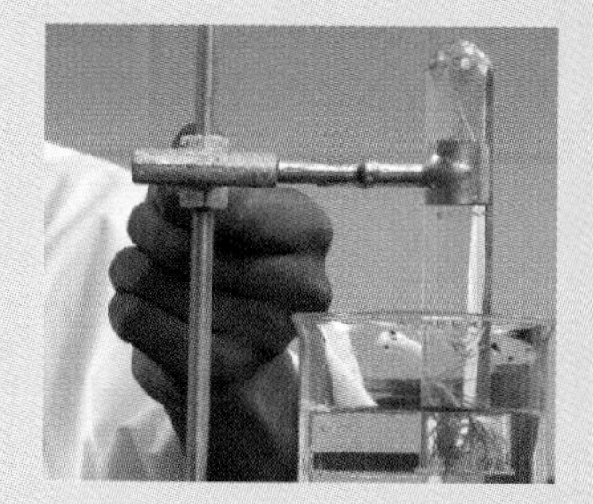

FIGURE 9.12 Secure the test tube in position with a clamp and retort stand.

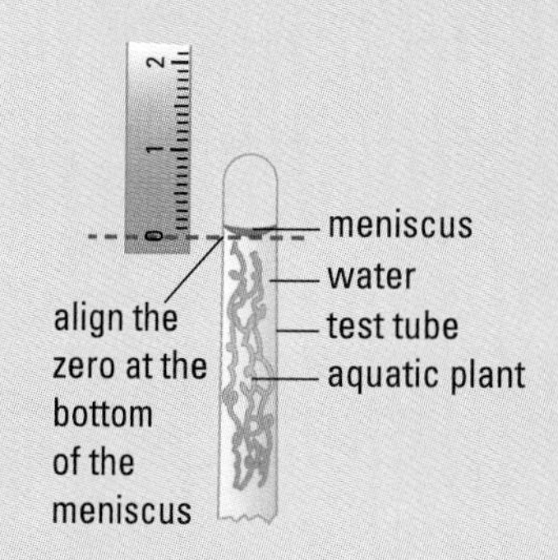

FIGURE 9.13 Measure the height of the gas with the ruler.

CONTINUED

Analysis and Conclusion

1. Create a line graph showing the change in gas levels over time for each of the test tubes, including the control.
2. Explain the purpose of the control test tube.
3. Identify the gas that was formed in the tube. Write an equation showing how this gas is formed.
4. Referring to your line graph, summarize the effect of each of the factors tested on the ability of Elodea to produce oxygen in this micro-environment.

Extension and Connection

5. When plants are producing large amounts of oxygen gas, they are also producing large amounts of sugars. Using this information, infer the effect of increasing the distance between the beaker and the light on the amount of sugars produced. If this plant were your only food supply, how would this affect you?
6. Is the aquatic plant in the beaker an example of a self-sustaining environment? Give reasons for your answer.

9.2 Systems to Sustain Life

A **system** is any set of connected parts that function together. For example, a bicycle is a system composed of wheels, pedals, brakes, a chain, a frame, handles, and a seat (Figure 9.14). One way of looking at Earth is as a series of systems called ecosystems. An **ecosystem** is all the living and non-living elements that interact in an environment. Just as the parts of a bicycle work together, each component of an ecosystem works with the others to maintain the environmental conditions that sustain life. An ecosystem keeps the inputs and outputs of the environment in balance.

FIGURE 9.14 A system is a set of interacting parts. How do the parts of a bicycle work together?

Earth contains many different ecosystems. Figure 9.15 shows three examples of ecosystems that can support human life: a forest, a classroom, and a farm.

A)

B)

C)

FIGURE 9.15 A) A forest ecosystem, B) a classroom, C) a dairy farm. How do these ecosystems provide an environment that can support human life?

Biotic and Abiotic Factors

An ecosystem is composed of biotic and abiotic factors that work together to sustain life. **Biotic factors** are living organisms or anything produced by living organisms, including the remains and wastes of those organisms. **Abiotic factors** are non-living factors in an ecosystem, such as water and air. Human life requires both biotic and abiotic factors. These factors can differ from ecosystem to ecosystem. For example, if you live in Canada's Arctic, the plants available for food in your ecosystem will be different from those available if you live in the fruit belt of the Niagara region. Figure 9.16 shows some examples of biotic and abiotic factors that help to support human life. How many of these are available where you live?

You may need to know how the systems in an environment work together for your project, Another Way of Life, at the end of Unit 5.

Biotic and Abiotic Factors that Support Human Life on Earth

Biotic factors	Abiotic factors
Plants	Water
Animals	Energy
Micro-organisms	Air

FIGURE 9.16 Biotic and abiotic factors help to sustain human life on Earth. What does each of these factors provide?

Matter and Energy

Matter is anything that occupies space and has mass. All the cells of your body are made of matter, and all the particles making up your desk are made of matter. Figure 9.17 shows some examples of matter. When you eat, some of the matter in the food is converted to the matter that makes up your cells.

FIGURE 9.17 Matter can be any shape or size. Give two examples of matter in your classroom.

Energy is a measure of the ability of a system to do work. For example, when an engine burns fuel, it releases the energy needed to do the work of moving a vehicle from one place to another (Figure 9.18). Similarly, the human body uses food energy to fuel its processes and movements.

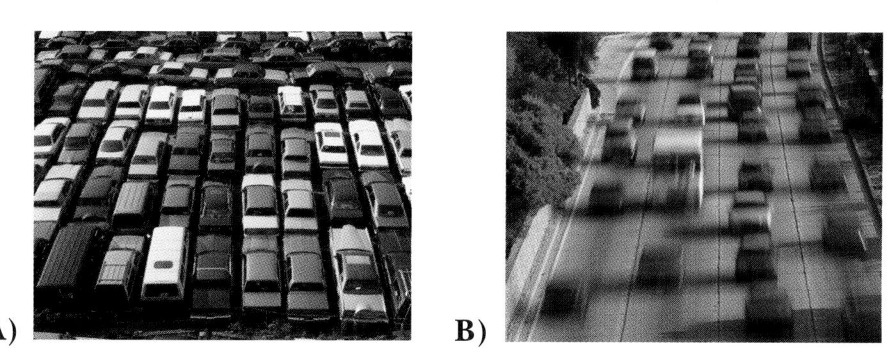

FIGURE 9.18 The engine of a passenger vehicle converts the energy stored in gasoline to energy of motion.

Most biotic and abiotic factors in an ecosystem are composed of both matter and energy. The exception is the Sun's rays, which are composed entirely of energy. Ecosystems function by transferring matter and energy between the biotic and abiotic factors they contain.

Matter Cycles

There is very little change in the amount of matter on Earth. Small changes occur occasionally, such as when a meteorite crashes to Earth's surface or a satellite is launched, but such events do not cause significant loss or gain of matter. However, matter is changed from one form to another, again and again. For example, the matter contained in a lettuce leaf in a sandwich can be converted to the matter making up the cells of your body. When these cells die, the matter they contain can be converted to other compounds by micro-organisms, and then taken up by plants. It may even become part of another lettuce leaf in another sandwich. On Earth, matter cycles between the biotic components (parts) and abiotic components of natural ecosystems. Most of the matter in any natural ecosystem is recycled through the carbon cycle, the nitrogen cycle, and the phosphorus cycle.

The Carbon Cycle

The **carbon cycle** is the continuous transfer of carbon atoms between biotic and abiotic factors in an ecosystem (Figure 9.19). Carbon is the main component of sugars and starches, which are used by living things for food energy. Sugars and starches are **carbohydrates**, which are organic compounds composed mainly of atoms of carbon, hydrogen, and oxygen. You can read more about organic compounds in Section 1.2. After an organism eats a carbohydrate, the chemical bonds in the carbohydrate are broken apart by the process of respiration. The broken chemical bonds release energy, which the organism uses to function and move. You can find out more information about chemical bonds in Section 1.1.

Some of the carbon that is released by respiration is used to make new compounds, such as those in cellular components. Every time a new compound is made, some chemical bonds are broken, releasing energy, and some new chemical bonds are formed. The energy in these new chemical bonds can then be used as a source of food energy by another organism. **Decomposers** are organisms that use the energy found in carbohydrates in wastes and dead tissues.

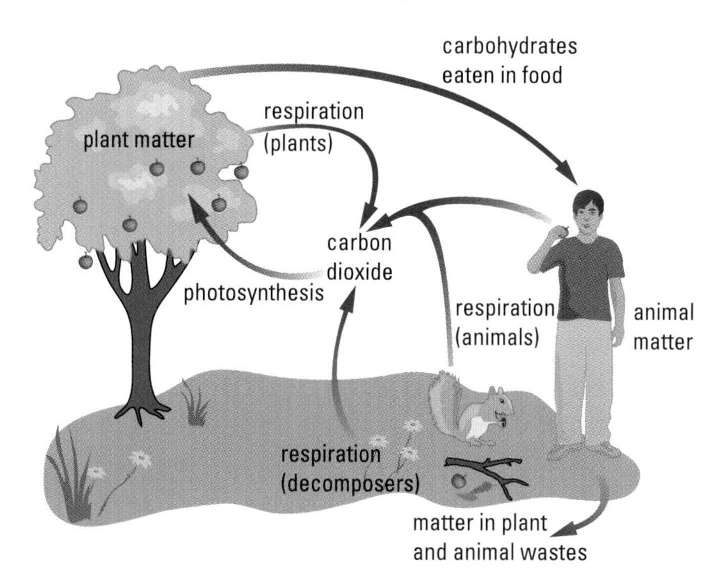

FIGURE 9.19 In the carbon cycle, carbon atoms are passed from one organism to another. Could this cycle operate without plants? Why or why not?

Not all the carbon and energy in food is passed between organisms, however. Some of it is released to the atmosphere, as carbon dioxide. If you compare the inputs and outputs of photosynthesis and respiration, you will see that photosynthesis may be thought of as respiration in reverse (Figure 9.20).

Photosynthesis Versus Respiration

Inputs		Outputs	
Photosynthesis	**Respiration**	**Photosynthesis**	**Respiration**
Energy	Sugar (glucose)	Sugars	Energy
Carbon dioxide	Oxygen	Oxygen	Carbon dioxide
Water			Water

FIGURE 9.20 The inputs of photosynthesis are the same as the outputs of respiration. Compare the outputs of these two processes.

On Earth, the rates of photosynthesis and respiration are roughly in balance. Without plants, carbon dioxide levels would build up and oxygen levels would decrease over time.

The Nitrogen Cycle

The **nitrogen cycle** is the continuous transfer of the nitrogen atoms in biotic and abiotic factors in an ecosystem (Figure 9.21). Nitrogen exists in the atmosphere of Earth as nitrogen gas. The cells of all living organisms contain nitrogen in the form of **proteins**, which are nitrogen-rich organic compounds. Humans and other animals obtain proteins by eating plants or other animals. Some of the nitrogen in these proteins is used to build and maintain cells. Excess nitrogen, along with any nitrogen released by cellular processes, is removed from the organism as waste. This waste is broken down by decomposers, which are micro-organisms such as bacteria and fungi, into nitrates and other compounds. **Nitrates** are nitrogen-containing substances that are soluble in water. You can learn more about solubility in Section 1.1. A small amount of nitrates is also formed directly from nitrogen gas in the atmosphere, through the action of nitrogen-fixing soil bacteria. Nitrates are taken up by plant roots (and by other soil bacteria), and used to build proteins. Section 7.2 contains more information about the importance of nitrogen to plants. Without the nitrogen cycle, wastes would build up in an ecosystem, and animals would quickly run out of protein.

The Nitrogen Cycle

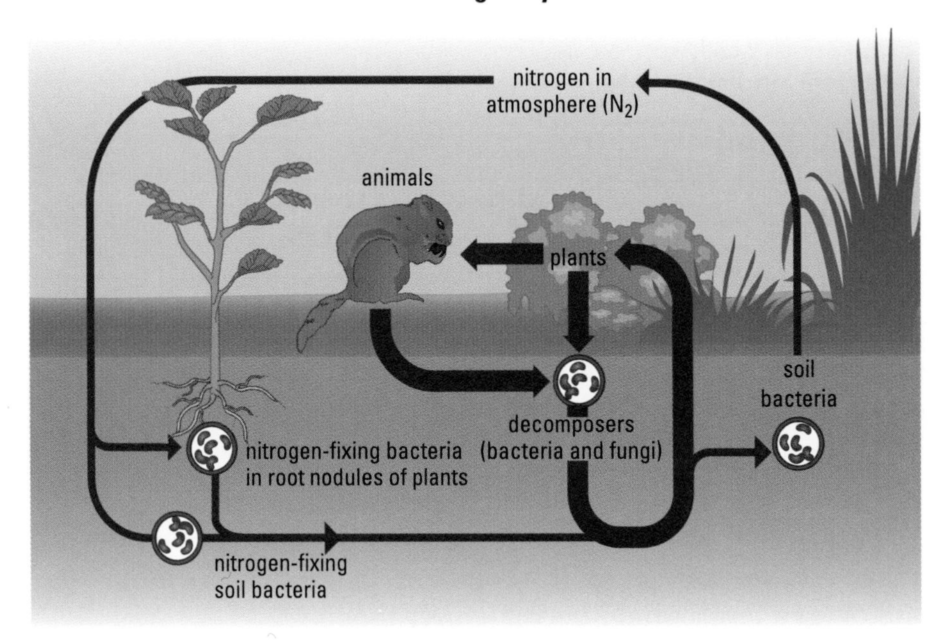

FIGURE 9.21
What would happen in an environment in which there were no decomposers? Could the nitrogen cycle exist?

The Phosphorus Cycle

The **phosphorus cycle** is the continuous transfer of the phosphorus atoms in biotic and abiotic factors in an ecosystem (Figure 9.22). Phosphorus is essential to all living things, since it is used to form DNA molecules. DNA is genetic material—you can find more information about DNA and its role in heredity in Section 6.1. Phosphorus is released from abiotic factors in the ecosystem through the action of water, such as when rain dissolves phosphorus contained in rocks. Some of the phosphorus is absorbed by certain species of fungi that produce phosphorus-containing substances. These substances may be taken up by plants and used to build and maintain plant cells. When other organisms eat the plants, the phosphorus in the plant cells is transferred to those organisms and then, in turn, to any organisms that eat them. Eventually, the phosphorus ends up in organic molecules produced as wastes by living organisms. These wastes are then broken down by decomposers.

Surf the Web

Terraforming means making the environment of another planet similar to that of Earth. This is one possible way of constructing alternative life-sustaining environments. Visit **www.science.nelson.com** and follow the links for ScienceWise Grade 12, Chapter 9, Section 9.2 to find out more about terraforming. In groups, discuss whether terraforming is a good idea.

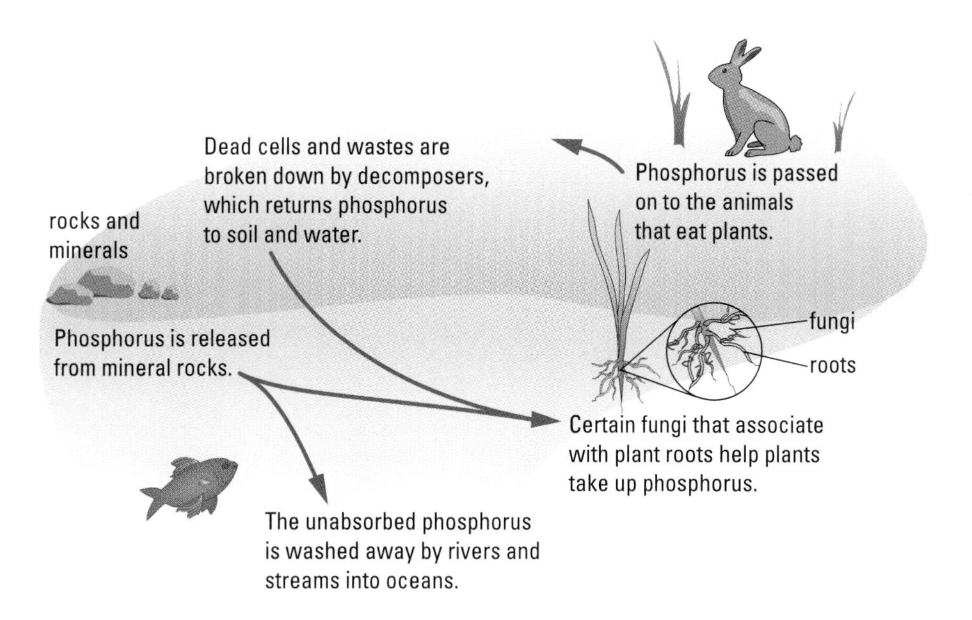

FIGURE 9.22 Name all the biotic and abiotic components of this ecosystem that are involved in the phosphorus cycle.

You may need to know about the water cycle for your project, Another Way of Life, at the end of Unit 5.

The Water Cycle

The **water cycle** refers to the circulation of water between the air, land, and oceans, lakes, and streams of Earth. Figure 9.23 shows the stages of the water cycle. As water passes through the water cycle, impurities are removed and the water is purified. Without the water cycle, Earth's water would gradually become less pure. Through precipitation, the water cycle also moves water from place to place. This helps to supply water to the other organisms we rely on, such as plants. Plants contribute to the water cycle by **transpiration**, a process in which liquid water is brought up from the soil through the roots and released through the leaves into the atmosphere as water vapour. Transpiration also helps to purify water.

The Water Cycle

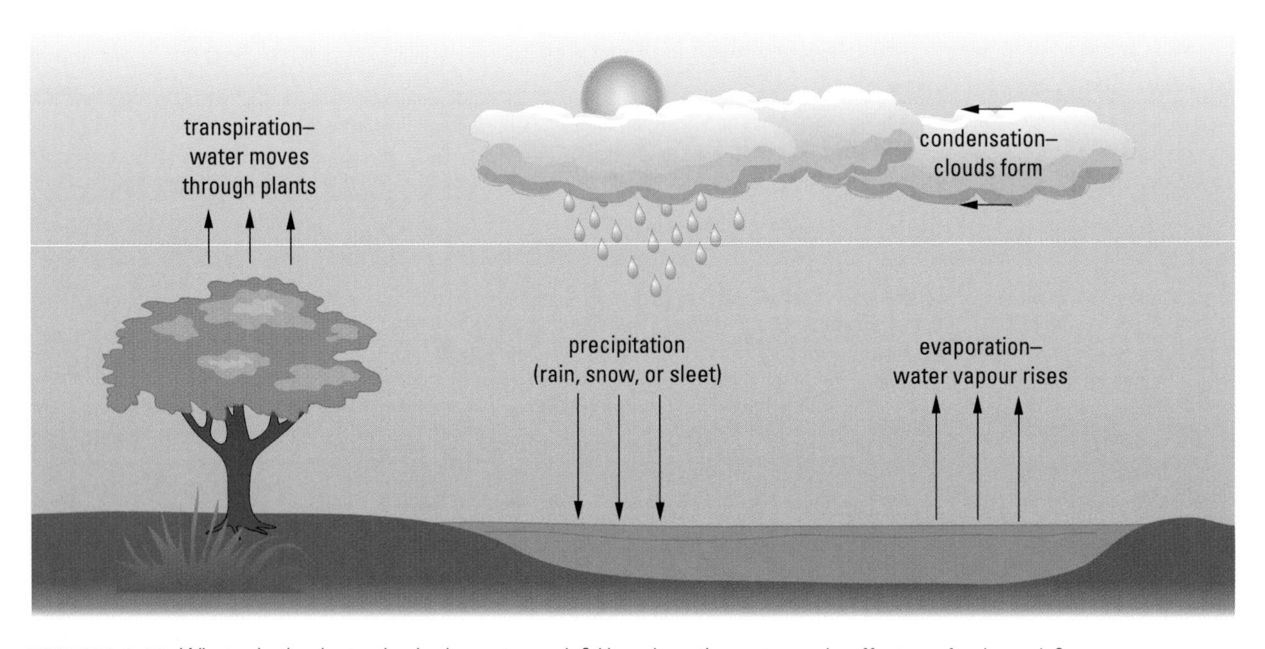

FIGURE 9.23 What role do plants play in the water cycle? How does the water cycle affect our food supply?

Energy Flows

As matter moves between biotic and abiotic factors in an ecosystem, the atoms in the matter remain unchanged, and so can be used again. For example, a carbon atom remains a carbon atom as it moves through the carbon cycle, although it may become part of many different chemical compounds. This is not true for energy. When energy is used, it is converted to another form that cannot be used in the same way. For example, when the energy in gasoline is converted to energy of motion of a passenger vehicle, that energy cannot be used to move the vehicle again. While the vehicle is moving, some energy is lost as heat energy and sound energy. Eventually, the driver must add more gasoline to keep the car running (Figure 9.24).

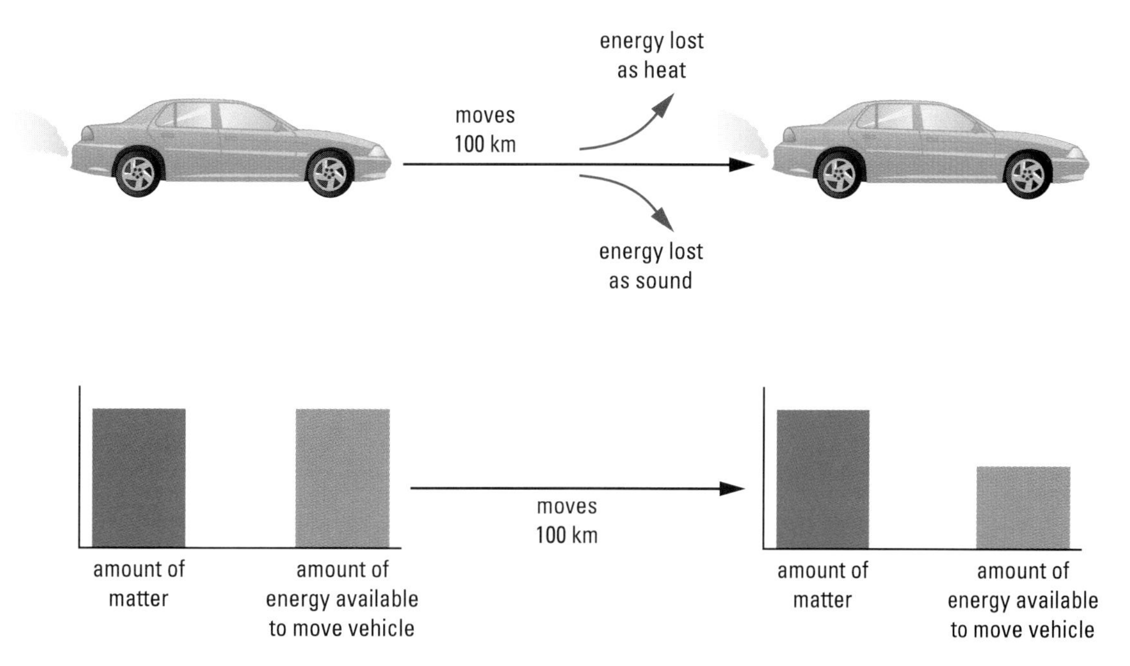

FIGURE 9.24 When a vehicle moves, the amount of matter in the vehicle remains the same, but the energy is converted to forms that cannot be used to continue to run the vehicle.

Similarly, when you eat a carrot, only part of the energy in the carrot is used to fuel your cells. Some of it will be lost as body heat, movement, or sound, depending on the activities you carry out. This energy cannot be recycled into another carrot. To be able to continue to support life, an environment therefore must have a constant supply of new energy. In Earth's ecosystems, the Sun provides this energy (Figure 9.25).

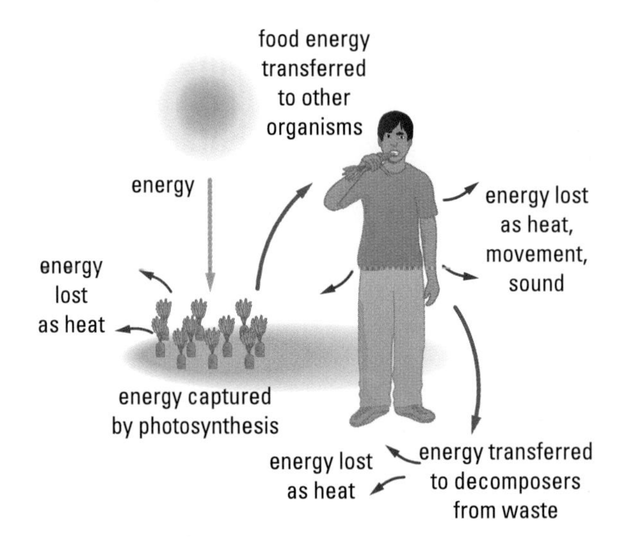

FIGURE 9.25 The Sun's energy is converted by plants into food energy, which is then passed on to the other organisms in an ecosystem. What would happen if this energy were not constantly renewed?

Matter and Energy in Alternative Environments

Earth's ecosystems are self-supporting because they recycle all matter, and depend on the Sun for their energy source. So far, humans have not been able to design a truly self-supporting alternative environment. For example, the Butterfly Conservatory in Niagara Falls, Ontario, is a huge greenhouse that mimics the ecosystem of a tropical rainforest. You can find out more about how greenhouses work in Section 8.1. The plants within the Conservatory are able to sustain the life of the butterflies, but matter and energy must be brought in and removed from the facility by human action in order to maintain the environment. For example, to keep the plants healthy, fertilizers and water must be brought in from outside the structure, and plant wastes must be removed. Similarly, the heating and cooling systems that help to maintain the temperature of the building are powered by gas or electricity brought from elewhere. You will find out about how matter and energy are handled in other alternative environments in the rest of this unit.

Review and Apply

1. Illustrate or describe two examples of ecosystems.
2. Identify each of the following as biotic factors or abiotic factors:
   a) bacteria
   b) dead leaves
   c) a pond
   d) snails
   e) sunlight
   f) wind
   g) a human being
   h) an apple tree
3. Decribe the role of carbohydrates in the carbon cycle.
4. Using a diagram, describe how the nitrogen in a salmon could end up in a pine tree.
5. Using an example, explain why energy must be replenished in a system.
6. You are a member of the first human colony on the Moon. One of the colony members has just developed a device that will remove all the micro-organisms from the colony buildings. Will this device help the colony to survive? Explain your answer.
7. Add the new concepts from this section to the graphic organizer you started in Section 9.1.
8. **Human Activity and the Carbon Cycle**

   Here on Earth, humans do much more than simply survive. For example, people might build houses or highways, manufacture paper or clothing, or travel in cars or planes. These activities require energy that is supplied by energy sources such as oil and gas. In a small group, use print and electronic resources to research the effects of the use of oil and gas on the levels of carbon dioxide in the atmosphere. Start your research by visiting www.science.nelson.com and following the links for ScienceWise Grade 12, Chapter 9, Section 9.2. When you have enough information, create a poster of the carbon cycle that includes this source of carbon dioxide. Compare this cycle to the one shown in Figure 9.19.

CASE STUDY

Biosphere 2

FIGURE 9.26 Insects were included among the biotic factors of Biosphere 2. Why do you think insects were included?

In 1991, four men and four women were sealed inside a huge glass and metal frame structure that sprawled across three acres of desert in the state of Arizona in the United States. This structure, Biosphere 2, was an ambitious attempt to simulate the systems of "Biosphere 1," Earth itself. Biosphere 2 is a huge, self-contained alternative environment, composed of four ecosystems similar to those found on Earth: a rainforest, a desert, an ocean, and an agriculture-forestry ecosystem. All the biotic and abiotic components known to be needed to maintain these ecosystems were included in the design of Biosphere 2: air, soil, water, plants, and animals (Figure 9.26). The temperature in the building was also controlled to mimic the natural ecosystems.

The eight people were to live in isolation in Biosphere 2 for two years, without any additional material being provided from outside. They were to obtain their food from the plants sharing Biosphere 2. The aim was to study how the biotic and abiotic components of Earth's ecosystems interact to sustain life. This information could assist the planning of alternative environments elsewhere, such as on the Moon and Mars. It was also hoped that Biosphere 2 could help us learn more about Earth's environment.

a) Choose one of the ecosystems that were built in Biosphere 2. Make a list of the biotic and abiotic factors in that environment. You may need to conduct additional research.

b) Using some of the biotic and abiotic factors in your list, draw a diagram of one possible carbon cycle that could exist in the ecosystem you chose.

Unexpected Troubles

Nine months after the Biosphere 2 experiment started, things began to go dreadfully wrong. Oxygen levels inside Biosphere 2 dropped, and levels of a toxic gas, nitrous oxide, rose. In an attempt to save the experiment, oxygen was pumped in from the outside. Then crops began to fail, so emergency food rations were secretly passed in through airlocks to the inhabitants of Biosphere 2. Tensions between the inhabitants rose. When it became evident they would soon have too little oxygen, the experiment was ended and the inhabitants left Biosphere 2.

FIGURE 9.27 Biosphere 2 was unable to sustain conditions suitable for human life.

c) What two processes that occur on Earth were expected to keep the oxygen levels in Biosphere 2 in a range that would support life?

Turning Trouble into New Knowledge

This failure of the Biosphere 2 project was an important step toward finding new knowledge about the interactions of the components of an alternative environment. The drop in oxygen levels was eventually traced to the action of soil bacteria. The respiration of these bacteria consumed oxygen far more quickly than the photosynthesis of the plants could replace it. This should have resulted in a rise in carbon dioxide levels, however, not nitrous oxide. It turned out that the concrete floor of Biosphere 2 was absorbing the carbon dioxide, further throwing off the balance of the ecosystem and causing the unexpected rise in nitrous oxide.

Although Biosphere 2 did not reach its original goals, its failure shows us how much more we need to learn about the interactions between the systems that support life. Researchers now use Biosphere 2 to study these interactions. That knowledge may help us to one day build a truly self-supporting alternative environment capable of supporting human life.

d) Discuss how the failure of the initial Biosphere 2 project was beneficial. Use specific examples in your answer.

CONTINUED

Analysis and Communication

1. What was the goal of the initial Biosphere 2 project? Describe the changes in the matter inside Biosphere 2 that would be necessary to achieve this goal.

2. Create a diagram to show what happened in the carbon cycle inside Biosphere 2. Include the abiotic factor that upset the balance of the cycle in your diagram.

3. Explain why removing the soil bacteria from Biosphere 2 would not have solved the problems that developed. Refer to a specific matter cycle in your answer.

Making Connections

4. Agree or disagree with the following statement: To successfully maintain an alternative life-supporting environment, we must first understand the interactions of ecosystems on Earth. Justify your answer by referring to the experience of Biosphere 2.

5. You are a researcher at Biosphere 2. Write a scientific question about alternative environments that you could investigate at Biosphere 2.

6. Sick building syndrome is a term used to describe health problems that arise whenever a person spends time in a particular building or part of a building. Symptoms of sick building syndrome include cough, chest tightness, fever, chills, and muscle aches. These symptoms disappear once the affected person has left the building. Conduct research to find out the causes of sick building syndrome. Summarize your research in a brief report. Explain why people who work in construction should be aware of sick building syndrome and its causes.

7. Name two occupations that might be involved in the construction or maintenance of Biosphere 2. Explain the potential role of each of these occupations. Conduct a survey of job advertisements for these occupations in your area. Summarize your research as a poster or electronic presentation.

9.3 Chapter Summary

Now You Can...

- Describe the inputs of food, energy, air, and water needed to maintain an alternative life-sustaining environment (9.1)
- Describe some of the outputs of an alternative life-sustaining environment (9.1)
- Identify the systems required to sustain human life in an environment (9.2)
- Relate information about sustaining life in an alternative environment to processes that sustain life in the natural environment (9.2)

Concept Connections

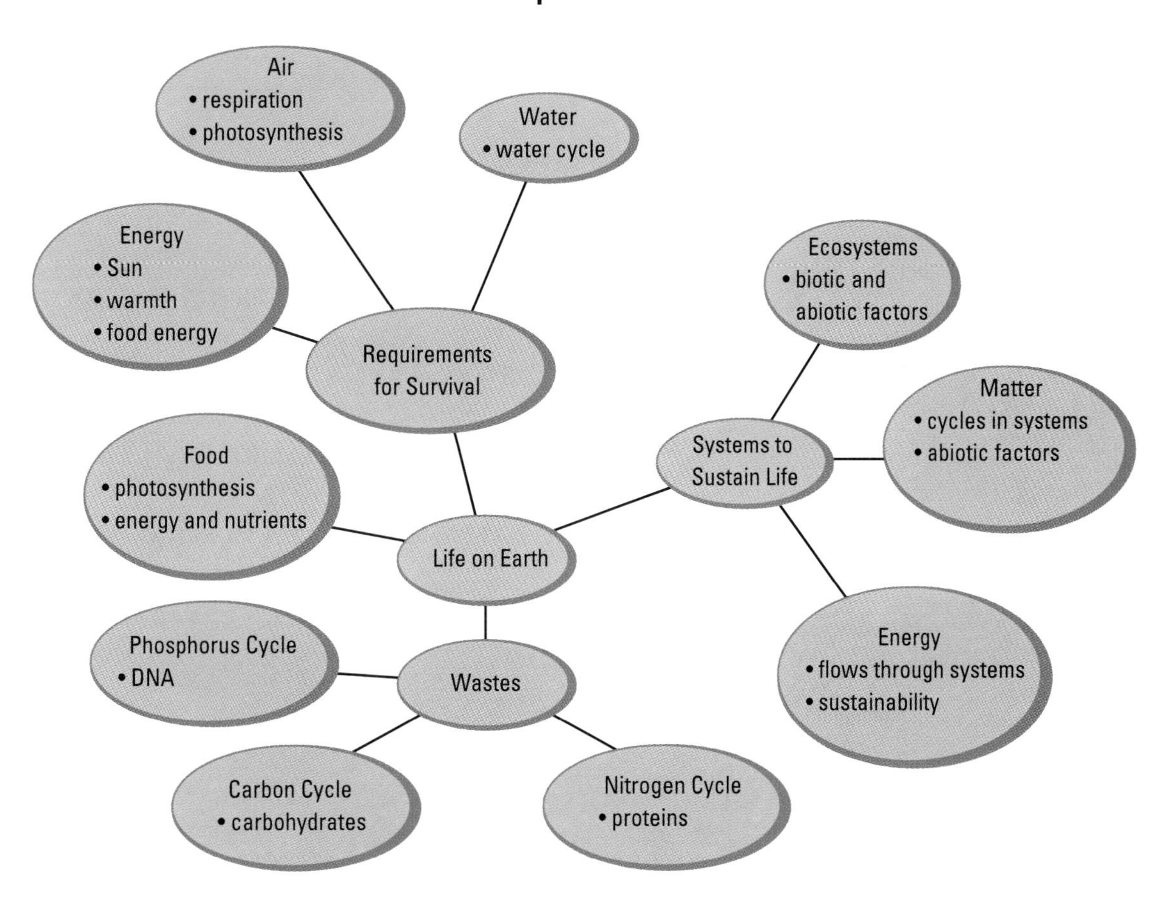

FIGURE 9.28 Compare your completed graphic organizer to the one on this page. How did you do? What new links can you add to your organizer?

CHAPTER 9 *review*

Knowledge and Understanding

1. In your notebook, state whether each of the following statements is true or false. Rewrite any false statements to make them true.
   - **a)** All the energy on Earth comes from photosynthesis.
   - **b)** To support human life, an environment must supply air, water, and food.
   - **c)** Plants help to support life on Earth by providing a source of food and maintaining the oxygen in air.
   - **d)** Plants contribute to the water cycle.
2. Why is Earth considered a sustainable environment? How might human activity affect this sustainability?
3. Examine the diagram shown in Figure 9.29. State the meaning of the diagram in your own words. Include specific examples in your answer.

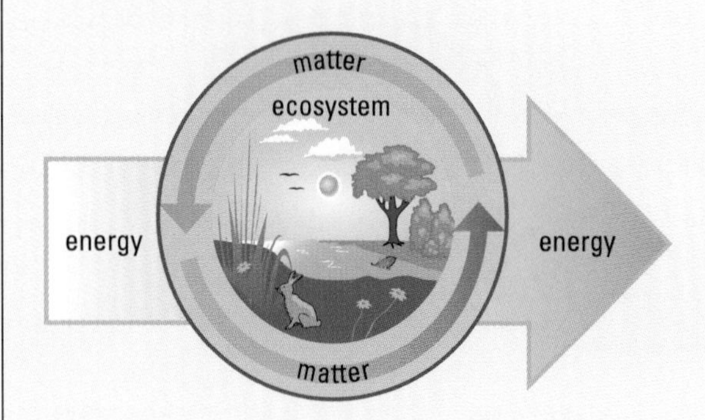

FIGURE 9.29

4. Describe how the processes of photosynthesis and respiration help to maintain the balance of carbon and oxygen in air. Include equations for these two processes in your answer.
5. Give at least two examples of carbohydrates. Explain why they are considered carbohydrates.
6. Why do plants need the nitrogen cycle to survive? Explain your answer.
7. Why is a television a system?
8. You work in a pet store that sells aquariums. A customer wants some information about how the system shown in Figure 9.30 functions. Write a short skit that outlines the conversation you might have with the customer. In your skit, you must identify all the biotic and abiotic components of the environment, and explain the role that each plays in maintaining this alternative environment. Note that only some components of the system are labelled in Figure 9.30.

FIGURE 9.30

9. Discuss the similarities and differences between the abiotic factors in the aquarium ecosystem shown in Figure 9.30 and the abiotic factors in the natural environment of the fish. How do the differences affect the sustainability of these two ecosystems?

Inquiry

10. For a school project, a group of students decided to test the effects of light intensity on the growth of plants. They set up two identical potted plants in the classroom, and placed one of the plants under a bright light bulb. Both plants were placed in the same type of soil and given the same amount of water. Besides light intensity, what factor is likely to vary between the environments of the two plants? How could the students change their experiment so that only light intensity will vary?

Making Connections

11. In the environment of a submarine, the human occupants are provided air, water, food, and energy. Is the submarine a self-supporting environment? Explain.

12. You are the owner of an animal shelter that takes in abandoned pets.
    a) In your notebook, draw a circle to represent the shelter. Using labelled arrows, show the general inputs and outputs that you will need to sustain the lives of the animals in your shelter.
    b) Suggest a way you could use plants to change the quantity of any of the inputs and outputs related to sustaining the shelter.

Communication

13. Using a poster or model, summarize the role of a tree in maintaining air quality, water quality, and the food supply on planet Earth. Include the source of energy that "runs" the tree in your answer.

14. Create a storyboard, skit, or short story based on the following scenario: A group of astronauts has accepted a mission to be the first people to live inside an alternative environment built on the planet Mars. They are to stay on Mars for six months. After only two weeks, the plants in the alternative environment begin to die. It takes four weeks for the Mars colony to receive supplies from Earth.

CHAPTER 10

New Worlds

Human life depends on four things: air, water, food, and energy. In Chapter 9, you learned about some of the ways that Earth's ecosystems provide us with these basic requirements. In this chapter, you will explore some human-built alternative environments that are capable of supporting human life. Look at the photographs in Figure 10.1. What systems would we need to produce food in an alternative environment? What do astronauts eat in space? What happens to their wastes? How can what we know about the systems on our home planet, Earth, help us to build alternative environments? What benefits would building an alternative environment off planet Earth have for you and your friends?

FIGURE 10.1

What You Will Learn

After completing this chapter, you will be able to:

- Describe the inputs of food, energy, air, and water needed to maintain a life-sustaining alternative environment (10.1, 10.2)
- Identify the components of an alternative environment, and describe how they must interact to be successful (10.1, 10.2)
- Describe the outputs of a life-sustaining alternative environment, and the systems required to handle them (10.1, 10.2)
- Describe the difficulties facing humans living in a weightless, self-supporting environment (10.2)
- Assess the Canadian contribution to the development of the International Space Station (10.2)
- Relate knowledge of how life is sustained in alternative environments to the processes that sustain life in the natural environment (10.1, 10.2)
- Analyze the costs and benefits to society, the economy, and the environment of constructing and operating an alternative environment capable of supporting human life (10.2)

What You Will Do

- Use flow charts to diagram the inputs, outputs, and interactions of the life-sustaining components of an alternative environment (Activity 10A)
- Analyze an existing life-sustaining environment in terms of sustainability, and make suggestions for its improvement (Activity 10A)
- Determine the effect of gravity on the growth of plants (Lab 10B)

Words to Know

biodome
gravity
microgravity
non-renewable energy source
offshore oil rig
renewable energy source
solar energy
weightlessness

A puzzle piece indicates knowledge or a skill that you will need for your project, Another Way of Life, at the end of Unit 5.

10.1 Alternative Environments on Earth

Earth's natural ecosystems provide us with air, water, and food, as well as a means to recycle our wastes. The Sun provides ecosystems with energy. To support human life, an alternative environment must also provide air, water, food, and energy, and some way to handle the waste we produce. In this section, you will consider two alternative life-sustaining environments here on Earth: offshore oil rigs and biodomes. You will explore how the systems of these alternative environments provide a life-sustaining environment, and to what degree they are sustainable.

Offshore Oil Rigs

An **offshore oil rig** is a structure that is designed and built to extract oil from undersea reserves. Since humans cannot survive in the natural environment of an ocean, offshore oil rigs must provide an alternative environment to sustain the lives of the people who work on them.

For example, the Hibernia oil rig off the Grand Banks of Newfoundland includes a structure called the Accommodations Module, which provides a life-sustaining environment for the approximately 185 people who live aboard year round (Figure 10.2). What inputs are needed to maintain the alternative environment of an offshore oil rig? How are the outputs handled?

FIGURE 10.2 The Hibernia offshore oil rig is designed to extract 135 000 to 150 000 barrels of oil daily.

Air

Since it is Earth-based, the air supply of an offshore oil rig is Earth's atmosphere. The air temperature is controlled by a heating and air-conditioning system, so the living quarters are sealed and well insulated. The air-handling systems must therefore remove the carbon dioxide produced by human respiration that would otherwise build up, and replace it with fresh air from outside.

Water

To survive, humans need fresh water, not the salt water that surrounds offshore oil rigs. Most of the fresh water is brought in by supply ships that regularly visit the structure. Some oil rigs, such as the Terra Nova platform off the coast of Newfoundland, contain facilities to produce fresh water from seawater. Such facilities can produce only a limited supply, however.

FIGURE 10.3 Most of the inputs and outputs required to support human life on an offshore oil rig must be transported from elsewhere, on supply ships such as this.

You will need to describe inputs and outputs of alternative environments for your project, Another Way of Life, at the end of Unit 5.

Food

Offshore oil rigs are entirely dependent on outside sources for their food supply (Figure 10.3). On site, refrigerated storage facilities allow the time between deliveries to be extended, but human life could not be sustained without the input of these food deliveries.

Energy

Most offshore oil rigs have oil and gas fuel brought in by tanker to supply energy. Some of the natural gas extracted by an oil rig can be used as fuel directly, for purposes such as heating and cooking. Some offshore rigs also have on-site generators that convert the energy in natural gas to electricity, which can power the oil rig's systems.

Waste

All the outputs of human life on an offshore oil rig are removed from the rig. The systems of the oil rig do not recycle or reuse human sewage or other wastes, such as plastics and food waste. Wastes are either removed by transport ships to be processed onshore, or else released into the ocean or atmosphere.

Sustainability of Offshore Oil Rigs

Any system that depends on a non-renewable energy source and does not cycle matter is not self-supporting, and cannot be sustained indefinitely. **Non-renewable energy sources** are energy sources that use up finite (limited) resources. In other words, non-renewable energy sources are those that can be completely used up. Earth's oil and gas resources are non-renewable, because if they were all used up, we would have no way of getting more. Since offshore oil rigs depend on non-renewable energy sources, they are not sustainable alternative environments. A **renewable energy source** is any energy source that does not use up finite resources. Renewable energy sources are those that cannot be used up, such as **solar energy** (energy from the Sun). There is also only limited, if any, recycling of matter on an offshore oil rig. Food and water are brought in from outside, and wastes are transported away (Figure 10.4).

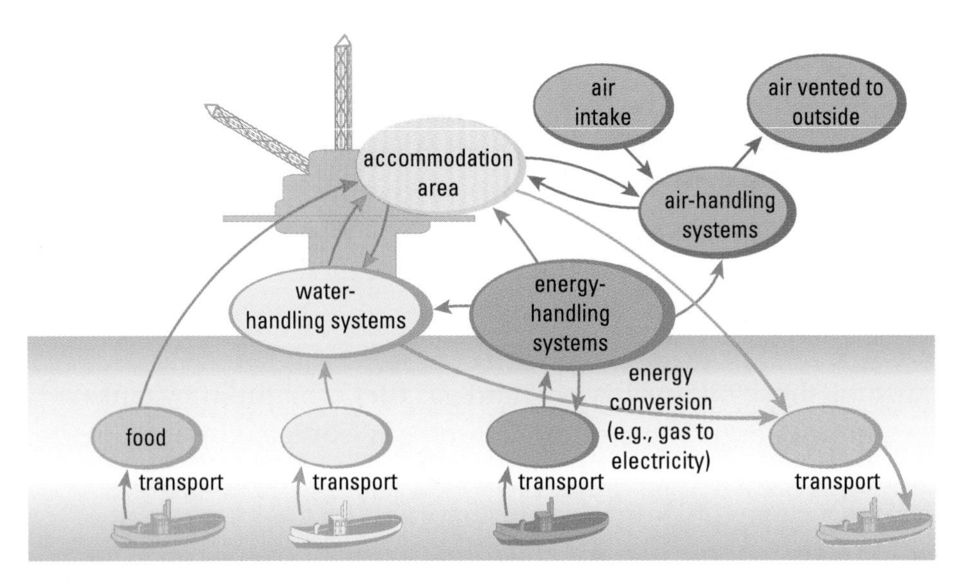

FIGURE 10.4 This flow chart shows the inputs and outputs of an offshore oil rig, and the systems required to handle them. In your notebook, supply the missing labels. Why is an offshore oil rig not a sustainable alternative environment?

Biodomes

FIGURE 10.5 The Montreal Biodome is designed to maintain environmental conditions that are different from the natural environment of the area.

A **biodome** is an alternative environment that is constructed for the purpose of sustaining life. Biodomes help us to study the interactions that support life here on Earth, and how we might construct sustainable alternative environments on space stations or other planets. There are a number of biodomes on Earth, including the Montreal Biodome in Quebec (Figure 10.5) and Biosphere 2 in Arizona, USA.

As you read in Section 9.2, Biosphere 2 was planned as a self-sustaining environment that could support human life. Due to some problems with its functioning, Biosphere 2 was converted to a research facility. Biosphere 2 is now the largest environmental laboratory in existence.

Food

The organisms that live in Biosphere 2 supply food for one another through the carbon cycle and nitrogen cycle. These cycles do not operate as well as they do in the natural environment of Earth, so additional nitrogen fertilizer must be added.

Wastes

Wastes and dead tissues are recycled in Biosphere 2 by carbon and nitrogen cycles that mimic those of Earth. However, these cycles do not recycle all the waste, so some materials must be removed to the outside to maintain the environment.

Air

Scientists can operate Biosphere 2 with a self-contained air supply, in which no air is allowed to enter or escape, or with air supplied from outside the facility, brought in by specially designed fans (Figure 10.6). When Biosphere 2 is operating with a self-contained air supply, it depends on an air-handling system known as "lungs." These "lungs" are flexible, which allow the building to expand and contract to maintain constant air pressure.

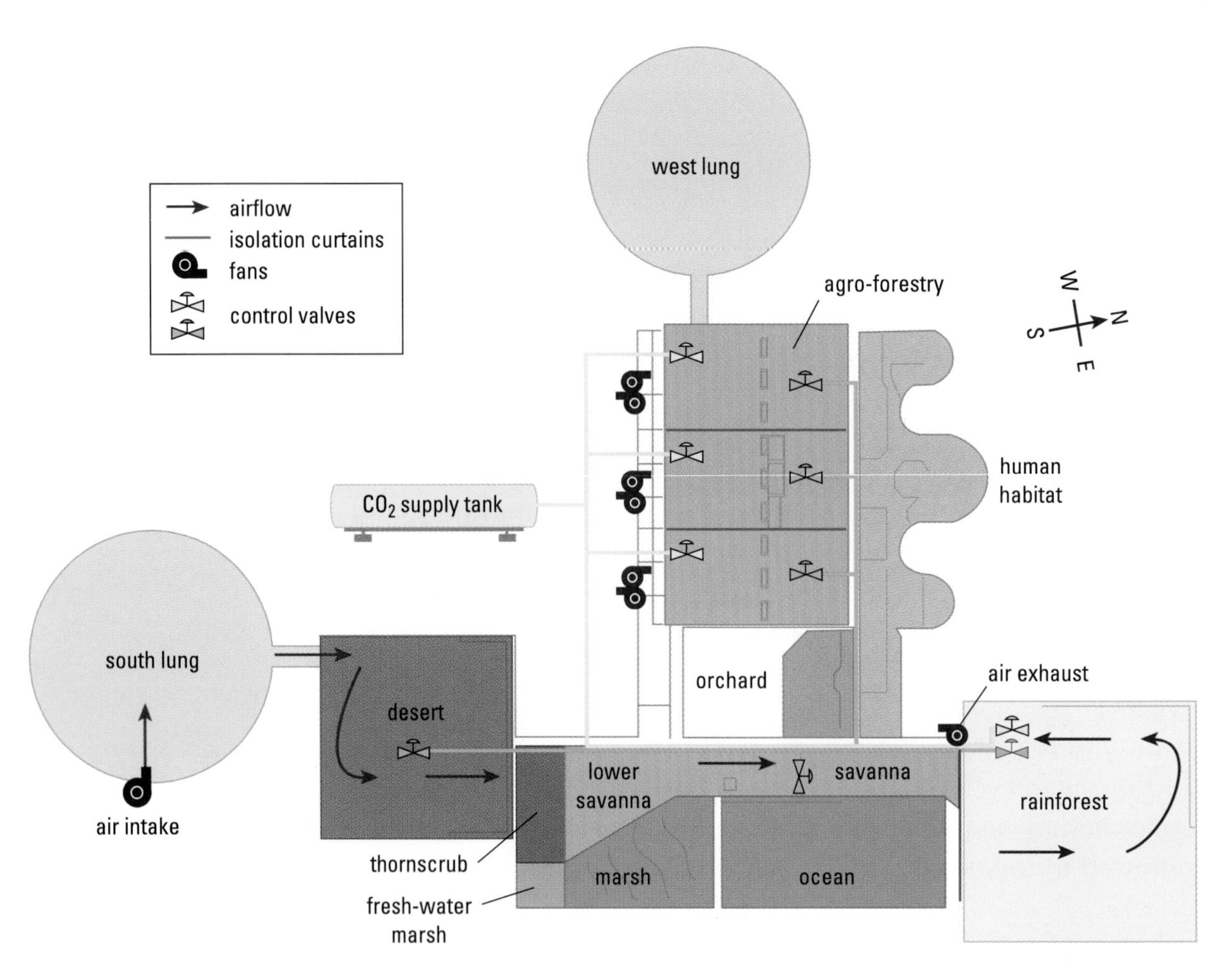

FIGURE 10.6 The air-handling system of Biosphere 2. How does this system differ from the way Earth's natural ecosystems maintain air quality? What are the similarities?

Water

Water in Biosphere 2 is recycled in a fashion similar to Earth's water cycle (Figure 10.7). Mechanical purification, storage, and water-movement systems are needed to keep the systems in balance. Since Biosphere 2 is not completely sealed, some water escapes to the outside, and must be replaced.

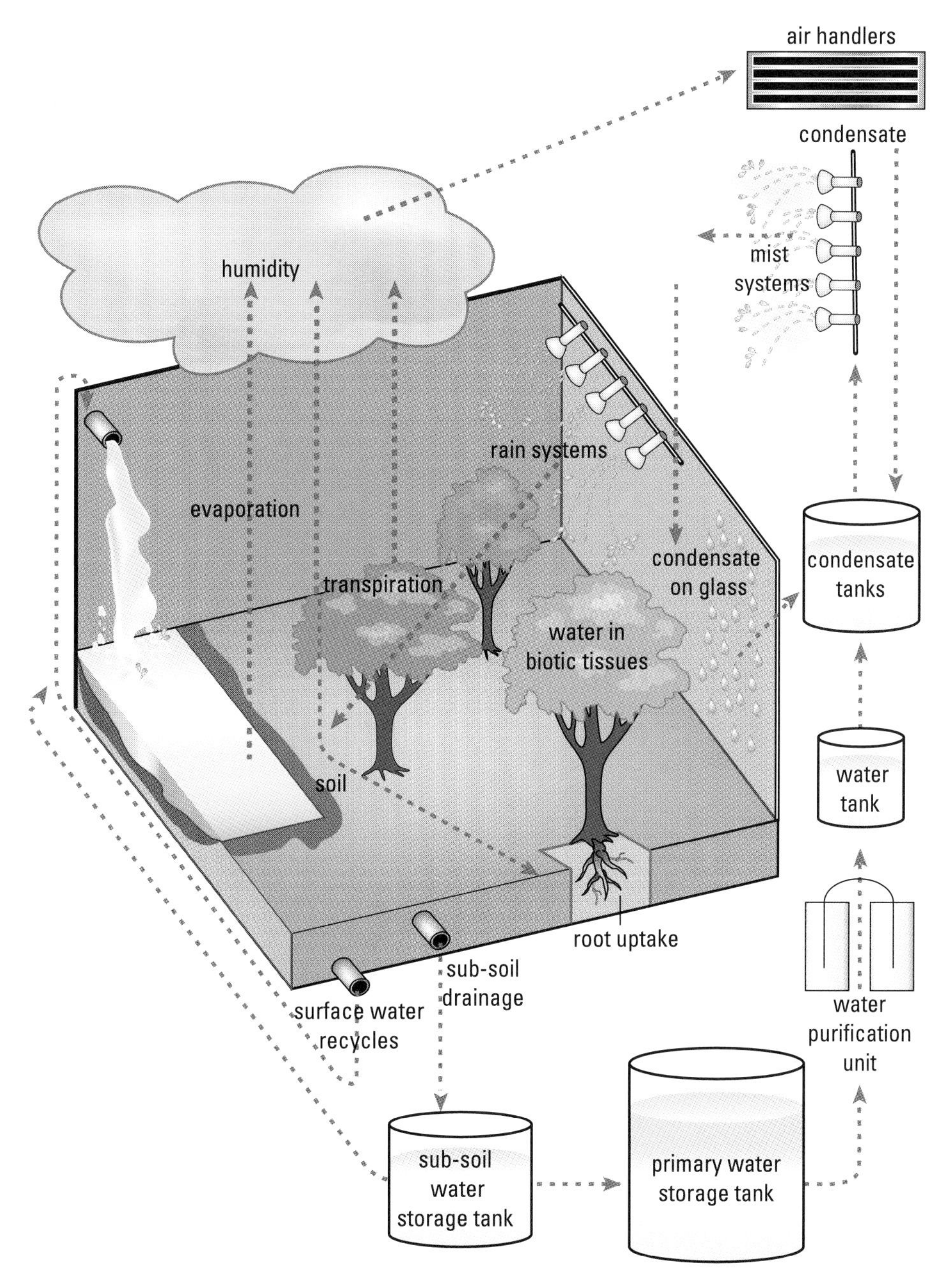

FIGURE 10.7 Water-cycling system in an area of Biosphere 2. Identify the biotic and abiotic factors in this water cycle. How do they compare to the biotic and abiotic factors of the water cycle on Earth?

FIGURE 10.8 The environment of Biosphere 2 includes a functioning water cycle, carbon cycle, and nitrogen cycle.

Energy

The energy for Biosphere 2 is provided in part by renewable solar energy, and in part by non-renewable sources, such as oil and gas. Sunlight provides the energy for the carbon cycle, and also some of the energy used in the water cycle. Sunlight also warms Biosphere 2. Electricity and gas-powered motors are used by the systems that maintain the temperature, water supply, and airflow, and to carry out maintenance activities such as excess waste removal.

Sustainability of Biodomes

Biodomes are designed to create the environmental conditions required to support the life of certain species. Maintaining these environmental conditions takes a lot of energy, some of which is supplied by non-renewable sources. None of the biodomes currently in existence is able to recreate the cycling of matter that exists in Earth's natural ecosystems. Biodomes are therefore more sustainable than offshore oil rigs, but less sustainable than Earth's ecosystems (Figure 10.9).

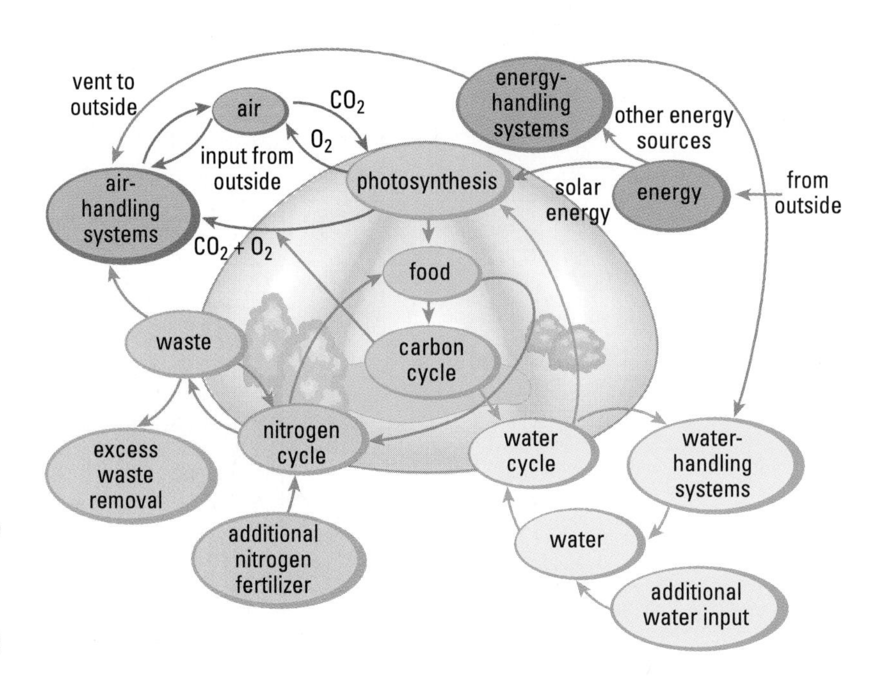

FIGURE 10.9 Flow chart of the inputs and outputs required to sustain life in a biodome. Compare the systems that maintain the life-sustaining environment of a biodome to those on Earth. What are the similarities and differences?

Job Link

Refrigeration and Air-Conditioning Technician

Refrigeration and air-conditioning technicians maintain air-handling equipment in indoor alternative environments here on Earth.

FIGURE 10.10 Refrigeration and air-conditioning technicians help to maintain the air quality in buildings.

Responsibilities of a Refrigeration and Air-Conditioning Technician

- install, maintain, repair, and overhaul refrigeration and air-conditioning systems, and combined heating and cooling systems
- troubleshoot and perform corrective action and minor repairs to prevent equipment or system failure
- may maintain a log of operational, maintenance, and safety activities, and communicate results in written reports

Where do they work?

- industrial plants and buildings
- large public facilities such as shopping malls, office complexes, or subways
- private homes

Skills for the Job

- ability to work with minimal supervision
- good problem-solving skills
- good written and verbal skills
- must be punctual, responsible, and well organized

Education

- completion of an apprenticeship training program and examination
- completion of secondary school may be required (must be at least 16 years old)

Surf the Web

To find related careers, visit **www.science.nelson.com** and follow the links for ScienceWise Grade 12, Chapter 10, Section 10.1.

ACTIVITY 10A

Inputs and Outputs of an Earth-Based Alternative Environment

You might not have realized it, but chances are that you have experienced life in an alternative environment. There are many places on Earth that can be thought of as alternative environments: airplane cabins, subway stations, even automobiles. How are these alternative environments maintained? What are the inputs and outputs needed to sustain life within them? In this activity, you will prepare a flow chart to diagram the inputs, outputs, and interactions of the various life-sustaining components of a common alternative environment here on Earth: a shopping mall.

You will need to be able to use a flow chart to diagram the inputs, outputs, and interactions between systems in an alternative environment for your project, Another Way of Life, at the end of Unit 5.

What You Will Need

- paper and pencils, or access to graphics software

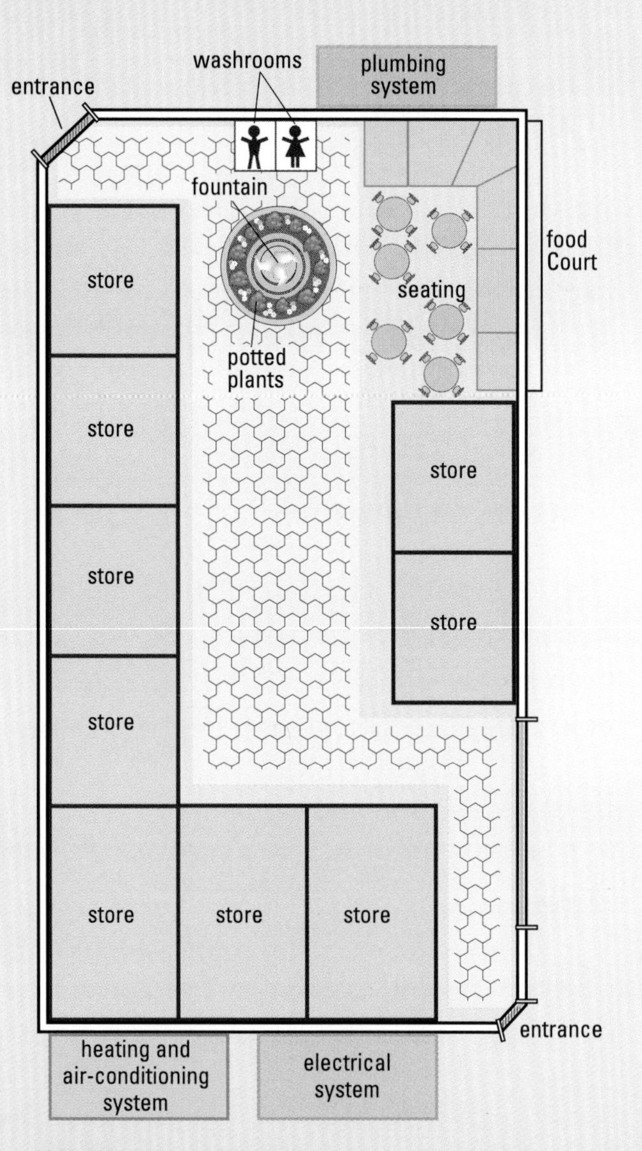

FIGURE 10.11 What systems support human life in a shopping mall? What are the inputs and outputs?

What You Will Do

1. Figure 10.11 shows some common components of a shopping mall. In point form, describe how air, water, food, and energy are provided by this alternative environment.
2. In point form, describe the outputs of this alternative environment, and the systems required to handle them.
3. Using the factors and systems you described in steps 1 and 2, create a flow chart that diagrams the inputs and outputs of the shopping mall. Indicate on your flow chart how the different systems in the shopping mall interact to maintain a suitable environment for human life.

What Did You Find Out?

1. Exchange your flow chart with a partner. After you have looked over your partner's flow chart, re-evaluate your own work. Make any improvements to your chart that you can.

Making Connections

2. Compare the life-sustaining components of the shopping mall to the processes through which the natural environment sustains life. For example, what does the heating and air-conditioning system have in common with the natural process of photosynthesis?
3. Briefly describe two occupations that are involved in building or maintaining the alternative environment of a shopping mall.
4. Suggest one or two ways that the alternative environment of a shopping mall could be made more sustainable.

Surf the Web

The Montreal Biodome is a Canadian-built, life-sustaining alternative environment where visitors can experience ecosystems from different parts of the world, all under one roof. Visit www.science.nelson.com and follow the links for ScienceWise Grade 12, Chapter 10, Section 10.1 to find out more about the Montreal Biodome. What are the systems that maintain the environments of the Montreal Biodome? What are the inputs and outputs of this alternative environment? Present your findings as a poster or web page.

■

Review and Apply

1. How are the inhabitants of an offshore oil rig supplied with food and water?
2. How are human wastes disposed of on an offshore rig?
3. Could an offshore rig be used for long-term human habitation? Give reasons for your answer.
4. Draw a diagram that shows how the airflow system and the water cycle in Biosphere 2 interact to maintain an environment that can support life.
5. What is the function of the "lungs" of Biosphere 2?
6. Explain why bacteria in the soil affect the amounts of both oxygen and carbon dioxide in Biosphere 2.
7. Give three examples of Earth-based alternative environments not mentioned in this textbook. Explain your choices.
8. Which alternative environment is more sustainable, an offshore oil rig or a biodome? Justify your answer by referring to the inputs and outputs of both environments. Suggest two ways that either of these environments could be made more sustainable.
9. Organize the concepts you have learned in this section in a graphic organizer.
10. **Design a Biodome**

    In small groups, create a design for a biodome that will support the life of a particular organism. For example, you might design a biodome to support the life of a lizard. Start by researching the requirements for life of the organism you choose. Your biodome design must use easily available materials, and have some means of supplying the organism with air, water, and food. It must also have a means of handling the outputs of the organism, such as carbon dioxide and wastes. Identify the energy source you will use to maintain the biodome. Present your completed biodome design as a poster or model, or by some other method of your choice.

10.2 Leaving Home

Humans are able to construct alternative environments off our planet, in space. These alternative environments include the many different spacecraft that have left Earth's atmosphere for a short time to enter orbit, and those that went beyond Earth to the Moon. Humans have also constructed orbiting space stations, which are alternative environments that allow us to live in space for long periods of time. How is human life sustained in these alternative environments?

ScienceWise Fact

Russian cosmonaut Leonid Kizim was the first human to live in space for more than one year. On December 21, 1987, Kizim left the space station Mir after living in space for 365 days, 22 hours, and 39 minutes.

Space Stations

The world's first space station was Mir, built by the former Soviet Union. Mir operated continuously from 1986 to 2000. In 1995, an American Space Shuttle docked with Mir (Figure 10.12) as the first step to operating a shared space station. This joint facility was occupied by American astronauts and Russian cosmonauts for 27 months. This shared space station laid the foundation for a structure designed for long-term human habitation in space, the International Space Station, or ISS.

FIGURE 10.12 This photograph shows Russian cosmonaut Vladimer N. Dezhurov and American astronaut Robert L. Gibson shaking hands through a linked hatch between Mir and the Space Shuttle. An important benefit of space research is to provide such opportunities for nations to work toward common goals.

Surf the Web

The International Space Station is usually visible in the night sky with the naked eye. Find out the position of the Space Station by visiting www.science.nelson.com, and following the links for ScienceWise Grade 12, Chapter 10, Section 10.2.

The International Space Station is constructed of specialized modules, each constructed on Earth (Figure 10.13). The completed modules are then brought to the station by the space shuttle, and final assembly takes place in space. Canada provided the Mobile Servicing System, which is made up of a 17-m, 125 000-kg capacity robotic arm called the "Space Station Remote Manipulator System," and a hand-like, 3.5-m "Special Purpose Dexterous Manipulator." This unit is also known as Canadarm 2. The first Canadarm design is part of the Space Shuttle. These robotic arms are used in fitting the modules together, and decrease the time astronauts must work outside a spacecraft.

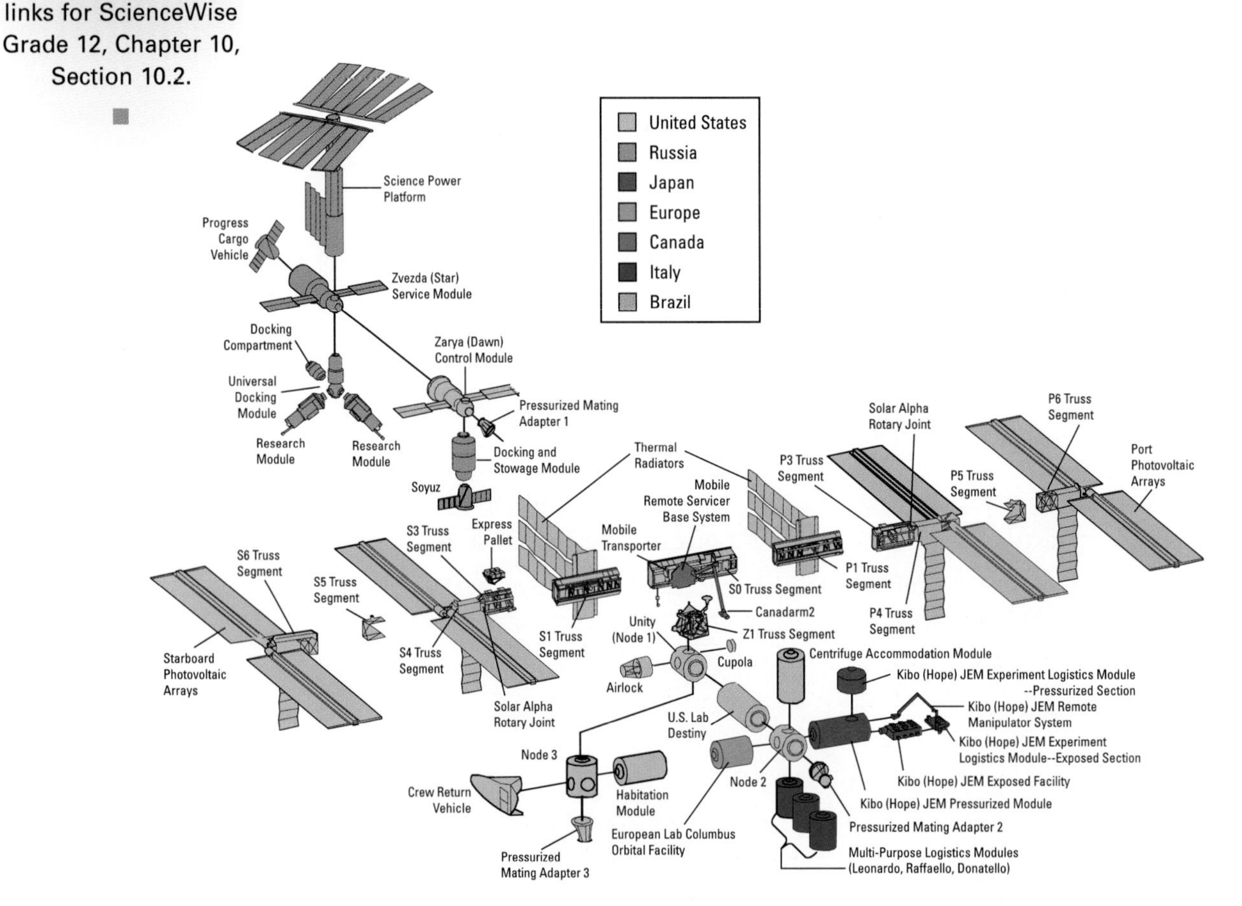

FIGURE 10.13 The International Space Station is constructed from a series of modules that fit together. The modules are assembled on Earth, then transported to space by the Space Shuttle.

Life in Low-Gravity Environments

A problem unique to life in space is low gravity. **Gravity** is the force of attraction between any two objects. Gravity tends to pull objects toward each other. On Earth, we feel gravity as a force pulling down, toward Earth's centre. Any object in orbit, including the International Space Station, is subject to microgravity. **Microgravity** is a gravitational force so small that objects appear to be weightless. **Weightlessness** means that an object is subject to no gravitational pull at all (Figure 10.14).

You may need to know how humans are affected by low-gravity environments for your project, Another Way of Life, at the end of Unit 5.

FIGURE 10.14 This photograph shows American astronaut Ellen Ochoa (left) and Canadian astronaut Julie Payette (right) handling supplies in the microgravity of the International Space Station. Although weightlessness can be fun, it can also cause problems. What difficulties might weightlessness cause when eating and drinking?

Weightlessness causes many problems in the Space Station. For example, liquids cannot be poured from container to container, food crumbs can float into sensitive instruments, and astronauts can lose their sense of up and down. The life support systems must also compensate for microgravity. For example, toilets aboard spacecraft cannot be flushed, because flushing depends on gravity pulling water and solids downward. On the International Space Station, toilets are cleared by jets of air that push material into the waste system.

The human body is adapted to life in Earth's gravity. Our muscles grow and function best under Earth's force of gravity, as do our bones and other body components. Our circulatory system also needs Earth's gravity to function. On Earth, gravity pulls body fluids, such as blood, down toward the feet. If you hang upside down, the blood is pulled to your head instead (Figure 10.15). Your body has a hard time moving the blood away from your head from this upside-down position.

FIGURE 10.15 The circulatory system is designed to compensate for the pull of gravity downward on our bodies.

ScienceWise Fact

One problem for humans living in a weightless environment is that their digestion works far less efficiently than on Earth. Elimination of the gases that accumulate in the body during digestion depends on gravity.

During prolonged space missions, microgravity can cause changes in the human body. In low gravity, the body's fluids are no longer pulled downward, so they tend to pool in the upper parts of the body. Fluid can build up in the brain, which then triggers calcium loss from bones. In a space mission lasting 6 to 12 months, an astronaut may lose one percent of his or her bone mass each month. The change in the distribution of the body's fluids also thickens the blood by about 10 percent, which can lead to circulatory problems. Astronauts therefore occasionally wear a lower-body negative-pressure suit to counteract the build-up of fluids. The pressure in these suits forces body fluids back down to the lower extremities (Figure 10.16).

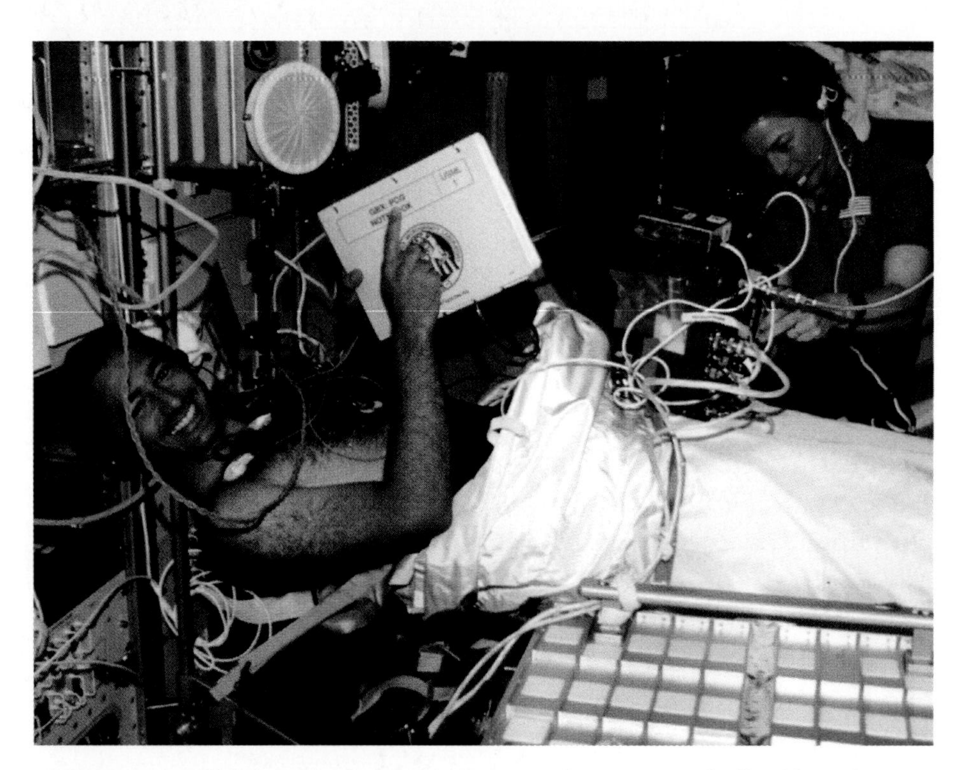

FIGURE 10.16 Why do astronauts need to wear negative pressure suits like this one?

Without gravity, the muscles of the body encounter little resistance and so work a lot less. Astronauts' muscles therefore shrink and become weak in microgravity environments, even though they exercise regularly. Astronauts returning to Earth's gravity after a prolonged time in space may be so weak that they find it difficult to walk.

Life Support Systems

The life support systems of the completed Space Station will be in a module called the Environmental Control and Life Support Systems. This module was designed to provide an environment where astronauts could live and work safely, and relatively comfortably, for prolonged periods. The module had to be extremely reliable and efficient, but compact enough for transport into space (Figure 10.17).

FIGURE 10.17 The Environmental Control and Life Support Systems module will help create an alternative environment capable of supporting human life in space for prolonged periods.

The Environmental Control and Life Support Systems module will perform the following functions:

- maintain the pressure, temperature, and amount of moisture in the air
- monitor and control the quality of air, including the amounts of oxygen and carbon dioxide gases
- recycle water
- manufacture oxygen gas from water
- detect and control any fires

Air

Providing a continuous supply of breathable air is critical to any environment in space. The systems on the International Space Station must replenish the oxygen used up and remove the carbon dioxide produced by respiration of the astronauts' respiration. It must also remove any other gases produced as a result of the operation of the Space Station.

The Space Station stores large amounts of oxygen in pressurized tanks. Since transporting and storing oxygen is not sustainable, other means of supplying oxygen are also needed. One device uses electricity to break the chemical bonds in water molecules to form oxygen and hydrogen gas (Figure 10.18). Section 1.1 has more information about chemical bonds. Currently, the hydrogen gas produced is vented into space. The Environmental Control and Life Support Systems module will supply oxygen by this process on a large scale.

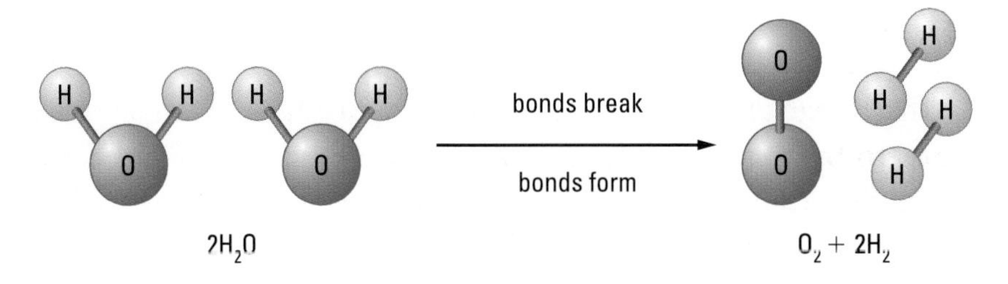

FIGURE 10.18 Electricity can be used to produce oxygen gas (O_2) and hydrogen gas (H_2) from water (H_2O).

Carbon dioxide is removed from the air by a machine that contains a substance called zeolite. Zeolite acts like a sieve; it captures the larger carbon dioxide molecules, but allows smaller molecules such as oxygen and nitrogen to pass through. The captured carbon dioxide is vented into space. Figure 10.19 is a flow chart that shows how the major components of the air-handling systems must work together to maintain the air quality inside the Space Station.

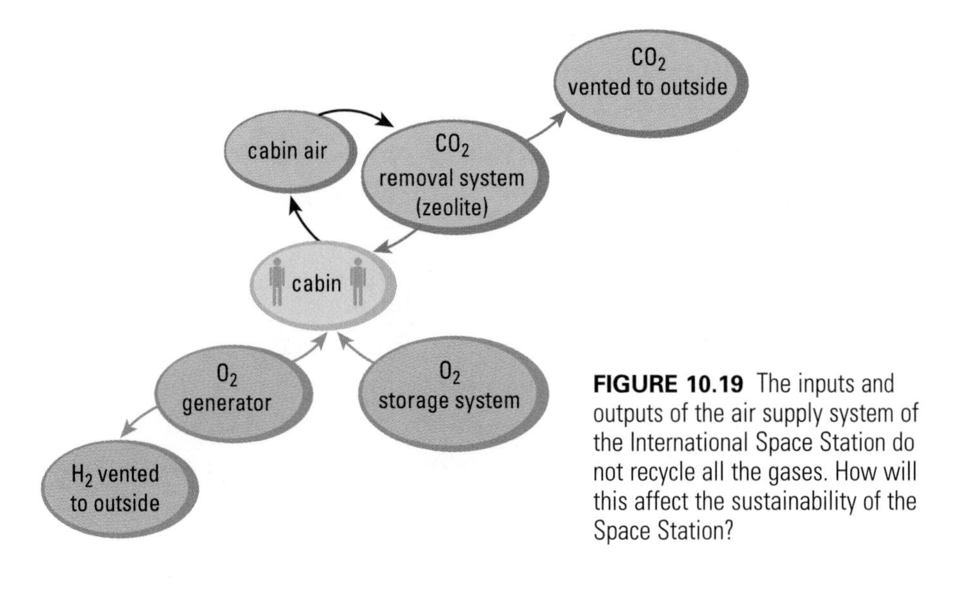

FIGURE 10.19 The inputs and outputs of the air supply system of the International Space Station do not recycle all the gases. How will this affect the sustainability of the Space Station?

Water

All the water on the Space Station was initially brought from Earth. Water takes up a lot of space and is quite heavy, so the Space Station uses this water very efficiently. All the water is captured, purified, and reused, including the water in the astronauts' sweat. The crew also use far less water than the average person on Earth. However, even with this recycling, the Space Station slowly loses water. This lost water must be replaced from Earth.

The water purification systems on the Space Station partially mimic the water cycle of Earth. The water is first filtered to remove particulate matter, similar to how water is filtered by soil on Earth. However, the Space Station does not have decomposers or plants to break down and use the trapped matter. Instead, a system of fine filters and chemicals purify the water, which is then passed through a machine that kills any micro-organisms.

Food

Currently, astronauts rely entirely on food supplies brought from Earth (Figure 10.20). Experiments are in progress to find ways of growing plants in space. If plants can be grown successfully on the Space Station, they will supply at least some food. Plants would also contribute to maintaining the air and water in the Space Station.

FIGURE 10.20 Food used in space is usually dried to make it lighter. It is also packaged in a way that prevents crumbs from escaping into the spacecraft. Why is this important?

Energy

Huge arrays of solar panels collect the Sun's energy to provide power to the International Space Station (Figure 10.21). However, during the orbit of the Space Station, there are periods in which Earth is situated between the Space Station and the Sun. At these times, the panels are in darkness and cannot collect energy. Therefore, the Space Station is also equipped with storage batteries that provide power during the dark phase of orbit.

FIGURE 10.21 How is the role of the solar panels of the International Space Station similar to the role of plants on Earth?

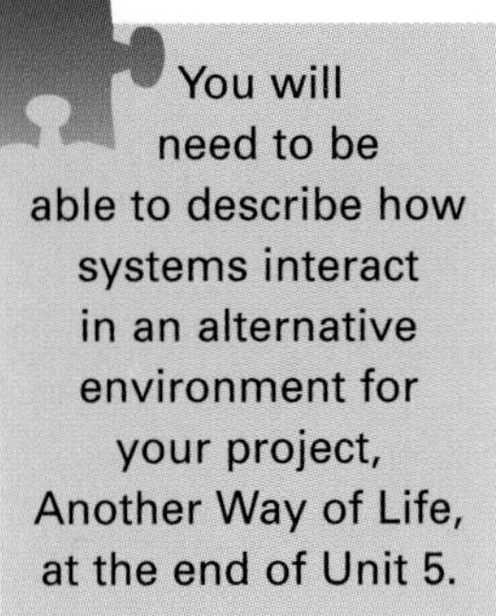

Wastes

Human wastes on the Space Station are only partially recycled. The water in urine is recovered and purified along with all the other water on the space station. Solid wastes and impurities from the water are transferred to a waste management system and eventually returned to Earth for disposal (Figure 10.22).

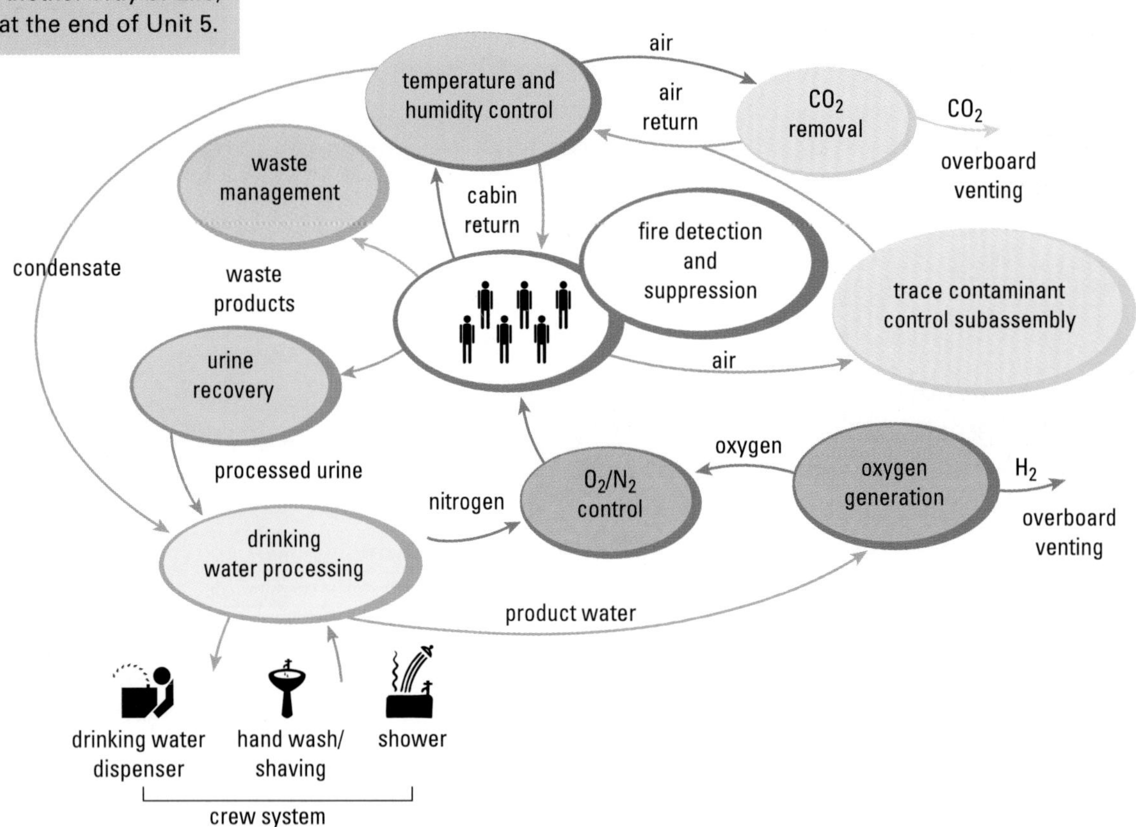

FIGURE 10.22 The systems for handling air, water, and wastes work together. Study this flow chart and describe these interactions.

Costs and Benefits of the Space Station

Building and operating the Space Station is an expensive venture. Many resources, including fuels, materials, ideas, and labour, are needed for every phase. How does this affect human life here on Earth?

The Space Station is an international facility that allows 16 nations to work together (Figure 10.23). Working together increases understanding between nations, which improves the chances for world peace. These nations represent a small proportion of the world's population, however, and tend to be the wealthier nations. This could result in an increase in the technological and knowledge gap between the world's richer and poorer nations.

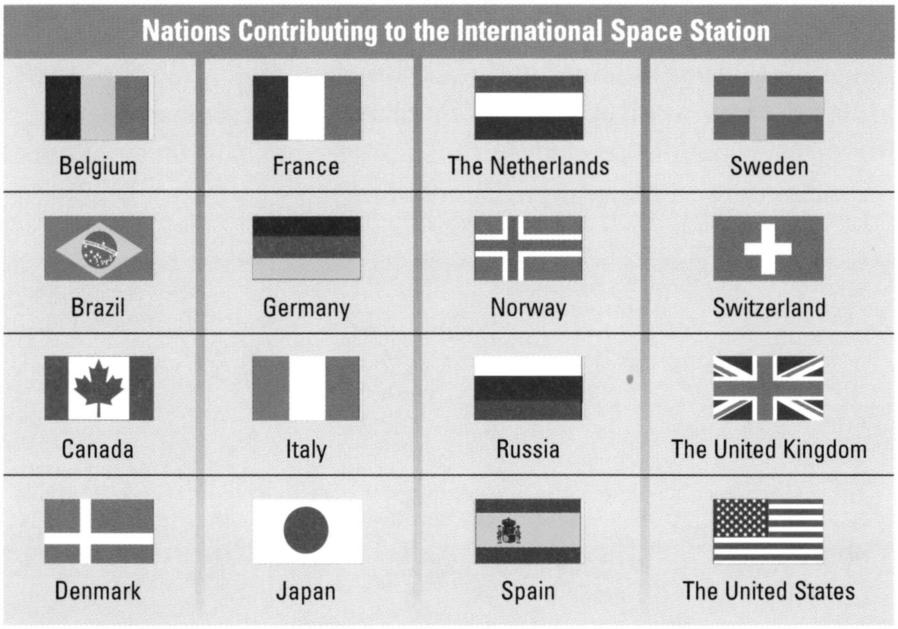

Nations Contributing to the International Space Station			
Belgium	France	The Netherlands	Sweden
Brazil	Germany	Norway	Switzerland
Canada	Italy	Russia	The United Kingdom
Denmark	Japan	Spain	The United States

FIGURE 10.23 The contributors to the International Space Station are wealthy nations. How will this affect nations that cannot afford to participate?

FIGURE 10.24 These products would not exist without the space program. How does the development of new products such as these affect Canada's economy?

Many materials and devices developed for use in space find their way into consumer products here on Earth. For example, the scratch-resistant coating on the lenses of eyewear was first developed for space applications (Figure 10.24), as were the materials used to produce lightweight protective equipment, such as helmets.

Maintaining the Space Station has environmental consequences. Some of the fuel used to launch the Space Shuttle comes from non-renewable energy sources. These fuels also releases pollutants into Earth's atmosphere. Some of the matter used in the Space Station is not recycled, such as the hydrogen and carbon dioxide that is vented from the air supply systems. This matter is no longer available to life on Earth. We have also left machinery from old satellites, rocket pieces, fuel, and other toxic fluids floating in orbit around our planet. These may eventually be recovered and recycled, but for now, we are causing a pollution problem beyond our planet (Figure 10.25).

FIGURE 10.25 Learning to live in space has environmental consequences, such as when we leave materials in orbit, as shown in this artist's representation.

Review and Apply

1. Describe how a weightless environment affects human life.
2. Write a question about creating or maintaining an alternative life-sustaining environment in space that you would like to ask an expert at the Canadian Space Agency.
3. Look at the flow chart of the life support systems on the Space Station in Figure 10.22. Describe what happens to the water vapour that is exhaled by the astronauts.
4. In a short essay, outline the costs and benefits of maintaining the Space Station to society, the economy, and the environment.
5. With a partner or in a group, brainstorm changes that could be made to the systems of the International Space Station so that it could run for a period of 10 years without new supplies from Earth.
6. Add the new concepts you learned in this section to the graphic organizer you started in Section 10.1.

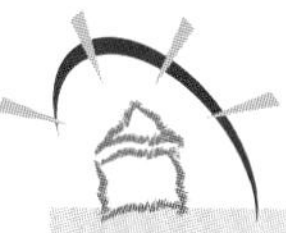

Try This at Home

Balancing Act

You can feel the force of gravity by trying this test. Stand with your back to the wall, feet together, and heels touching the wall. Place an object about 30 cm away from your toes (Figure 10.26). Try to pick it up without bending your knees or moving your feet. What happened? Explain your experience using the concept of gravity. Do you think you could do this in space? Explain.

FIGURE 10.26 How does gravity affect your ability to move?

LAB 10B

Gravity and Plant Growth

Photosynthesis by plants helps to support life on Earth. Plants remove carbon dioxide from the atmosphere and replace it with oxygen. Ultimately they supply food for all life on Earth. Along with decomposers, plants also play an important role in recycling the nutrients in wastes. Plants also help to keep the water supply clean. Scientists are therefore investigating the use of plants in alternative environments in space. These environments will have little or no gravity. How do plants respond to gravity?

Purpose

To study the effect of gravity on the growth of plants

Materials

- two small seedlings.
- moistened potting soil
- two transparent flat containers, such as an empty cassette tape case or a compact disc case
- modelling clay
- masking tape
- black construction paper

Procedure

1. Separate the top and bottom of the cassette tape or compact disc case so you have two sections. Add a small piece of modelling clay to each corner of one of the pieces, as shown in Figure 10.27.

FIGURE 10.27 Step 1

2. Gently remove one seedling from its container, being careful not to break the roots. Place the seedling on the case section with the modelling clay, so that the roots are on the plastic (Figure 10.28).

3. Carefully place soil around the roots of the plant. Place the other side of the cassette tape or compact disc container on the modelling clay. You should now have a container composed of two flat pieces with a narrow space in between them, with a plant and soil in the space (Figure 10.29).

4. Clean the edges of the container. Apply tape around the three sides of the container to seal them (Figure 10.30). If necessary, add a small amount of water to the soil.

5. Repeat steps 1–4 for a second seedling.

6. Cover the sides of the containers with black construction paper to block out light. Use masking tape to secure the paper. Label the first container "1" and the second "2."

FIGURE 10.28 Step 2

FIGURE 10.29 Step 3

FIGURE 10.30 Step 4

CONTINUED

7. Place both containers where they will receive plenty of light, such as on a windowsill. Stand container "1" upright, and place the sealed container "2" on its side (Figure 10.31).

8. After about two days or more, remove the paper and observe the roots and plant. Draw labelled diagrams to show any differences between the two seedlings.

FIGURE 10.31 Step 7

Analysis and Conclusion

1. Describe the direction of growth of the foliage of plants 1 and 2.

2. Describe the direction of growth of the roots of plants 1 and 2.

3. To what force do the roots respond? To what force does the foliage respond?

Extension and Connections

4. How would the lack of gravity affect the growth of plants on a Space Station? Suggest a solution to this problem.

5. Predict the results if you were to conduct this experiment (Lab 10B) on the Moon, where the force of gravity is only $\frac{1}{6}$ as strong as on Earth.

10.3 Chapter Summary

Now You Can...

- Identify the various life-sustaining components in an alternative environment and the ways in which they interact (10.1, 10.2)
- Describe the difficulties of living in microgravity (10.2)
- Describe Canadian contributions to alternative environments (10.2)
- Discuss the sustainability of existing alternative environments (10.1, 10.2)
- Compare the sustainability of alternative environments to the sustainability of the natural environment on Earth (10.1, 10.2)
- Describe the inputs, outputs, and interactions of a life support system for an alternative environment, and the systems in place to handle them (10.1, 10.2)

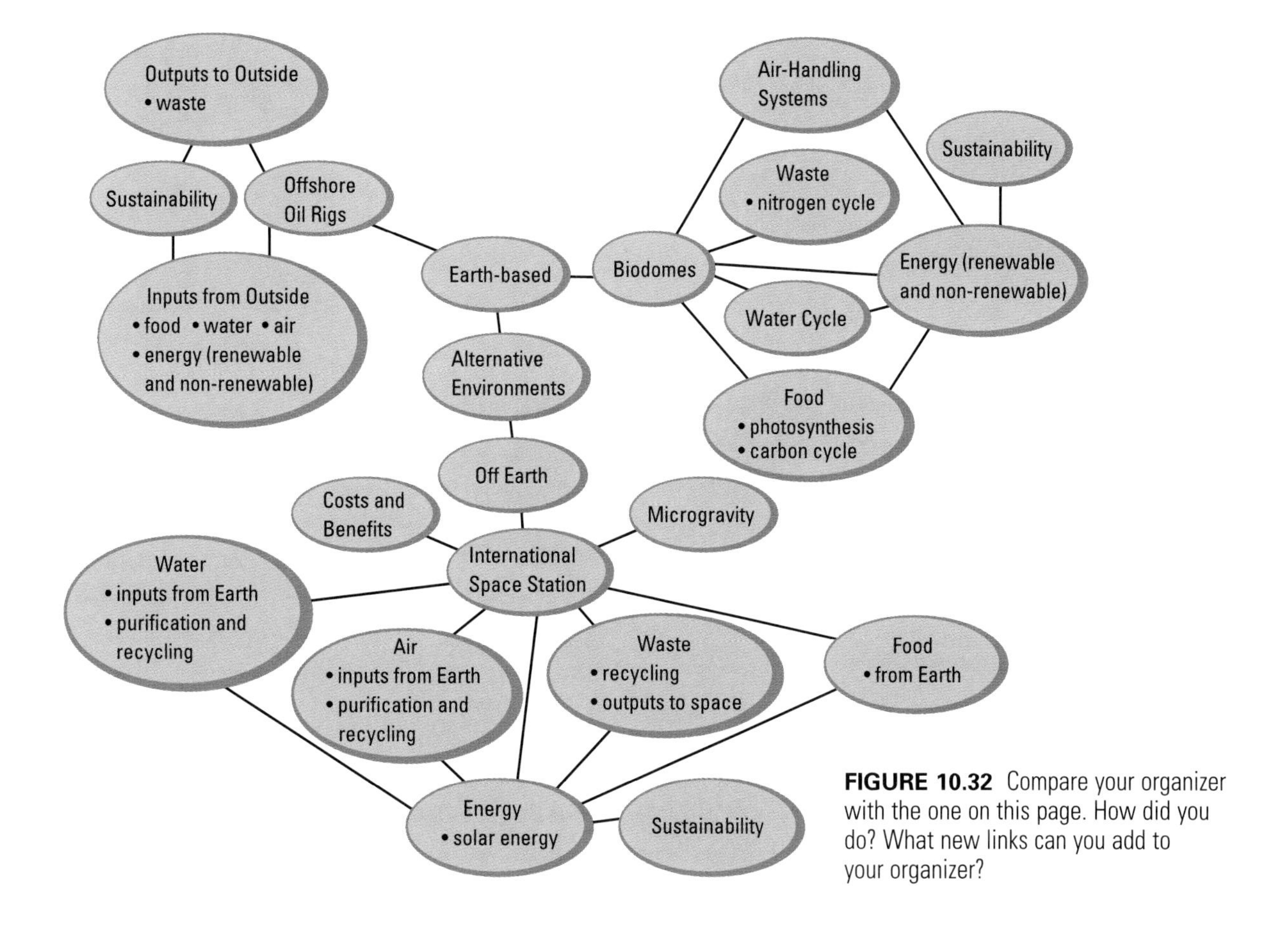

FIGURE 10.32 Compare your organizer with the one on this page. How did you do? What new links can you add to your organizer?

CHAPTER 10 *review*

Knowledge and Understanding

1. List the inputs and outputs needed to support life in an alternative environment.
2. Why is an offshore oil rig not a sustainable alternative environment?
3. What components of Biosphere 2 function most like natural ecosystems on Earth?
4. Is Biosphere 2 a self-sustaining environment? Support your answer with examples.
5. Describe the interaction(s) between the air and water cycles in a closed biodome.
6. In your own words, explain the difference between gravity and microgravity.
7. How could a spacesuit be considered a alternative life-sustaining environment? Give an example of an alternative life-sustaining environment that was not described in this textbook.
8. State two products that are available on Earth that were developed as a result of efforts to build alternative life-sustaining environments in space.
9. Could an alternative environment be sustainable if it relied on a non-renewable energy source? Explain your answer.

Inquiry

10. You are an astronaut living in a weightless environment.
    a) What would happen if you tried to boil water in a pot?
    b) How would you make the water uniformly hot?
    c) What difficulties might you encounter when you try to pour the water into a cup?
    d) How might you overcome this problem?
11. How would low gravity affect an astronaut's ability to water plants growing in a Space Station?

Making Connections

12. List the factors that affect the ability of an aquarium to support life.

13. You work on the tenth floor of an office building. Explain how an office building is an example of an alternative environment. What are the systems that support your life in this building?

14. Firefighters work in conditions that are hazardous to life. How might the development of the Space Station benefit a firefighter?

Communication

15. Create a skit showing the difficulties that a person would face living in a weightless environment. Include in your skit some suggestions for overcoming these difficulties.

16. Use print and electronic resources to research the history of Canada's contribution to the development of the International Space Station. Include information on any of the components of the Space Station, as well as information on Canada's astronauts. Create a graphic timeline to display the results of your research.

PUTTING IT ALL TOGETHER

Another Way of Life

Alternative life-sustaining environments must be able to provide all the necessities for life. To be sustainable over the long term, an alternative environment must recycle all matter and use only renewable energy sources. So far, we have been unable to create an alternative environment that is as sustainable as the natural environment of Earth. How far have we gotten toward this goal?

The Plan

In this activity, you will use your knowledge of what an alternative life-sustaining environment needs to be able to support human life to analyze the sustainability of an existing alternative environment. You will work in a group to suggest ways in which the alternative environment might be improved or further developed. You will also create a model or display of the alternative life-sustaining environment that summarizes your work.

What You May Need

- Internet access and library resources
- materials to make a model or poster presentation
- coloured pens, markers, pencils, or a computer and colour printer

What You Will Do

1. As a group, conduct research on existing alternative environments, both on and off planet Earth. Based on your initial research, decide as a group on one alternative environment that you would like to explore further. Here are some examples of alternative environments you might explore:
   - the Sudbury Neutrino Observatory
   - a nuclear submarine
   - NASA's Lunar-Mars Support Project (LMLSTP) Phase IIA
   - NASA's Lunar-Mars Support Project Phase III
   - NASA's Lunar-Mars Support Project BIO-Plex

FIGURE E1 Begin your research at **www.science.nelson.com.**

2. As a group, conduct research to answer the following questions about the alternative environment you chose. Divide the work among the members of your group so that each person is responsible for finding at least one answer.
   a) What are the biotic and abiotic components needed to support human life in this environment?
   b) What are the sources of air, water, energy, and food in this environment?
   c) What are the systems that maintain the supplies of energy, air, water, and food?
   d) How do the different systems interact in this environment?
   e) What are the outputs of the alternative environment, and what are the systems that handle them?
   f) How sustainable is this alternative environment?
   g) What are the similarities and differences between the life support systems of the alternative environment and the natural environment of Earth?

3. When research is complete, meet as a group and share the information. Decide as a group how you can best present this information.

4. Prepare your presentation.

CONTINUED

What Did You Find Out?

1. When the presentation is complete, review the inputs and outputs of the alternative environment. Identify any areas in which matter is not recycled, or in which non-renewable energy sources are used. Focusing on these areas, make suggestions on how this alternative environment might be made more sustainable.

2. Add your suggestion for improving the alternative environment to your presentation.

Assessment

3. Display your project. Elect at least one person in your group to remain with your project to answer any questions. The rest of the group will visit the presentations of your classmates. Make notes of what worked well and what did not work well in other presentations.

4. Which of the presentations did you find most interesting? Why?

5. How did the alternative life-sustaining environments compare to the environment on Earth?

6. Were there any similar problems among the environments? What were they?

Glossary

A

abiotic factor: a non-living factor in an ecosystem, such as water and air

addition reaction: a chemical reaction in which compounds react when they are combined; occurs when a double or triple bond breaks and new single bonds are formed

air: the mixture of gases that makes up Earth's atmosphere; made mostly of nitrogen gas, but it also contains carbon dioxide, oxygen, and water vapour

allele: each different form of a specific gene

alternative environment: any environment other than the natural environment of Earth

amniocentesis: a procedure in which a doctor removes fluid from around a developing baby by inserting a needle into the mother's uterus; the removed cells can be used to diagnose genetic abnormalities

amphiphilic: refers to molecules that consist of two distinct parts—a hydrophilic ("water loving") head and a hydrophobic ("water hating") tail (*see* hydrophilic and hydrophobic)

amplifier: a device that strengthens the signal produced by an input transducer so the signal can travel farther

amplitude: the horizontal distance between the rest position and an extreme of a longitudinal or transverse wave

analog system: a system where the movement of the signal (e.g., sound) is comparable, or analogous, to the movement of a physical variable (e.g., electrical current)

annual: a plant that lives for one year only

anode: the positive terminal inside a picture tube that attracts electrons from the cathode (*see* cathode)

atom: the smallest unit of matter found on Earth

B

balance: the visual impact of each feature in a landscape in relation to other features on a site

base pair: a pair of chemicals on opposite sides of the DNA double helix that make up genes

beat: difference in sound intensity caused by lines of constructive and destructive interference passing the ear from sources of different pitches

biennial: a plant that lives for two years only

biodome: an alternative environment that is constructed for the purpose of sustaining life; helps us to study the interactions that support life on Earth, and how we might construct sustainable alternative environments on space stations or other planets

biotechnology: the application of biological science facts to solve human problems

biotic factor: a living organism or anything produced by a living organism in an ecosystem, including remains and wastes

bulb: an underground structure that stores energy and nutrients for the next year's plant growth

C

carbohydrates: organic compounds composed mainly of atoms of carbon, hydrogen, and oxygen

carbon cycle: the continuous transfer of carbon atoms between the biotic and abiotic factors of an ecosystem (*see* biotic and abiotic factors)

carbon microphone: an input transducer that uses carbon instead of a magnet to generate an analog signal (*see* input transducer and analog signal)

cathode: the negative terminal inside a picture tube that produces electrons

cathode ray tube: a picture tube

chemical bond: the attraction that holds atoms together to make a molecule

chemotherapy: the treatment of a disease, usually a cancer, with chemical medications

chromosome: a piece of DNA containing many different genes

clay: small particles of minerals called silicates

clear-cutting: harvesting forests by removing all of the trees from a selected area, regardless of their age or type

climate: the average conditions of temperature, precipitation (rain and snow), and sunlight of an area

cloning: the process of making a genetic duplicate of an organism

cochlear implant: a small electronic device, consisting of a microphone within a headset, which picks up sound and stimulates the nerves involved in hearing

colloid: a mixture in which invisible pieces of solid or drops of liquid are suspended that will not settle out

colour: the quality of the sound of a musical note; distinctive character or timbre (*see also* timbre); an element of landscape design where features, such as plants, rocks, and buildings, are organized by colour

commercial thinning: the selection and removal of smaller trees from a forest so that the remaining trees will be better able to reach the desired size and shape

composite video signal: a signal for a silent, black-and-white television consisting of beam intensity information and horizontal and vertical retrace signals

compound: a substance made by bonding atoms together in a particular way

compression: the place in a longitudinal wave where the particles are close together

computed axial tomography (CAT): a diagnostic technology that uses computers to combine many X-ray images together and generate cross-sectional (cut-away) views of the body

condensation reaction: a chemical reaction in which two molecules combine to form a single organic molecule and another small molecule, usually water

constructive interference: the increase in the amplitude of a wave that occurs when two crests or rarefactions or two troughs or compressions pass through each other

contrast: an element of landscape design where features, such as shape, size, colour, or texture, are organized to show their differences or contrasts

corrective lenses: glasses or contact lenses

covalent bond: a chemical bond in which atoms share electrons to form single, double, or triple bonds; weaker than an ionic bond

crest: the high point of a transverse wave

cultivated plants: plants produced by humans for their specific qualities

cutting: a section of a plant, usually about 10 to 15 cm long, comprising a section of stem and at least one leaf, used for plant propagation

cycle: a complete sequence of motion that repeats itself

D

decibel (dB): the unit of measure for the loudness of a sound (i.e., the amount of sound energy entering the ears)

decomposers: organisms that use the energy found in carbohydrates in wastes and dead tissues

defibrillator: a device used to apply an electric shock to the heart; used when a patient is experiencing irregular heartbeats

deoxyribonucleic acid (DNA): a molecule that contains all the genetic material for a particular individual, written in a chemical code; like a blueprint for a particular living thing

destructive interference: the decrease in the amplitude of a wave that occurs when a crest and trough or a rarefaction and compression pass through each other

diagnose: determine the specific cause of

dialysis: the process of separating large particles in a solution, such as cells and proteins, from small particles, such as salts; requires a filter that only allows particles below a certain size to pass through

dominant allele: the allele that appears physically in an individual with two different alleles for a gene

dormant: the state of a plant when it stops growing; occurs when the temperature falls below a certain level

dynamic microphone: a microphone that contains a permanent magnet

E

echocardiogram: an ultrasound scan of the heart (*see* ultrasound)

ecosystem: all the living and non-living elements that interact in an environment

electromagnetic waves: transverse waves that do not need a medium to travel through; includes X-rays, microwaves, radio waves, and visible light waves

electronegativity: the electron-attracting ability of an element

emulsifying agent: any substance that helps keep other substances from separating out of the mixture

emulsion: a type of colloid in which tiny liquid droplets are mixed in another liquid; the two liquids are immiscible (*see* colloid and immiscible)

energy: a measure of the ability of a system to do work

environment: the specific conditions of an area in which an organism lives

F

focal point: a point where waves meet after reflection and become amplified

forestry: the science and practice of planting, caring for, and managing forests

frequency: the number of waves that pass by a certain point every second; measured in hertz (Hz)

G

gametes: egg and sperm cells; produced by meiosis

gene: a piece of DNA that contains the code for a particular genetic trait; usually hundreds of base pairs in length

genetic disease: *see* genetic disorder

genetic disorder: a medical condition that can be passed on from parent to child

genetic engineering: biotechnology that involves adding or changing specific genes in an organism

genetic trait: a trait that is inherited and can be passed on to the next generation

genome: all the genes on all the organism's chromosomes (*see* gene and chromosome)

germination: the first stage of growth of a plant, from a seed or spore to a seedling

gravity: the force of attraction between any two objects

greenhouse: a structure designed specifically to grow plants

H

habitat: an area with environmental conditions that allow a particular organism to survive

hard materials: non-living objects used in the design of gardens, such as walls, paving stones, rocks, decks, and gazebos

hard water: water that contains dissolved minerals/ ions such as calcium, magnesium, and iron

hardiness: the relative ability of a plant to survive extreme climate conditions

harmony: the effect of matching features in a landscape, such as colour, size, and shape

hearing aid: a medical device that is placed in the ear to amplify sound

hemodialysis: a medical technology in which a machine purifies the blood by using dialysis to mimic the functioning of the kidneys

hertz (Hz): the unit of measure for frequency, measured per second (*see* frequency)

heterogeneous: visibly made up of different substances

homogeneous: looks like one substance

horticulturalist: a person who produces flowers

humidity: moisture content of the air

humus: partially decomposed material from living organisms; the main source of nutrients in soil, helps soil hold water

hydrolysis reaction: a chemical reaction that uses the addition of a molecule of water to break apart large molecules; the reverse of a condensation reaction

hydrophilic: refers to the polar, water-soluble end of an amphiphilic molecule (*see* amphiphilic)

hydrophobic: refers to the non-polar, non-water-soluble end of an amphiphilic molecule (*see* amphiphilic)

hydroponics: the cultivation of plants in nutrient-enriched water

I

immiscible: refers to a solute that does not mix with a solvent (*see* insoluble)

in phase: refers to two waves that are identical in every way, including direction

input transducer: a device that converts non-electrical energy (e.g., sound) to electrical energy

insoluble: refers to a solute that does not mix with a solvent (*see* immiscible)

interference: the interaction of two or more waves

ionic bond: a chemical bond in which electrons are not shared among atoms but completely transferred from one atom to another

J

joint replacement: a prosthesis that replaces joints (e.g., knees, wrists, or elbows) damaged by accident or disease

K

karyotype: a photograph of all the chromosomes in the nucleus of a single cell; used to determine the number, size, and shape of all the chromosomes in an organism

L

landscape gardening: the activity of laying out or designing gardens

landscaping: *see* landscape gardening

longitudinal wave: a wave that moves parallel to the movement of the particles of the substance in motion

M

magnetic resonance imaging (MRI): a diagnostic medical technology that uses magnetism, radio waves, and a computer to produce detailed images of different parts of the body

matter: anything that occupies space and has mass

medication: any substance used for medical treatment; can cure a disease, treat disease symptoms, or prevent a disease

medium: the substance that a wave travels through

meiosis: a form of cell division that gives rise to daughter cells with half the amount of DNA as the parent cell

microgravity: a gravitational force so small that objects appear to be weightless

miscible: refers to a solute than mixes with a solvent (*see* soluble)

mitosis: a form of cell division that gives rise to daughter cells with the same amount of DNA as the parent, or starting, cells

mixture: a combination of two or more different types of particles

molecule: a group of atoms held together that could be considered one particle

monomers: simple molecules that are combined to form polymers (*see* polymers)

mutation: a change in the DNA of an individual, causing a genetic disorder

N

native plants: plants that grow naturally

natural: occurs in nature

nitrates: nitrogen-containing substances that are soluble in water

nitrogen cycle: the continuous transfer of the nitrogen atoms in the biotic and abiotic factors of an ecosystem (*see* biotic and abiotic factor)

non-polar molecule: a symmetrical molecule in which the bonding electrons are evenly distributed so that both ends of the molecule have the same amount of negative or positive charge

non-renewable energy source: an energy source that uses up finite (limited) resources (i.e., can be completely used up)

O

offshore oil rig: a structure that is designed and built to extract oil from undersea reserves, and to sustain the lives of the people who work on it

optimal growth conditions: conditions under which a plant will grow the strongest and healthiest possible, and produce the desired type of growth

organic compounds: all carbon-containing compounds except carbon monoxide, carbon dioxide, and ionic carbon compounds

oscilloscope: an instrument that shows the wave forms fed into it

out of phase: refers to waves that are not completely identical

output transducer: a device that converts electrical energy into non-electrical energy by reversing the process of the input transducer (*see* input transducer)

overtones: other pitches or tones that overlap the basic, or fundamental, pitch

P

pacemaker: an electronic device that sends timed electrical signals to the muscles of the heart to keep the heart beating regularly; usually implanted directly in the body

pedigree: a graphical representation that shows to whom a person is related and how; traces the presence and absence of a particular trait in each individual within a group of related people

perennial: a plant that lives for more than two years

period: the amount of time it takes for a complete oscillation or cycle; measured in seconds

petrochemicals: synthetic substances made mostly from petroleum (crude oil)

phosphor: a chemical that glows when hit with electrons

phosphorus cycle: the continuous transfer of the phosphorus atoms in the biotic and abiotic factors of an ecosystem (*see* biotic and abiotic factors)

photoelectric effect: a property of certain metals that causes them to emit electrons when light is shone on them

photosites: sites on a video camera screen that are made of materials sensitive to light; absorb varying amounts of electric charge depending on the intensity of light that hits them

photosynthesis: the process by which green plants convert light energy and carbon dioxide into sugars

pitch: the frequency of a sound wave; in music, how high or low the note is

pixels: dots on a video or television screen that make up the image we see

plant propagation: the production of new individuals of a particular plant

polar covalent bond: an unequal sharing of electrons between two atoms

polar molecule: an asymmetrical molecule with polar bonds (i.e., one side of the molecule is more negative than the other due to a higher concentration of electrons)

polymer: an extremely large molecule made by combining hundreds or thousands of identical simple molecules

polymerization: the process that forms polymers from monomers (*see* polymers and monomers)

prosthesis: an artificial body part that replaces damaged or missing body parts

proteins: nitrogen-rich organic compounds found in the cells of all living organisms

R

radar: an instrument that uses radio waves to identify the presence or speed of something; includes a transmitter and a receiver

radiation therapy: a cancer treatment that uses high-energy radiation to kill cancer cells

rarefaction: the place in a longitudinal wave where the particles are stretched apart

receiver: a device that detects waves

recessive allele: the allele that does not appear physically in an individual with two different alleles for a gene

reflection: occurs when a wave hits a medium different from the one in which it has been travelling

refraction: the bending of a wave when it passes from one medium into another that has a different optical density

renewable energy source: any energy source that does not use up finite resources (i.e., cannot be used up)

repetition: an element of landscape design where features, such as shapes, colours, or patterns, are repeated

respiration: the process by which glucose is broken down to release energy that is used in our cells

S

sand: very small particles of rock

saponification: the hydrolysis of a special type of ester (fats and oil) to produce soap (*see* hydrolysis reaction)

scale: a deposit of calcium carbonate caused by boiling hard water that contains hydrogen carbonate

seedling: a young plant with only its first few leaves

selective cutting: removing some trees in all size classes, either singly or in groups

self-supporting environment: an environment that provides all the requirements for life without any need for people to bring any material in or out of the environment (e.g., Earth)

silt: a mixture of sand and clay particles that have been carried by water and then deposited

soft materials: plants such as trees, shrubs, bushes, and bedding plants used in landscaping

soft water: water that contains few or no dissolved minerals/ions

solar energy: energy from the Sun

soluble: refers to a solute that mixes with a solvent (*see* miscible)

solute: the substance being dissolved in a solution

solvent: the substance that causes the solute to dissolve in a solution (*see* solute)

sonogram: a picture of the pattern of sound waves produced during an ultrasound scan

sound: a wave that can be detected by the ear

suspension: a mixture containing suspended solids that cause a cloudy appearance but that will eventually settle out

sustainable development: development that can be continued over the long term; ensures that we use the resources we have today in a manner that will not damage the environment or economy tomorrow

synthetic: artificially made, does not exist in nature; manufactured

system: any set of connected parts that function together

T

technology: the application of scientific facts to solve human problems

texture: a characteristic of the surface of an object, such as softness, roughness, and smoothness

timbre: the quality of the sound of a musical note as a function of overtones (*see also* colour and overtones)

topography: the shape and size of land features, such as slopes

total internal reflection: reflection of a wave off the medium boundary; occurs if the angle at which the wave hits the medium boundary is too large for refraction to occur (*see* refraction)

trait: a characteristic of an organism

transducer: a device that converts energy from one form to another

transmitter: a device that produces waves

transpiration: a process in which liquid water is brought up from the soil through the roots of plants and released through the leaves into the atmosphere as water vapour; helps to purify water

transverse wave: a wave that moves at right angles to the movement of the particles of the substance in motion

trough: the low point of a transverse wave

U

ultrasound: sound waves that are very high in frequency, and cannot be heard by the human ear; used to create images of the internal parts of the human body, including the soft tissues

V

variable: an experimental condition that is purposely changed during an experiment

vegetative propagation: propagation of plants using any part of the plant (root, stem, or leaf) other than the reproductive structures (seeds or spores)

vibrating: moving back and forth

W

water cycle: the circulation of water between the air, land, and oceans, lakes, and streams of Earth

wave: a moving disturbance

wavelength: the distance between successive crests or successive troughs of a transverse wave, or the distance between successive rarefactions or successive compressions of a longitudinal wave

weightlessness: the state of an object that is not subject to gravitational pull

wilt: to droop (a plant) if given too little water

X

X-ray radiation: a diagnostic technology in which high-energy beams (X-rays) are directed on a part of the body; passes through soft tissues, such as muscle, but is stopped (absorbed) by the bones of the skeleton

Index

N

O

P

R

Photo Credits

Every effort has been made to track ownership of all copyrighted material, secure permission from copyright holders, and acknowledge correctly the sources of the material reproduced in this book. The publisher welcomes any information that will enable it to rectify, in subsequent editions, any errors or omissions.

Cover: (left) Image 100/Royalty-Free/Corbis/Magma, (right) EyeWire/Getty Images, (bottom) FK Photo/Corbis/Magma.

Contents: p. v Alan Marsh/firstlight.ca; p. vi John Colletti/MaXx Images/Index Stock; p. vii Bill Tice/MaXx Images; p. viii PhotoDisc/Getty Images; p. ix NASA.

Unit 1 opener: p. 1 Alan Marsh/firstlight.ca. **Chapter 1** opener: p. 2 (left) Owen Franken/Corbis/Magma, (top centre) Jeff Greenberg/Visuals Unlimited, (right) Brooklyn Productions/The Image Bank/Getty Images; p. 7 Barry Cohen; p. 8 Jo Prater/Visuals Unlimited; p. 9 Jack Ballard/Visuals Unlimited; p. 10 (top to bottom) VU/George Herben/Visuals Unlimited, Amoz Eckerson/Visuals Unlimited, Jeff Greenberg/Visuals Unlimited; p. 11 Barry Cohen; p. 14 (top) Richard Kellaway/PC Services, (bottom left) Jeff Greenberg/Visuals Unlimited, (bottom centre) Steve Callahan/Visuals Unlimited, (bottom right) K9 Storm Inc. www.k9storm.com; p. 15 (top left and right) Richard Kellaway/PC Services, (centre left) Kevin and Betty Collins/Visuals Unlimited, (centre right) EyeWire/Getty Images, (bottom left) Jean-Francois Causse/Stone/Getty Images, (bottom right) Eyewire/Getty Images; p. 16 (top left) UUMC/Visuals Unlimited, (top right) SIU/Visuals Unlimited, (bottom) Inga Spence/Visuals Unlimited; p. 17–19 Richard Kellaway/PC Services; p. 20 FK Photo/Corbis/Magma; p. 33 (left) Hulton-Deutsch Collection/Corbis/Magma, (right) Jim Cummins/Taxi/Getty Images. **Chapter 2** opener: p. 34 (top left and centre) Barry Cohen, (bottom left and right) PhotoDisc/Getty Images; p. 37, 40 Barry Cohen; p. 41 PhotoDisc/Getty Images; p. 42 (left) E.S. Ross/Visuals Unlimited, (right) Barry Cohen; p. 44 (left) Barry Cohen, (right) Richard Kellaway/PC Services; p. 45–47 Barry Cohen; p. 50 (left) Barry Cohen, (right) Martyn F. Chillmaid/SPL/Publiphoto; p. 51 (left) Kim Fennema/Visuals Unlimited, (right) Barry Cohen; p. 53, 62 Barry Cohen; p. 63 Jeff Greenberg/Visuals Unlimited.

Unit 2 opener: p. 67 John Colletti/MaXx Images/Index Stock. **Chapter 3** opener: p. 68 (top left) Barros & Barros/The Image Bank/Getty Images, (bottom left) Barry Cohen, (right) Don Smetzer/Stone/Getty Images; p. 70 David Jett/Corbis/Magma; p. 72 Jeff Clark; p. 78 (left) PhotoDisc/Getty Images, (right) Ed Lallo/MaXx Images/Index Stock; p. 80 (top left) Rob Lewine/Corbis/Magma, (top right) Jon Feingersh/Corbis/Magma, (bottom left) G&M David de Lossy/The Image Bank/Getty Images, (bottom centre) Agefotostock/firstlight.ca, (bottom right) John Kelly/The Image Bank/Getty Images; p. 82 (top) ©Canada Post Corporation, 1999. Reproduced with permission., (bottom left) Science Photo Library, (bottom right) Bettmann/Corbis/Magma; p. 83 (top left) Lucent Technologies, (top right) Bettmann/Corbis/Magma, (bottom) Blaise Edwards/CP Picture Archive; p. 86 (top) David Nunuk/firstlight.ca, (bottom) Stewart Cohen/MaXx Images/Index Stock; p. 89 (top) Ron Watts/firstlight.ca; p. 90 Neal Preston/Corbis/Magma; p. 91 Loren Winters/Visuals Unlimited; p. 96 Courtesy of Bill Wilkinson. **Chapter 4** opener: p. 100 (top left and bottom left) EyeWire/Getty Images, (right) Jeff Greenberg/Visuals Unlimited; p. 109 Richard Kellaway/PC Services; p. 110 Ed Young/SPL/Publiphoto; p. 113 Tek Image/SPL/Publiphoto; p. 116 Tom Uhlman/Visuals Unlimited; p. 121 Barry Cohen; p. 125 Nathalie Lemaire; p. 126–127 Barry Cohen; p. 129 Francois Sauze/SPL/Publiphoto; p. 130 Richard Kellaway/PC Services; p. 131 Roger Ressmeyer/Corbis/Magma; p. 138 Rudi Van Briel/Photo Edit, Inc.; p. 139 (top row, left to right, first four images) EyeWire/Getty Images, (fifth image) Richard Kellaway/PC Services, (sixth image) Reuters NewMedia Inc./Corbis/Magma, (bottom) John Sohlden/Visuals Unlimited.

Unit 3 opener: p. 141 Bill Tice/MaXx Images. **Chapter 5** opener: p. 142 (left) Eric Larrayadieu/Stone/Getty Images, (centre) Jeff McIntosh/CP Picture Archive, (right) Spike & Ethel/MaXx Images/Index Stock; p. 144 (top) Curtis Lantinga/firstlight.ca, (bottom) G.K. & Vikki Hart/The Image Bank/Getty Images; p. 153 (top left and right, and bottom right) Dept. of Clinical Cytogenetics, Addenbrookes Hospital/Science Photo Library, (bottom left) L. Willatt, East Anglian Regional Genetics Service/Science Photo Library; p. 155

Dr. P. Marazzi/Science Photo Library; p. 157 MugShots/firstlight.ca; p. 158 (top left) Barry Cohen, (bottom right) CP Picture Archive; p. 161 Chris Knapton/Science Photo Library. **Chapter 6** opener: p. 168 (top left) Tim Flach/Stone/Getty Images, (bottom left) G. Tompkinson/SPL/Publiphoto, (right) EyeWire/Getty Images; p. 170 David Wrobel/Visuals Unlimited; p. 171 Saturn Stills/SPL/Publiphoto; p. 172 (top and bottom left) AlphaPresse/BSIP, (bottom right) Will & Deni McIntyre/Photo Researchers, Inc.; p. 173 (top) photo by Paul Barette, courtesy of Canada Diagnostic Centres, (bottom left) EyeWire/Getty Images, (bottom right) SIU/Visuals Unlimited; p. 174 Eric Schremp/Photo Researchers, Inc.; p. 175 Michael Newman/Photo Edit, Inc.; p. 176 (left) David Wrobel/ Visuals Unlimited, (centre) Saturn Stills/SPL/ Publiphoto, (right) EyeWire/Getty Images; p. 177 Custom Medical Stock Photo; p. 179 EyeWire/Getty Images; p. 180 (top) Kenneth Green/Visuals Unlimited, (bottom) Custom Medical Stock Photo; p. 186 Bettmann/Corbis/Magma; p. 187 (top left) Bettmann/ Corbis/Magma, (top centre) SIU/Visuals Unlimited, (top right) AFP/Corbis/Magma, (bottom centre) SIU/Visuals Unlimited, (bottom right) Science VU/ Visuals Unlimited; p. 188 James King-Holmes/SPL/ Publiphoto; p. 189 Brad Nelson/Custom Medical Stock Photo; p. 190 Richard Kellaway/PC Services; p. 191 Paul A. Souders/Corbis/Magma; p. 196 (left) James A. Sugar/Corbis/Magma, (right) Jeff Greenberg/ Visuals Unlimited.

Unit 4 opener: p. 199 PhotoDisc/Getty Images. **Chapter 7** opener: p. 200 (top left) Image 100/ Royalty-Free/Corbis /Magma, (bottom left) Jeff Greenberg/Visuals Unlimited, (right) Perron/Visuals Unlimited; p. 202 (top row, left to right) PhotoDisc/ Getty Images, Adam Jones/Visuals Unlimited, Corbis/ Magma, Bill Beatty/Visuals Unlimited, (bottom row, left to right) Wally Eberhart/Visuals Unlimited, Steve Callahan/Visuals Unlimited, David Sieren/Visuals Unlimited, Fritz Polking/Visuals Unlimited; p. 203 (top row, left to right) Wally Eberhart/Visuals Unlimited, EyeWire/Getty Images, D. Cavagnaro/ Visuals Unlimited, Inga Spence/Visuals Unlimited, (bottom row, left to right, first two images) John Sohlden/Visuals Unlimited, (third image) Inga Spence/Visuals Unlimited, (fourth image) PhotoDisc/ Getty Images; p. 205 (left and centre) PhotoDisc/ Getty Images, (right) Wally Eberhart/Visuals Unlimited; p. 206 Inga Spence/Visuals Unlimited; p. 207 (top) PhotoDisc/Getty Images, (bottom left and centre) International Bloembollen Centrum Hillegom, Holland, (bottom right) Richard Kellaway/PC Services; p. 212 Jerome Wexler/Visuals Unlimited; p. 213 (left) Douglas Peebles/Corbis/Magma, (right) Patrick Johns/Corbis/Magma; p. 214 (top) International Bloembollen Centrum Hillegom, Holland, (bottom left) Michael Boys/Corbis/Magma, (right) PhotoDisc/ Getty Images; p. 215 PhotoDisc/Getty Images; p. 219 (top to bottom) Wally Eberhart/Visuals Unlimited, Science Vu/Visuals Unlimited, Wally Eberhart/Visuals Unlimited, Inga Spence/Visuals Unlimited; p. 220 (top to bottom) George D. Lepp/Corbis/Magma, Richard Kellaway/PC Services, Ken Wagner/Visuals Unlimited, R.F. Ashley/Visuals Unlimited, Ken Wagner/Visuals Unlimited; p. 221 (top to bottom, first two images) E. Webber/Visuals Unlimited, (third image) Nigel Cattlin/Holt Studios International/Photo Researchers, Inc., (fourth image) Science Vu/Visuals Unlimited, (fifth image) David Newman/Visuals Unlimited; p. 223 Richard Kellaway/PC Services; p. 228 (top left) Wally Eberhart/Visuals Unlimited, (bottom left) E. Webber/Visuals Unlimited, (right) Brad Mogen/ Visuals Unlimited. **Chapter 8** opener: p. 230 (top left) Kirtley-Perkins/Visuals Unlimited, (bottom left) Bernd Wittich/Visuals Unlimited, (right) Mark E. Gibson/ Visuals Unlimited; p. 232 David Sieren/Visuals Unlimited; p. 234 (top) EyeWire/Getty Images, (bottom left to right, first image) Paul Gier/Visuals Unlimited, (second image) John Sohlden/Visuals Unlimited, (third and fourth images) Richard Kellaway/PC Services; p. 235 (top right) Tom Edwards/Visuals Unlimited, (centre) Glenn Oliver/Visuals Unlimited, (bottom left) Patrick Johns/Corbis/Magma, (bottom centre) Louise Tanguay/gardenIMAGE, (bottom right) William J. Weber/Visuals Unlimited; p. 236 (top left) Mark E. Gibson/Visuals Unlimited, (top right) Rob & Ann Simpson/Visuals Unlimited, (bottom) Charles Gupton/ Corbis/Magma; p. 237 (top left to right) Science Vu/Visuals Unlimited, Bill Banaszewski/Visuals Unlimited, Ron Austing, Adam Jones/Visuals Unlimited, (bottom) Barry Cohen; p. 238 (left) John Heseltine/ Corbis/Magma, (right) Richard Kellaway/PC Services; p. 239 John D. Cunningham/Visuals Unlimited; p. 241 Barry Cohen; p. 243 (top left) EyeWire/Getty Images, (top right) John D. Cunningham/Visuals Unlimited, (bottom left) Richard Kellaway/PC Services, (bottom right) Jeff Greenberg/Visuals Unlimited; p. 244 (top left) Jeff Greenberg/Visuals Unlimited, (top right and centre left) Adam Jones/Visuals Unlimited, (centre right) Link/Visuals Unlimited, (bottom left) Adam Jones/Visuals Unlimited, (bottom right) Max Hunn/Visuals Unlimited; p. 245 (top left) Michael S. Yamashita/Corbis/Magma, (top right, and bottom left and right) Richard Kellaway/PC Services; p. 248 John D. Cunningham/Visuals Unlimited; p. 249 Al Davis; p. 250 (top left) Ned Therrien/Visuals Unlimited, (top right) Jeff Greenberg/Visuals Unlimited, (bottom left) Lester Lefkowitz/Corbis/Magma, (bottom right) Barry Cohen; p. 251 (top left) Gerald & Buff Corsi/Visuals Unlimited, (top right) Ron Austing, (bottom left) Ray Coleman/Visuals Unlimited, (bottom right) Jana R. Jirak/Visuals Unlimited; p. 252 (left) Bert Krages/Visuals Unlimited, (right) D. Long/Visuals Unlimited; p. 253 Tom Edwards/Visuals Unlimited; p. 255 Mark E. Gibson/Visuals Unlimited; p. 256 Patrick J. Endres/Visuals Unlimited; p. 260 (left) John Sohlden/Visuals Unlimited, (right) Richard Kellaway/PC Services.

Unit 5 opener: p. 263 NASA. **Chapter 9** opener: p. 264 (top left) Patrick J. Endres/Visuals Unlimited, (top centre) Michael DeMocker/Visuals Unlimited, (bottom left) G&M David de Lossy/The Image Bank/Getty Images, (right) Tibor Bognar/Corbis/Magma; p. 266 (top) Terje Rakke/The Image Bank/Getty Images, (bottom) Richard Kellaway/PC Services and courtesy of Chris Duller/Primarily Pets; p. 267 The SOHO-EIT Consortium: SOHO is an ESA-NASA programme of international cooperation; p. 269 Winston Fraser/Ivy Images; p. 271 Deb Fairchild; p. 275 (top right and centre right) PhotoDisc/Getty Images, (centre left) AbleStock, (bottom) David Barr/Photobar Agricultural Stock; p. 276 (left, top to bottom) D. Cavagnaro/Visuals Unlimited, Steve McCutcheon/Visuals Unlimited, Ron Dengler/Visuals Unlimited, (right, top to bottom, all three images) EyeWire/Getty Images; p. 277 (top left to right) Image Club/EyeWire, EyeWire/Getty Images, AbleStock, (bottom left and right) EyeWire/Getty Images; p. 286 Pascal Goetgheluck/Science Photo Library; p. 287 G. Prance/Visuals Unlimited; p. 290 Mark E. Gibson/Visuals Unlimited. **Chapter 10** opener: p. 292 (top left) Inga Spence/Visuals Unlimited, (bottom centre) Jeff Greenberg/Visuals Unlimited, (right top and bottom) NASA; p. 294–295 Hibernia Management and Development Company Limited; p. 297 Sean O'Neill and courtesy of Biodôme de Montréal; p. 300 G. Prance/Visuals Unlimited; p. 301 Deep Light Productions/SPL/Publiphoto; p. 305 NASA; p. 307 (top) NASA, (bottom) EyeWire/Getty Images; p. 308–311 NASA; p. 314 (top) Warren Morgan/Corbis/Magma, (bottom) Chris Butler/SPL/Publiphoto.

The publisher also wishes to thank the following sources. Page 53 (logo artwork) Environmental Choice Program; p. 298–299 (diagram reference) Columbia University Biosphere 2 Center; p. 306 and p. 312 (diagram information) NASA; p. 323 (art of Laurentian Forest map, on computer screen) Courtesy of Biodôme de Montréal.